# A Dictionary of
# **Weather**



SEE WEB LINKS

This is a web-linked dictionary. There is a list of recommended web links at the end of the book on p. 308. To acces the websites, go to the dictionary's web page at http://www.oup.com/uk/reference/resources/weather, click on Web links in the Resources section and click straight throught to the relevant websites.

**Storm Dunlop** is an experienced writer on meteorology and astronomy. He is a Fellow of both the Royal Astronomical Society and the Royal Meteorological Society, and past photographic editor of the journal, *Weather*. He is also the past president of the British Astronomical Association and has given many lectures and talks on all aspects of meteorology and astronomy. His previous works include *Collins Gem Weather* (1999), *Wild Guide Night Sky* (1999), *How to Identify Night Sky* (2002), *How to Identify Weather* (2002), *Wild Guide Weather* (2003), *Practical Astronomy* (2003), *Collins Atlas of the Night Sky* (2005), *Weather* (2006 & 2007), *Photographing Weather* (2007), and *Guide to Weather Forecasting* (2008).

A Dictionary of

# Weather

SECOND EDITION

STORM DUNLOP



OXFORD
UNIVERSITY PRESS

OXFORD
UNIVERSITY PRESS

Great Clarendon Street, Oxford OX2 6DP

Oxford University Press is a department of the University of Oxford.
It furthers the University's objective of excellence in research, scholarship,
and education by publishing worldwide in

Oxford New York

Auckland Cape Town Dar es Salaam Hong Kong Karachi
Kuala Lumpur Madrid Melbourne Mexico City Nairobi
New Delhi Shanghai Taipei Toronto

With offices in
Argentina Austria Brazil Chile Czech Republic France Greece
Guatemala Hungary Italy Japan Poland Portugal Singapore
South Korea Switzerland Thailand Turkey Ukraine Vietnam



Published in the United States
by Oxford University Press Inc., New York



First edition published 2001

Reissued 2005

Second edition published 2008



British Library Cataloguing in Publication Data
Data available

Library of Congress Cataloging in Publication Data
Data available

ISBN 978-0-19-954144-7
1

Typeset by SPI Publisher Services, Pondicherry, India
Printed in Great Britain by
Clays Ltd, St Ives plc

# Contents

# Foreword

My whole life has been connected with the weather, from the early days as an observer at the Christmas Island H-bomb tests to leading a team of broadcast meteorologists doing upwards of 100 television and radio broadcasts every day on the BBC. One of the first topics of conversation—even between people that have never met before—is the weather. It holds a strange fascination for most of us, probably because deep down we all see ourselves as being able to foretell the future, especially the weather. Many people can, such as farmers and fishermen whose livelihoods and even lives depend on them being able to understand the signs in the sky. Most of the weather folklore came from these people, and many of the sayings are correct some of the time.

Through the last century we saw enormous strides in the advancement of science, including weather forecasts, and it was L F Richardson who devised a numerical way of predicting the weather in the early part of the 20th century, long before computers had been invented. A lot of work was done at that time and between the wars to develop the skill of forecasting, including the Norwegian Frontal theory which is still used very widely today. With the invention and rapid expansion of the computer, weather forecasting has taken on a different role. There is still the need for the professional forecaster who can almost 'smell' the weather, but more and more reliance is placed on the super computers to help the forecaster look up to ten days ahead. Even with the bigger and faster computers there is still a great need for men and women who are passionate about the weather to come into the profession: those people who always look up into the sky first thing in the morning; those who read the thermometers and rain gauges in their gardens every day; and those who can remember when the last heavy snowfall in southern England was.

I have spent a lifetime looking and worrying about the weather, and one of the books I have missed has been a comprehensive reference book with simple explanations of the more common and rarer weather phenomena. There are, of course, many weather glossaries on the market but this new dictionary of weather surpasses them all by a long way. It is the definitive reference for all things meteorological, with an easy-to-understand explanation on such diverse topics as absolute instability through the ozone layer to the World Meteorological Organization.

As we move into the 21st century, the weather will take on even greater significance than it has done in the past. Through the greenhouse gases, which we are pumping into our atmosphere in ever greater amounts, we are rapidly changing the climate of all the parts of the world, and the forecasters will have to rely on even more powerful computers to brief the politicians on the likely outcome if we continue to pollute our environment as we are doing at the moment.

How I wish that I had had the *Dictionary of Weather* beside me over the last forty years of forecasting our weather.

Bill Giles OBE

*FORMER HEAD OF BBC WEATHER SERVICES*

# Preface

This second edition of the *Dictionary of Weather* largely follows the criteria used for the first edition; namely that the terms should be of use to students, amateur meteorologists, and the general public, rather than concentrate on the more specialized vocabulary required by professional meteorologists. The same criteria have been used in determining the level of coverage, which deliberately avoids highly technical descriptions, but which, it is hoped, still provides clear definitions of the various concepts.

All headwords have been checked and extensively revised where necessary. About 200 new terms have been included. As before, apart from terms in the fields of climatology, hydrology, and oceanography, additional headwords from less obviously related fields such as astronomy, geomagnetism, and geology are also incorporated. These help to ensure that both the current interactions of the Earth with its environment, and also the reasons for past climatic change, are adequately covered.

Although dictionaries document usage, rather than define it, some scientists appear to introduce terms to cause deliberate confusion. Some such terms have been omitted here, most notably 'misocyclone', which everyone, without exception, assumes to be a mis-spelling of 'mesocyclone'. This compiler, at least, hopes that it does not enter common usage, and that some term such as 'mini-cyclone' or even 'microcyclone' eventually becomes current.

The almost explosive growth in the number of abbreviations and acronyms has continued. There are now so many that although I have tried to be selective, and include only those that are most likely to be encountered, or that are the most important, reasons of space and balance mean that many have been omitted. Note that the abbreviations, acronyms, and capitalization of the names of experiments and instrumentation (particularly in the field of satellites) follow the general usage among the scientists and engineers who have developed and use the equipment.

At the time of writing, the Bali Conference on climate change has just taken place. It remains to be seen how successful the 'Bali roadmap' will be in obtaining agreement on the successor to the Kyoto agreement. One positive outcome is that, at last, all governments appear to accept the reality of global warming, even though those countries that have failed to meet their Kyoto obligations (most notably Australia, Canada, and the United States) will probably escape the sanctions to which they are liable.

In preparing both editions, I am extremely grateful to the many individuals who have not only provided me with extensive lists of terms for possible inclusion—many of which could not be incorporated for one reason or another—but also patiently discussed various concepts with me. I hope that they will forgive any errors and omissions, which are the result of my own misunderstanding rather than of their explanations.

Storm Dunlop

*CHICHESTER, WEST SUSSEX*
*JANUARY 2008*

# Acknowledgements

The author gratefully acknowledges the invaluable help provided for both first and second editions by the following:

Jay Anderson, University of Manitoba
Graham Bartlett, National Meteorological Library, Exeter
Andrew Best, Meteorological Service of New Zealand, Wellington
Grant Bigg, University of East Anglia, Norwich
Roger Brugge, University of Reading
Jim Dudhia, National Center for Atmospheric Research, Boulder, Colorado
Steve Edberg, La Cañada, California
Durham Garbutt, St Augustine, Florida
Dave Gavine, Edinburgh
Nick Grahame, Meteorological Office, Exeter
Les Hamilton, Aberdeen
Peter Hingley, Royal Astronomical Society, London
Steve & Donna O'Meara, Volcano, Hawaii
Jean Palutikof, University of East Anglia, Norwich
Richard Pettifer, Royal Meteorological Society, Reading
Bill Pike, Woodlands St Mary, Berkshire
Jonathan Shanklin, British Antarctic Survey, Cambridge
John Snow, National Severe Storms Laboratory, Norman, Oklahoma
Ian Strangeways, TerraData, Wallingford, Oxon.
Dick Tabony, Glasgow
Malcolm Walker, Royal Meteorological Society, Reading
Bob Wilderspin, Meteorological Office, Exeter
Austin Woods, European Centre for Medium-Range Weather Forecasts, Reading
Andrew Young, San Diego State University, San Diego, California

# List of Plates

# A

**AATSR** *See* ADVANCED ALONG-TRACK SCANNING RADIOMETER.

**ablation** The removal of surface snow or ice by *sublimation, melting, or *evaporation. The term is sometimes extended to include snow removed by the wind (*deflation) and also the calving of *glaciers.

**ablation zone** The zone of a *glacier in which losses through various processes, such as calving, *deflation, melting, etc., exceed any addition through snowfall or accumulation of *rime ice. *Compare* ACCUMULATION ZONE.

**abroholos (abrolhos)** A violent *squall, particularly prevalent in summer, that occurs off the south-eastern coast of Brazil.

**absolute humidity** The amount of *water vapour in air, determined as the mass of vapour per unit mass of air, and normally given in grams per cubic metre. *See also* HUMIDITY; SPECIFIC HUMIDITY.

**absolute instability** The state of a column of air that has a *lapse rate that everywhere exceeds the *dry adiabatic lapse rate. The column is therefore unstable throughout its height.

**absolute stability** The state of a column of air that has a *lapse rate that is always less than the *saturated adiabatic lapse rate and thus remains stable at all levels. *See* STABILITY.

**absolute zero** The theoretical temperature at which molecules of any substance have the least energy, all molecular motion ceases, and a perfect gas has zero volume. It is the zero point for the *Kelvin temperature scale, and is equivalent to −273.16 °C.

**absorption** **1.** The process by which a substance, such as a gas or *water vapour, reacts chemically with another. *Compare* ADSORPTION. **2.** The removal of certain wavelengths from incident radiation by a material such as glass or air by *scattering or by excitation of an atom or molecule through absorption of an energetic photon. Depending on the substance, the energy thus gained may be either reradiated almost immediately at a different wavelength or converted into thermal motion of the atoms and molecules within the material.

**absorption, atmospheric** The various components of the Earth's atmosphere act to reduce or completely block certain wavelengths of the radiation incident at upper levels. The principal active atoms are *oxygen and *nitrogen, and molecules of oxygen, *ozone (in the ultraviolet), and *water vapour (in the infrared). The overall spectrum at ground level is complex, but there are major windows where a significant amount of radiation reaches the surface. The most important are in the visible spectrum between about 0.3 and 0.9 μm (300 and 900 nm); multiple windows in the infrared between about 1.25 and 30 μm, and also beyond 300 μm; and at radio wavelengths between about 1 cm and 30 m.

**accessory cloud** A form of cloud that occurs only as an adjunct to certain of the ten principal *cloud genera (*see* CLOUD CLASSIFICATION). It is sometimes a distinct cloud mass (as with *pannus, *pileus, or *velum) and sometimes an extension of the main cloud (as with *arcus, *incus, *mamma, *tuba, or *virga). All the latter (and *praecipitatio) are more correctly described as *supplementary cloud features.

**acclimatization** The process by which animals (including humans) and plants adjust to changes in their environment. Such alterations may be the result of climatic change (*see feature*) or arise from variations in the range of a particular species.

**accretion** **1.** A term used in *cloud physics for the growth of a frozen *precipitation particle through the collision of an ice crystal or *snowflake with *supercooled water droplets, which freeze instantly on contact. *Compare* COALESCENCE. **2.** In a more general sense, the coalescence or *agglomeration of either liquid or frozen particles.

**accumulation zone** The zone of a *glacier in which the amount of snow and ice that is deposited exceeds that lost by *deflation, *evaporation, melting, or other means. *Compare* ABLATION ZONE.

**acid rain (acid precipitation)** A popular term for *precipitation that has a low pH (potential of hydrogen) value as a result of atmospheric (particularly anthropogenic) pollutants, and which causes environmental damage. Although a neutral solution has a pH value of 7, even in pre-industrial times natural rainwater was slightly acid with a pH value of 5.6, most probably because of dissolved *carbon dioxide, together with *nitrogen oxides created by *lightning, and other substances produced by volcanoes and forest fires. At the present day, the principal components responsible for increased acidity are sulphuric acid ($H_2SO_4$) and nitric acid ($HNO_3$). To allow for natural variation, some scientists use a slightly lower limit of 5.0 in defining acid rain.

**acoustic sounding** A method of using high-intensity sound waves to investigate the detailed structure of the *troposphere. The sound is emitted by one or more loudspeakers and the return signal detected by a suitable receiver.

The technique is capable of mapping *inversions, determining *turbulence, and, through the *Doppler effect, of detecting variations in the wind field.

**Acoustic Thermometry of Ocean Climate (ATOC)** An experiment to measure oceanic temperatures by determining the time for sound to travel from acoustic sources off California and Kauai (in the Hawaiian Islands) to hydrophones at various sites across the Pacific Ocean. The velocity of sound increases with water temperature, which the method should determine to within 0.05–0.1 degC. It is potentially a sensitive method of monitoring global warming (*see feature*).

**ACSYS** *See* ARCTIC CLIMATE SYSTEM STUDY.

**active surface** The surface layer that is in contact with the atmosphere and which undergoes the greatest diurnal temperature changes, absorbing heat by day and radiating it to the atmosphere at night. Examples are: bare soil, the top of a tree canopy, and the uppermost layer of the oceans.

**ACW** *See* ANTARCTIC CIRCUMPOLAR WAVE.

**adiabat (occasionally also known as an adiabatic)** A line on an *aerological diagram that illustrates the behaviour of dry or saturated air lifted under *adiabatic conditions. It represents the *dry, *pseudo-, or *saturated adiabatic lapse rate.

**adiabatic** [Greek: 'impassable'] A process in which heat does not enter or leave the system. Under most circumstances, a parcel of air that is not in contact with the surface may be assumed to behave adiabatically, enabling its behaviour to be calculated with a considerable degree of accuracy. A parcel of air that moves upwards is subject to expansion and **adiabatic cooling**, whereas one moving downwards is compressed and undergoes **adiabatic warming**. The rate of change in temperature with altitude is known as the *lapse rate, and this varies depending on whether the air is unsaturated (the *dry adiabatic lapse rate) or saturated (the *saturated adiabatic lapse rate). *See also* PSEUDO-ADIABATIC LAPSE RATE.

**adiabatic atmosphere** An atmosphere that has a *dry adiabatic lapse rate.

**adiabatic lapse rate** The theoretical *lapse rate for a parcel of air that moves vertically under *adiabatic conditions, i.e. that does not gain or lose heat to its surroundings. *See* LAPSE RATE.

**adsorption** The process by which a thin film of gas or liquid forms on a solid surface. The film does not react chemically with the material on which it is deposited. *Compare* ABSORPTION (1).

**Advanced Along-Track Scanning Radiometer (AATSR)** A *radiometer carried on the European *Envisat *polar-orbiting satellite, with seven channels, four in the visible/near-infrared and three at longer, thermal-infrared wavelengths. Its principal purpose is to obtain highly accurate *sea-surface temperatures for climate modelling.

**Advanced Microwave Sounding Unit (AMSU)** An improved version of an earlier instrument known as the Microwave Sounding Unit (MSU). A satellite-borne instrument specifically designed to determine the *humidity at different heights in the atmosphere regardless of cloud cover. The instrument monitors three microwave bands to determine water content and temperature, and hence humidity. It was first flown on the *NOAA 15 *polar-orbiting satellite, launched in May 1998. Legacy units are being fitted to *Eumetsat *MetOp satellites (*see* INITIAL JOINT POLAR SYSTEM).

**Advanced Scatterometer (ASCAT)** An instrument carried by the *MetOp satellites, and which is an advanced version of the *scatterometers employed on the *European Remote Sensing Satellites. It uses six antennae for simultaneous coverage of two swathes on either side of the ground track. Apart from wave height and wind direction, its observations are used in studies of land and sea ice, snow cover and thawing, snow properties, and soil moisture.

**Advanced Very High Resolution Radiometer (AVHRR)** The *radiometer carried by the latest *NOAA *polar-orbiting satellites. It operates at two wavebands in the visible region, and three in the infrared, with a resolution of 1.1 km. Legacy units are being fitted to *Eumetsat *MetOp satellites (*see* INITIAL JOINT POLAR SYSTEM).

**Advanced Weather Interactive Processing System (AWIPS)** A system installed by the *National Oceanic and Atmospheric Administration which integrates *Doppler radar (i.e. the *NEXRAD system), data from automatic surface stations, satellite imagery, and forecast models. The aim is to provide locally specific forecasting, particularly for floods and tornadoes (*see feature*), which has been impractical or difficult with technically distinct systems.

**advection** The transport of a substance (such as air) or of some specific property (such as *humidity) from one point to another. In meteorology, the term is generally applied to horizontal motion, although vertical advection may be important in certain cases, such as the hydraulic jump (*see* BORE) that accompanies the *morning glory.

**advection fog** **1.** *Fog that forms through the *advection of moist stable air that has been carried over a cold surface, which causes it to cool below its *dew point. **2.** Pre-existing fog that has been carried over a neighbouring area, e.g. sea fog that has invaded the land.

**aerobiology** The study of airborne organisms, including bacteria, fungal spores, and pollen.

**aerological diagram** A graph of the values of *pressure, *temperature, and *humidity obtained in a vertical sounding of the atmosphere, typically from a *radiosonde ascent. Reference lines incorporated to assist the plotting and interpretation of the diagram are *isobars, *isotherms, and dry and saturated *adiabats. Three versions are commonly used, each with its own advantages: the *tephigram, emagram, and Stüve diagram. A specific meteorological form of a **thermodynamic diagram.**

**aerology** A general term for the study of the atmosphere or any of its constituent layers. *Compare* AERONOMY; METEOROLOGY.

**aeronomy** The branch of science that studies the physical and chemical properties of the atmosphere. Although sometimes used for studies of the whole atmosphere (*compare* AEROLOGY), it is generally restricted to describing the investigation of the upper layers (above the lower *stratosphere), where *dissociation and *ionization are dominant processes.

**aerosol** Finely divided liquid droplets or solid particles that are uniformly dispersed and able to remain suspended in the atmosphere rather than settling out. Sizes are typically in the range 0.001–1.0 μm. Although not strictly correct, the term is also frequently applied to larger particles and droplets (<50–100 μm), such as those found in *fog or *haze. (Even larger particles may remain in suspension under conditions of extreme *turbulence, such as those found in vigorous *cumulonimbus clouds.) Aerosols are important in atmospheric processes, in particular as *condensation and *freezing nuclei, and for the part they play in *atmospheric electricity and various chemical reactions.

**aerosonde** A small robotic aircraft designed to carry out long-duration flights and obtain meteorological data from inaccessible regions, thus complementing data from *radiosondes, *ultra-long duration balloons, and *meteorological satellites. Two types are under development, one for use over oceanic areas, and the second for flights at altitudes over 25 km (and possibly above 35 km). In August 1998, a prototype low-altitude aerosonde crossed the Atlantic Ocean from Newfoundland to Scotland (3200 km) in 26 hours. Proposals exist for solar-powered high-altitude forms that might remain aloft for as long as a year.

**afterglow** *See* ALPINE GLOW; TWILIGHT ARCH.

**Afternoon-Train (Afternoon-Train constellation, A-Train)** A group of *polar-orbiting satellites flying in formation (known as a **constellation**), and following one another in orbit, thus monitoring a single swathe of the Earth with a sequence of different instruments at essentially the same time. The orbit has a nominal altitude of 705 km, an inclination of 98°, and crosses the Equator at *c.*13:30 UT (hence 'Afternoon'). The interval between the leading and trailing satellites is maintained to be less than 15 minutes. Currently, the satellites (in order) are: *Aqua, *Cloudsat, *Calipso, *Parasol, and *Aura. The *Orbiting

Carbon Observatory (OCO) will be inserted ahead of Aqua on launch, which is expected to be in 2008.

**ageostrophic wind** An important theoretical wind, representing the vector difference between the *geostrophic wind and an actual wind. It is related to pressure changes with place and time. There is, for example, an ageostrophic component between the surface wind and the *gradient wind.

**agglomeration** A term used in *cloud physics for any of the processes by which *precipitation grows through the collision and amalgamation of various cloud or precipitation particles. *See also* ACCRETION; AGGREGATION; COALESCENCE.

**aggregation** A term used in *cloud physics for the process by which ice crystals or *snowflakes grow by colliding and adhering to one another.

**Agulhas Current (Mozambique–Agulhas Current)** The narrow, warm current, an extension of the Indian Ocean's *South Equatorial Current, that flows south along the eastern coast of Africa. One of the fastest ocean currents, it may reach as much as 5 *knots off South Africa. Most of its flow then turns eastward and merges with the *Antarctic Circumpolar Current, but some may round the Cape of Good Hope into the South Atlantic.

**air** **1.** The mixture of gases that forms the Earth's *atmosphere, i.e. excluding *aerosols, *dust particles, *precipitation (in any form), or *pollutants. An extremely important variable component is *water vapour, but if this is excluded, dry air has a relatively constant composition (within the *homosphere), which is given in the table below. The water-vapour content may vary between zero and about four per cent. Other gases present in variable amounts include *ammonia, *carbon monoxide, *ozone, and *sulphur dioxide.

| GAS | ATMOSPHERIC ABUNDANCE (% BY VOLUME) |
|---|---|
| *nitrogen ($N_2$) | 78.09 |
| *oxygen ($O_2$) | 20.95 |
| *argon (Ar) | 0.94 |
| *carbon dioxide ($CO_2$) | 0.03 |
| *neon (Ne) | $1.8 \times 10^{-3}$ |
| *helium (He) | $5.2 \times 10^{-4}$ |
| *methane ($CH_4$) | $2.0 \times 10^{-4}$ |
| *krypton (K) | $1.0 \times 10^{-4}$ |
| *hydrogen (H) | $5.0 \times 10^{-5}$ |
| *nitrous oxide ($N_2O$) | $5.0 \times 10^{-5}$ |
| *xenon (Xe) | $8.0 \times 10^{-6}$ |

**2.** A gentle breeze; a **light air** being defined as force 1 on the *Beaufort scale.

**air frost** A temperature below 0 °C, as measured in a standard *thermometer screen.

**airglow** The weak ultraviolet, visible, or infrared emission produced in the *ionosphere primarily by the *recombination of ionized atoms and molecules. These are initially ionized by solar radiation or by collision with energetic particles emitted by the Sun (the *solar wind). The three forms of airglow (*dayglow, *twilight glow, and *nightglow) have different characteristics.

**air mass** A large body of air, whose properties (*temperature, *humidity, and *lapse rate) are essentially homogeneous over an area several hundred kilometres across. *Fronts form the boundaries between air masses with differing properties.

The nature of air masses is determined by three factors: the properties of the *source region; the age of the air mass; and the changes that occur as it moves across the globe. The principal source regions are the semi-permanent high-pressure zones (the *subtropical and *polar anticyclones), together with the continental anticyclones that develop in winter. Primary classification is based on temperature, giving *Arctic (A), *polar (P), or *tropical air (T); and on the nature of the surface in the source region: maritime (m), or continental (c), the latter symbols being used as a prefix:

| | |
|---|---|
| mA | *maritime Arctic air |
| cA | *continental Arctic air |
| mP | *maritime polar air |
| cP | *continental polar air |
| mT | *maritime tropical air |
| cT | *continental tropical air |

Various additional and secondary types are recognized, in particular *equatorial air (E), *Antarctic air (A or occasionally AA), and *Mediterranean air. Air masses may be modified as they move away from their source regions. If continental Arctic (cA) air crosses the sea, for example, it becomes maritime Arctic (mA) air.

Additional classifications are sometimes used to indicate how in the northern hemisphere (in particular) polar air may move towards the Equator and acquire a particular character and then subsequently be drawn back towards the pole. A suffix indicates whether the air is warmer (w) or colder (k) than the surface. The former becomes more stable, and the latter more unstable. Older works may use the concept of *returning air (r) and these also often reverse the order of the symbols, with Tc for continental tropical or rPm for returning maritime polar air.

**air-mass analysis** In *synoptic meteorology, the identification of a specific *air mass and the location of any *fronts that form along its boundaries.

**air-mass climatology** A method of describing the *climate of a particular region in terms of the *type and frequency of the *air masses that affect the region during a representative period.

**air-mass thunderstorm** A *thunderstorm that develops, generally late in the day, within a warm, moist *air mass through convective heating from the surface. Neither widespread uplift nor vertical motion at a *front is involved in its formation, and it does not normally become severe or persist for very long.

**air meter** A portable instrument for measuring air flow in buildings, ventilation systems and similar situations. Generally consists of a set of vanes that rotate about a horizontal axis and with some form of mechanical or digital system of recording the air's velocity. *See also* ANEMOMETER.

**Aitken nucleus** *See* NUCLEUS.

**Alaska Current** A relatively warm ocean current that flows anticlockwise around the Bay of Alaska. A branch of the *Aleutian Current.

**albedo** The ratio (usually expressed as a percentage) between the quantity of radiation falling on to a surface, and the amount that is reflected. For certain surfaces, notably water, the albedo varies greatly depending on the angle of incidence. Some typical values are:

| SURFACE | ALBEDO (%) |
|---|---|
| water (normal incidence) | 5 |
| forest | 5–10 |
| wet earth | 10 |
| rock | 10–15 |
| dry earth | 10–25 |
| grass | 25 |
| sand | 20–30 |
| old snow | 55 |
| clouds | 50–65 |
| water (grazing incidence) | 70 |
| fresh snow | 80 |

**Aleutian Current** The northern, relatively cold branch of the *Kuro Shio Extension that flows eastwards across the northern Pacific Ocean. On encountering the western coast of North America, it divides into the *Alaska Current and the *California Current.

**Aleutian low** The semi-permanent low-pressure centre located at about 50° N over the Aleutian Islands. It is an important *centre of action in the northern hemisphere. *See also* PACIFIC DECADAL OSCILLATION.

**Alexander's dark band** *See feature* RAINBOW.

**along-track scanning radiometer (ATSR)** A *radiometer that scans parallel to the ground track of a *polar-orbiting satellite, often in a fixed direction, perpendicular to the surface, i.e. looking towards the *nadir.

**alpine glow** A sequence of colours seen on snow-covered mountain tops at sunset, sometimes called **alpenglow** or (German) *alpenglüh*. The true alpine glow begins when the Sun is still above the horizon. Mountain tops that are illuminated by the Sun become yellow, pink, reddish, and finally purple, before being overtaken by the *shadow of the Earth. Subsequently, when the Sun is at least 3–4° below the horizon, and if the *purple light is present, the mountains may be illuminated by a purple glow, known as the **afterglow**. A reversed sequence may be seen at sunrise. *See also* SUNRISE/SUNSET COLOURS.

**altimeter setting** The atmospheric *pressure reading for a particular airport and time that is to be set on an aircraft altimeter. The latter will then indicate the correct altitude of the airport when the aircraft is on the ground (assuming the appropriate allowance has been made for the instrument's height above ground as installed in the aircraft cockpit).

**altitude** **1.** In the horizontal coordinate system, the vertical angle between the horizon and an object, sometimes known as **elevation**. A negative value implies that the object is below the horizon. **2.** The height (in metres or feet) of an aircraft or atmospheric layer above a reference level, usually mean sea level.

**altocumulus (Ac)** One of the main *cloud genera: a middle-level cloud (*see* ÉTAGE) that occurs as a layer or patch of more or less separate cloudlets in the form of heaps, rolls, or pancakes (Plate 1). The cloud elements have an apparent width of 1–5° (by which they are distinguished from the higher *cirrocumulus and the lower *stratocumulus). They are white, or white and grey, normally with darker shading. Altocumulus clouds predominantly consist of water droplets, but ice crystals are often present. They display *diffraction phenomena such as *coronae and *iridescence. Altocumulus clouds are extremely varied, and occur in the *cloud species *castellanus, *floccus, *lenticularis (Plate 2), and *stratiformis; and the *cloud varieties *duplicatus, *lacunosus, *opacus, *perlucidus, *radiatus, *translucidus, and *undulatus.

**altostratus (As)** One of the main *cloud genera: a middle-level cloud (*see* ÉTAGE) that occurs as a grey or bluish-grey sheet that may be completely uniform or appear fibrous or striated (Plate 3). It may be thin enough for the disk of the Sun to be visible, but does not exhibit *halo phenomena. The edges of a layer may sometimes show a *corona or *iridescence. Altostratus may be either a water droplet or a *mixed cloud. It frequently produces *precipitation, but this may not reach the ground. The main *cloud varieties are: *duplicatus, *opacus, *radiatus, *translucidus, and *undulatus.

**ambient pressure/temperature** The pressure or temperature that applies in the atmosphere immediately surrounding the parcel of air, cloud, or other phenomenon that is being considered.

**ammonia ($NH_4$)** An atmospheric pollutant that may react with combustion gases, such as *nitric oxide, to produce tiny particles of ammonium nitrate, which aggravate respiratory and coronary problems. Often described as arising primarily from livestock wastes, but comparable (or greater) quantities are found in urban areas, produced by certain vehicles, apparently as a result of some action of catalytic converters.

**AMO** *See* ATLANTIC MULTIDECADAL OSCILLATION.

**AMOC** *See* ATLANTIC MERIDIONAL OVERTURNING CIRCULATION.

**amorphous cloud** A term sometimes applied to a low layer of featureless cloud, often *nimbostratus or *stratus *fractus. There are no distinct undulations or breaks, nor even a well-defined *cloud base.

**amplitude** For a variable quantity that exhibits a sine wave, the greatest divergence from the mean value. With an asymmetrical variation, half the mean difference between the extreme values of the fluctuation. Note that although this is the generally accepted definition in physics, in some fields, notably astronomy, amplitude is taken to mean the overall range of variation.

**AMSU** *See* ADVANCED MICROWAVE SOUNDING UNIT.

**anabatic wind** A localized wind that blows upslope when a mountainside is heated by the Sun, normally in the form of a *valley wind. Such winds tend to be slightly less common and weaker than the descending counterpart, the *katabatic wind (*mountain wind).

**anafront** A warm or cold *front above which the warm air is ascending relative to the adjacent cold air. Such a front is generally more active and gives greater precipitation than a corresponding *katafront.

**analogue method** A method of forecasting based on finding past occasions that are analogous to the current *synoptic situation, and also on the assumption that the weather will subsequently develop in a similar way. For various reasons, including those advanced by *chaos theory, such forecasting has limited success.

**analysis** Detailed examination of the state of the atmosphere based on *synoptic observations, in particular for forecasting or research purposes. This usually involves the determination of the values of the individual components, the comparison of observations from adjacent sites, the preparation of the appropriate charts (such as *isobaric or *isothermal charts), and their subsequent interpretation.

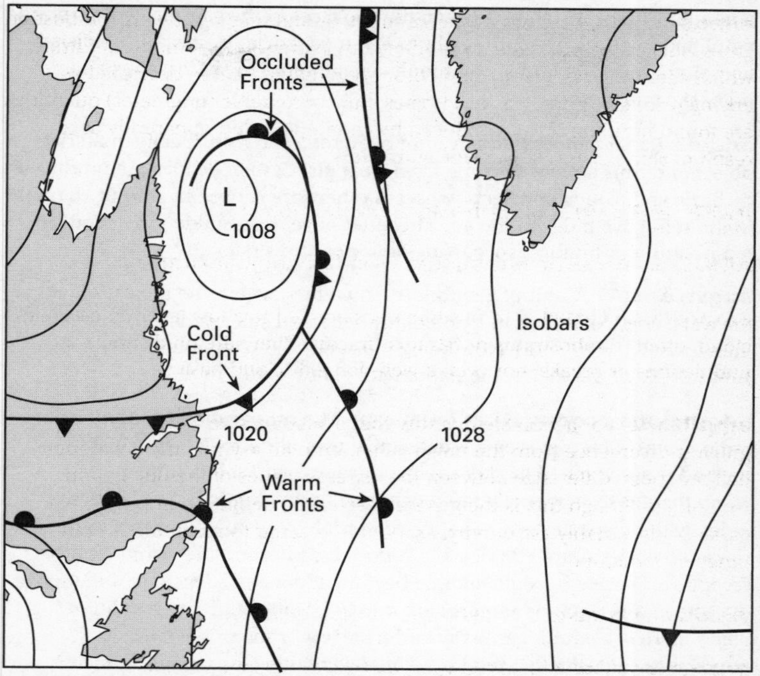


**Analysis** A surface analysis involves plotting *isobars, and assessing the location and type of *fronts.

**andhi** A violent *squall or *dust storm occurring in late spring in north-western India.

**anemogram** A recording of wind speed, normally in the form of a paper chart.

**anemograph** An *anemometer that provides a continuous record of wind speed, and sometimes direction, usually by means of a paper chart.

**anemometer** Any instrument for measuring wind speed. There are four broad categories, based on rotation, pressure, heat loss, and variation in the transmission of sound. Significant types within each group are: *cup- and *propeller-type, *drag-force and *pressure-plate, *thermal, and *ultrasonic anemometers.

**aneroid barometer (sometimes known as an elastic barometer)** [Greek: 'not wet'] A *barometer that does not use a liquid, i.e. without a mercury column. A corrugated metal capsule is partially evacuated of air, and prevented from collapsing by an internal spring. Changes in air pressure cause alterations in the distance between the two faces of the capsule, which may be amplified to give a pressure reading. The mechanism is used in *precision aneroid barometers and is commonly encountered in *barographs and domestic barometers.

**angels** Radar echoes that have not been returned by a specific material object and thus appear to come from clear air. Certain echoes are produced by birds and clouds of insects, whereas others are caused by abrupt changes in the refractive index of the air. The latter arise from sudden variations in temperature or humidity occurring at or near *inversions.

**ångström (Å)** A unit of length formerly often used in the measurement of the wavelength of radiation, particularly light, and still extensively found in the literature ($1\,\text{Å} = 10^{-10}\,\text{m}$). Now being replaced by the internationally approved submultiple of the metre, the nanometre ($1\,\text{nm} = 10^{-9}\,\text{m} = 10\,\text{Å}$).

**anomalous propagation (anaprop)** The propagation of sound or radio waves such that they are detectable over abnormally large distances, beyond those reached by the usual surface waves. Such conditions arise when there is exceptionally rapid decrease in the atmosphere's refractive index with height (known as **super-refraction**), causing the waves to be refracted back towards the Earth's surface well beyond the normal limit of reception. Sounds become audible beyond a *zone of silence, and the range of radio waves (including radar) is extended. Under extreme conditions, when there is a strong temperature *inversion or major decrease in atmospheric density in a relatively thin layer fairly close to the surface, a **radio duct** may be formed, channelling (**ducting**) radio and television transmissions to great distances. This often gives rise to what broadcasters call **co-channel interference**. Such conditions frequently occur in an *anticyclone when there is subsidence from higher levels.

**anomaly** **1.** The deviation of a property at a particular location from the long-term mean value. Anomalies, such as *sea-surface temperature anomalies, are often an extremely significant factor in *teleconnections such as those found in the El Niño Southern Oscillation (*see feature*) and the *North Atlantic Oscillation. **2.** The difference between a forecast and the actual conditions that pertain at the predicted time.

**Antarctic air (A)** A *continental-type *air mass (cA) that originates over the Antarctic ice-cap and the surrounding *ice shelves and *pack ice. It is always extremely cold and dry, but becomes increasingly unstable as it moves north over the warmer waters around the continent. It is always colder than *Arctic air at the surface, and at all levels in the southern autumn and winter. The symbol AA is used by some meteorologists.

**Antarctic anticyclone** The essentially permanent high-pressure region that exists over the Antarctic continent. There is inflow at upper-tropospheric levels, *subsidence, and an *anticyclonic circulation at the surface, giving rise to polar easterlies.

**Antarctic Bottom Water** A bottom current of extremely cold, dense, and saline (34.66 parts per thousand) water that flows north from its source in the Weddell Sea and the South Atlantic. The densest water found in any of the world's oceans. Initially its temperature is around −0.4 °C. It forms a significant contribution to the *Great Ocean Conveyor Belt.

**Antarctic Circumpolar Current (ACC; sometimes known as the West Wind Drift)** The dominant ocean current in the southern hemisphere, which flows in an easterly direction, driven by the strong prevailing westerlies. Unimpeded by any major land masses, it encircles the globe, and has no counterpart in the northern hemisphere. *See* ANTARCTIC CIRCUMPOLAR WAVE.

**Antarctic Circumpolar Wave (ACW)** A set of *anomalies in atmospheric *pressure, sea-ice cover, *sea-surface temperature, and *wind stress that propagate eastwards around the Antarctic continent. The anomalies originate from oceanic *Rossby waves within the *Antarctic Circumpolar Current, approximately between latitudes 40° S and 70° S. These waves are most noticeable as four alternating regions of warm tropical water and cold polar water. They take *c.*8–10 years to circle the globe, and may be directly linked through a *teleconnection with El Niño Southern Oscillation activity (*see feature*). Distinct correlations exist with weather over South America, South Africa, Australia, and New Zealand.

**Antarctic Coastal Current (sometimes known as the East Wind Drift)** A current that flows west around Antarctica, driven by the polar easterlies blowing outwards from the *Antarctic anticyclone.

**Antarctic Convergence** A line of *convergence in the ocean surrounding Antarctica between latitudes 50° S and 60° S, where cold surface water from Antarctica (**Antarctic Surface Water**) sinks below warmer, less saline water from the middle latitudes (**Subantarctic Surface Water**), becoming *Antarctic Intermediate Water. Often known as the **Antarctic Polar Front**, but this term must not be confused with the atmospheric *Antarctic Front.

**Antarctic Divergence** An oceanic boundary, lying at *c.*60–65° S, between the eastward-flowing *Antarctic Circumpolar Current and the westward-flowing *Antarctic Coastal Current.

**Antarctic Front** A persistent atmospheric *front that develops between the extremely cold polar *easterlies flowing out from the Antarctic ice sheet and the *polar maritime air found farther north. It generally lies at *c.*60–65° S and may encircle a considerable portion of the Antarctic continent at any one time.

**Antarctic Intermediate Water** Cold and moderately saline water that has formed around Antarctica as **Antarctic Surface Water**, and flowed north to the *Antarctic Convergence, where it sinks below warmer, less saline **Subantarctic Surface Water**. It may subsequently be traced at a depth of about 900m beneath the South Atlantic. *See also* ANTARCTIC BOTTOM WATER.

**Antarctic Oscillation** *See* ARCTIC OSCILLATION.

**anthelic arc** A *halo phenomenon consisting of a white arc that is visible on the opposite side of the sky to the Sun, and which is probably created by a single internal reflection within ice crystals. Pairs of anthelic arcs are normally seen, intersecting to give a diagonal cross, centred on the *parhelic circle, 180° from the Sun, where they may enhance an *anthelion (if present).

**anthelion** A *halo phenomenon (also known as a **counter sun**) that resembles a *parhelion (or **mock sun**). It lies on the *parhelic circle at the same altitude, but 180° in *azimuth from the true Sun. This bright spot is thought to arise from double reflection within ice crystals. It cannot occur when the solar elevation is greater than 46°, and below about 15° it takes the form of an **anthelic pillar**. The brightness may be enhanced by the intersection of *anthelic arcs with the *parhelic circle at this point.

**anticorona** *See* GLORY.

**anticrepuscular rays** *Crepuscular rays that appear on the opposite side of the sky to the Sun, and which, because of perspective, appear to converge at the *antisolar point.

**anticyclogenesis** The formation of an *anticyclonic circulation, or the strengthening of the airflow around an existing *anticyclone. The opposite of *anticyclolysis.

**anticyclolysis** The weakening or complete disappearance of the circulation around an existing *anticyclone.

**anticyclone** A region of relatively high atmospheric *pressure, also known as a **high**. On a *synoptic chart, it appears as a set of closed, approximately circular or elliptical *isobars. Air descends from the upper *troposphere in its core, causing it to become the source of outflowing air, which exhibits *anticyclonic circulation. Generally more persistent and slower moving than the corresponding low-pressure systems (*cyclones, *depressions, or *lows), anticyclones normally bring calm weather that changes slowly.

The descending air is warmed by compression, its *humidity decreases, and it becomes increasingly *stable, frequently giving clear, cloudless skies, which may, however, allow temperatures to drop dramatically at night, sometimes producing *radiation fog. An *inversion is often present and moist air trapped beneath this may produce conditions of *anticyclonic gloom.

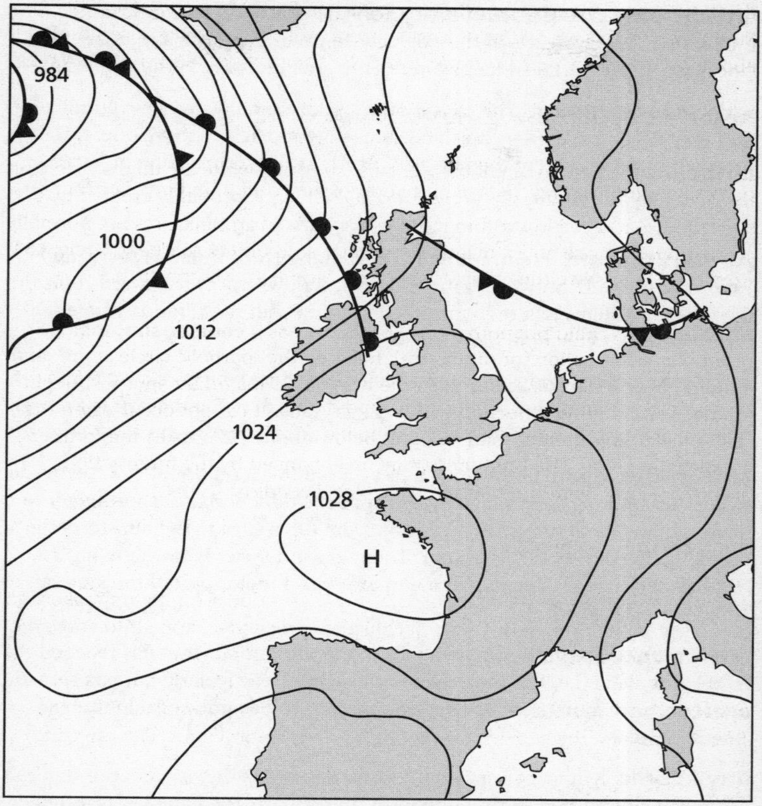


**Anticyclone** A high-pressure area, here centred over the Bay of Biscay, tends to force *depressions to move around it.

Anticyclones may be of two types: 'warm' and 'cold'. Warm anticyclones are typical of the semi-permanent subtropical high pressure areas (the *subtropical anticyclones), such as the *Azores and *Bermuda highs. They are deep, with warm temperatures throughout the troposphere (except sometimes at the very surface), with a high *tropopause and cold *stratosphere above them. They are slow moving and often extend their influence in the form of *ridges of high pressure. *Blocking anticyclones at high latitudes are also of this type. Cold anticyclones have low temperatures, and are shallow with a low tropopause and warm stratosphere. They are more mobile and shorter-lived than the warm variety. Semi-permanent cold anticyclones occur in winter over continental land masses, particularly over Siberia. The greatest atmospheric air pressure (1085.7 hPa) occurred at Tosontsengel, Mongolia, on 19 December 2001.

**anticyclonic circulation (anticyclonic curvature)** Airflow around the local vertical which is in the opposite sense to the Earth's rotation, i.e. clockwise in the northern hemisphere, and anticlockwise in the southern.

**anticyclonic gloom** The conditions of poor visibility and low illumination that may occur within a well-established, slow-moving *anticyclone. It is usually a combination of the effects of overcast skies, consisting of *stratus or *stratocumulus, and the accumulation of dust and pollutants beneath an *inversion.

**Antilles Current** A current that is part of the North Atlantic *gyre, and an extension of the *North Equatorial Current. It flows north-westwards outside the Greater Antilles and then north, combining with the *Florida Current off Cape Hatteras to become the *Gulf Stream.

**antisolar point** The point on the *celestial sphere which, relative to the observer (or camera), is diametrically opposite the Sun. Rainbows (*see feature*), *glories, and certain other optical phenomena are centred on the antisolar point.

**anti-trades** Upper-level winds that occur at low latitudes above the *trade winds and blow in approximately the opposite direction, transporting air to higher latitudes. The vertical transition from the trades to the anti-trades is normally marked by an *inversion. The two wind systems form the basic *Hadley cells in each hemisphere and are thus a major part of the general circulation (*see feature*).

**anvil cloud** The common name for a *cumulonimbus that has reached the *capillatus stage (i.e. has undergone *glaciation), particularly if it has spread out at an *inversion (such as the *tropopause) to become cumulonimbus *incus (Plate 9).

**anvil dome** A large or persistent *overshooting top that is often present with a *multicell or *supercell thunderstorm (particularly the latter). It is a sign of vigorous *thermals within the cloud mass.

**aphelion** The point of an astronomical body's orbit at which it is farthest from the Sun and the incident radiation is at a minimum. For the Earth, this occurs around 1 July, when the distance from the Sun is *c.*154 636 160 km, or 101.67 per cent of its mean value. *Compare* PERIHELION.

**apogee** The point of the orbit of the Moon or an artificial satellite at which it is farthest from the Earth. *Compare* PERIGEE.

**Appleton, Sir Edward Victor** (b. Bradford, England, 6 September 1892; d. Edinburgh, 21 April 1965) British physicist who carried out fundamental, pioneering research into the Earth's upper atmosphere. He established the existence of various reflecting layers in the *ionosphere and was able to link variations in their intensity with *solar activity. Awarded the Nobel Prize for Physics in 1947 for this work.

**Appleton layer** *See* IONOSPHERE.

**applied meteorology** The application of meteorological knowledge, current weather data, analyses, or forecasts to specific problems, such as those relating to transport, industrial processes, agriculture, etc. **Applied climatology** is the similar application of climatological data and knowledge.

**APS** *See* ATMOSPHERIC PRESSURE SENSOR.

**APT** *See* AUTOMATIC PICTURE TRANSMISSION.

**Aqua** A NASA *polar-orbiting satellite, part of the Earth Observing System, the second to be launched (after the *Terra satellite) on 4 May 2002. It is primarily designed to monitor the Earth's *hydrological cycle, but is also monitoring *aerosols, energy flux, phytoplankton, air, surface, and water temperatures, and vegetation cover. It is currently first in the *Afternoon-Train satellite constellation. *See also* AURA; CALIPSO; CLOUDSAT; ORBITING CARBON OBSERVATORY; PARASOL.

**Arago's point** *See* POLARIZATION OF THE SKY.

**arc, auroral** An auroral form (*see* AURORA) that appears as an arch across the sky, normally with a fairly well-defined lower edge and a more diffuse upper border. The highest point of the arc is towards the magnetic north. An arc is often the first distinct form to become visible during an auroral display, although it may be preceded by an overall **veil**, a horizon **glow**, or (occasionally) by diffuse **patches**. Two varieties are commonly seen: a **homogeneous arc**, which is largely featureless, and a **rayed arc**, with distinct vertical striations. *See* RAY, AURORAL.

**arc of contact** A *halo phenomenon that consists of an arc (or arcs) tangent to the 22° halo. The form varies with solar elevation. At low elevations, both upper and lower contact arcs are convex towards the Sun. The lower arc is then normally hidden below the horizon but may be visible from an aircraft. At higher elevations the upper arc becomes concave. The two arcs bend and, at an elevation of about 40°, merge into a single arc, known as the **circumscribed halo**. At *c.*70° elevation, the arc can no longer be distinguished from the 22° halo.

**Arctic air (A)** An *air mass that originates over the Arctic Ocean. During winter, when high pressure dominates, it is very dry and of *continental type (cA), but differs only slightly from the continental *polar (cP) *air masses that develop over Siberia and northern Canada. (It is generally colder at all levels.) In summer, the pool of cold air is shallow, and the air takes on more *maritime characteristics (mA).

**Arctic Climate System Study (ACSYS)** A *World Climate Research Programme project to obtain data on the significance of the Arctic for global *climate, including long-term monitoring of the overall energy budget and obtaining hydrographic data on the deep oceanic basins.

**Arctic continental air** *See* ARCTIC AIR.

**Arctic Front** A shallow *front that separates an outbreak of frigid, very dense, *Arctic air to the north from the warmer *continental polar or *maritime polar air to the south. Such a front is often present in the early part of the year, lying between Greenland and Scandinavia.

**Arctic frost smoke (Arctic sea smoke)** *See* STEAM FOG.

**Arctic maritime air** *See* MARITIME ARCTIC AIR.

**Arctic Oscillation** An *oscillation, approximately decadal in length, in the atmospheric parameters over the Arctic and those in a zone centred at approximately 40° N. When the *polar vortex is strong and extends down to the surface, the Arctic surface air pressure decreases, the westerly *zonal flow increases in strength, moves northwards, and transports more heat to northern Europe and the polar region causing a decrease in Arctic ice cover. In the opposite phase, there is a *sudden warming, the strength of the vortex declines, and the westerlies track farther south. It is essentially the same phenomenon as the *North Atlantic Oscillation. A similar feature, the **Antarctic Oscillation**, occurs in the southern hemisphere.

**Arctic Polar Front** An oceanic frontal zone that separates the subtropical and subpolar *gyres, which is both a temperature and salinity front. In the North Atlantic, where it is also known as the **cold wall** or **North Wall**, it lies between the warm and saline *Gulf Stream and *North Atlantic currents, and the cold, relatively fresh, *Labrador Current. In the North Pacific, it is in two portions, divided by the islands of Japan. The larger frontal zone lies between the warm, saline *Kuro Shio and the cold, fresh Oyashio current.

**arcus (arc)** [Latin: 'arch'] A *supplementary cloud feature in the form of a long, dense shelf or roll of cloud, often with ragged edges, that lies along the lower leading edge of the main cloud. When well developed it may appear as a dark, threatening arch, and it is sometimes known as an **arch cloud**. It is generally associated with *cumulonimbus, but occasionally appears beneath *cumulus. Specific features are present in the types of arcus known as *roll clouds and *shelf clouds.

**Argo** A project that has deployed 3000 *profiling floats throughout the world's oceans, aiming to obtain data on temperature and salinity profiles from even the most remote areas.

**argon (Ar)** An *inert gas, atomic number 18: the third most abundant atmospheric component after *nitrogen and *oxygen, amounting to 0.94 per cent by volume (1.3 per cent by weight). Three *isotopes exist: $^{36}$Ar, $^{38}$Ar, and $^{40}$Ar. The last forms the major portion of terrestrial argon (99.6 per cent) and has arisen from the radioactive decay of primordial potassium-40 ($^{40}K$). The rate of decay is used in potassium–argon dating to determine the age of the Earth ($4.55 \times 10^9$ years).

**arid climate (also known as a desert climate)** A climate where rainfall is insufficient to support appreciable amounts of vegetation. Arid climates occur in regions subject to persistent high pressure, such as those beneath the *subtropical anticyclones, or where topography blocks the passage of moisture-bearing systems, as with the Gobi Desert in Central Asia. Technically, the high Antarctic plateau has an arid climate, because of its extremely low *precipitation.

**Aristotle** (b. Stagira, Chalcis, Greece, 384 BC; d. Athens, 322 BC) Greek logician, philosopher, and early scientist. Aristotle produced the earliest comprehensive discussion of meteorology in his work *Meteorologikon,* which was based on two primary theories: (a) that the universe was spherical and that motions of the celestial bodies were governed by a system of concentric, crystalline (and unchanging) spheres as described by Eudoxus; (b) that (as suggested by Empedocles) the terrestrial region consisted of four elements—earth, water, air, and fire—also in concentric spheres, with Earth at the centre. These theoretical concepts influenced all Aristotle's generally erroneous descriptions of the causes of meteorological phenomena. His writings remained at the heart of all works on meteorology until the rise of experimental science in the 17th century.

**Arrhenius, Svante August** (b. Vik, near Uppsala, Sweden, 19 February 1859; d. Stockholm, 2 October 1927) Swedish physical chemist and Nobel laureate. Although his main contributions to science lay outside meteorology and climatology, Arrhenius noted Joseph *Fourier's suggestions of the significance of *carbon dioxide and the greenhouse effect (*see feature*), and he developed a theory that attempted to explain the *ice ages by variations in the carbon-dioxide and water-vapour content of the atmosphere. This theory found some acceptance, but it was soon superseded by the Milankovich theory (*see feature*) of changes in the Earth's orbit as the primary mechanism. Carbon dioxide is now thought to form a significant, amplifying feedback mechanism, and is particularly important in the *Snowball Earth theory. Although the data Arrhenius used were inaccurate, his calculation that the effect of a doubling of $CO_2$ content would lead to an increase of 5–6 deg C may be compared with the latest estimate from the *Intergovernmental Panel on Climate Change of an increase of 2–4.5 deg C.

**Asiatic low** A low-pressure *trough that lies over southern Asia in early summer, located approximately over India and the head of the Bay of Bengal. It is a major *centre of action for the northern hemisphere during that period of the year.

**aspect** The direction in which sloping ground faces. Technically, it is the compass direction (or *azimuth) indicated by a line perpendicular to the steepest contours.

**Assmann psychrometer** A *psychrometer in which a consistent rate of ventilation is provided by a motor-driven fan.

**ASTER** The Advanced Spaceborne Thermal Emission and Reflection radiometer carried on the *Terra satellite. A high-resolution instrument for monitoring *desertification, floods, ice cover, etc., as well as having the capability of pointing at specific special targets such as erupting volcanoes.

**astronomical theory** *See feature* MILANKOVICH THEORY.

**ATD (arrival time difference)** *See* SFERICS FIX.

**Atlantic Meridional Overturning Circulation (AMOC; North Atlantic thermohaline circulation)** The *thermohaline oceanic circulation system in the North Atlantic, in which warm water is carried north and east by the *Gulf Stream and *North Atlantic Current, with cold water returning south at depth. The circulation, which forms part of the *Great Ocean Conveyor Belt, is driven by the intense evaporation off Labrador and Greenland, with a consequent increase in density and salinity, the water sinking to become North Atlantic Bottom Water. *See also* RAPID.

**Atlantic Multidecadal Oscillation (AMO)** A series of long-term changes in *sea-surface temperatures involving almost the whole of the North Atlantic, with a periodicity of 20–40 years. It is determined from annual ocean temperature *anomalies, averaged across the North Atlantic from 0° to 70° N. Note that it differs from the *North Atlantic Oscillation, which is defined by the pressure pattern, not the temperature distribution.

Its effects are found in ice-core and tree-ring records, covering at least the last 1000 years. Recent warm phases occurred during 1860–80 and 1940–60, and cool phases during 1905–25 and 1970–90. Although of low amplitude (~0.5 deg C), it involves changes in the north–south and overturning circulations, affecting air temperatures and precipitation over Europe and North America, and is positively correlated with a similar oscillation in the northern Pacific (the *Pacific Decadal Oscillation). It alternately masks and increases the effects of global warming (*see feature*) in surrounding areas. In its warm (positive) phase it appears to increase drought in the American Midwest and Southwest, and to increase the number of major Atlantic *hurricanes, possibly doubling their frequency, but does not affect the numbers of *tropical storms, nor of weak hurricanes. The link with hurricanes is, however, controversial. Rainfall in Florida, the Pacific Northwest, and Europe increases during a warm phase, the latest switch to which occurred around 1995. *See also* ARCTIC OSCILLATION; NORTH ATLANTIC OSCILLATION.

**atmosphere** **1.** The gaseous envelope of the Earth and other planets. It may be of two basic types. A primary atmosphere such as that found on the four gas giants (Jupiter, Saturn, Uranus, and Neptune) largely consists of the gases that were present when the planet accreted from the primordial solar nebula. A secondary atmosphere, such as those of Venus, Earth, and Mars, arises through the loss of a substantial fraction of the primary atmosphere to space through various mechanisms, or through significant chemical evolution (or both).

In the case of the Earth, the hot, primordial atmosphere probably consisted of *carbon dioxide, *nitrogen, and smaller quantities of more complex molecules such as sulphuric acid ($H_2SO_4$) and *methane. Most of these gases were lost through the action of the violent *solar wind produced by the Sun early in its lifetime. As the planet cooled, water vapour (perhaps derived from cometary impacts) began to condense and fell as torrential rain, estimated by some scientists to have persisted for some 10 million years. The extreme acidity of this primordial rain dissolved calcium from the early crust. This reacted with atmospheric carbon dioxide to produce calcium carbonate and locked most of it away in carbonate sediments. With the rise of lifeforms, substantial quantities of *oxygen were released by photosynthesis, giving rise to the predominantly nitrogen–oxygen secondary atmosphere present today. For the composition of this atmosphere, *see* AIR. For its structure, *see* ATMOSPHERE, STRUCTURE OF.

**2.** A unit of pressure, equivalent to 101 325 Pa (1013.25 hPa) or 760 mm Hg, and normally used only for expressing high pressures such as those encountered in the atmospheres of Venus and Jupiter, or in chemical processes.

**atmosphere, structure of** The Earth's atmosphere consists of various layers, the nomenclature and altitudes of which depend upon the criteria used for their definition. Based upon the changes of temperature with height (i.e. the *lapse rate), four layers may be distinguished. Closest to the surface is the *troposphere, in which most of the significant weather occurs, where temperature generally declines with height, and which extends up to the *tropopause at *c.*7 km over the poles and 14–18 km over the Equator. The lowest region of the overlying *stratosphere is often *isothermal, but the temperature then increases with height to the *stratopause at *c.*50 km. In the *mesosphere, temperature again falls with height to the *mesopause at *c.*86–100 km. The outermost layer is the *thermosphere, in which atoms and molecules may attain extremely high temperatures (velocities), and which extends into interplanetary space. The region above 700 km, at which height atoms may begin to escape into space, is sometimes termed the *exosphere.

On the basis of chemical composition, the atmosphere consists of just two layers, the *homosphere (largely identical with the troposphere, stratosphere, and mesosphere), in which the composition is essentially constant, and the overlying *heterosphere.

Specific ionization and photochemical processes occur in the *ionosphere (encompassing part of the upper mesosphere and thermosphere) and the *chemosphere and *ozonosphere (both part of the upper stratosphere).

**atmospheric absorption** *See* ABSORPTION, ATMOSPHERIC.

**atmospheric chemistry** The study of the chemical composition of the atmosphere, of *aerosols, *pollutants, and *rainfall, and the various chemical reactions in which they are involved. It includes the processes of *dissociation, *ionization, and *recombination, and the general group of reactions known as *photochemistry. Certain chemical cycles are of particular importance, especially the *carbon cycle, the *hydrological cycle, and the *nitrogen cycle.

**atmospheric composition** *See* AIR (1).

**atmospheric electricity** Electrical phenomena that occur in the atmosphere, excluding the phenomena involved in the *ionosphere. Even under normal conditions (i.e. in the absence of *thunderstorm activity) the upper atmosphere has a substantial positive *space charge relative to the Earth's surface. Various mechanisms (such as *cosmic rays) act to ionize atmospheric atoms and molecules, and in regions of fine weather, positive *ions move down towards the negatively charged surface, whereas negative ions migrate upwards. There is also a substantial flow of current upwards that accompanies the downward transfer of charge in thunderstorms. *See also* BLUE JET; ELF; GIGANTIC JET; LIGHTNING; ST ELMO'S FIRE; SPRITE.

**atmospheric optics** The various optical phenomena that occur in the atmosphere and arise through *diffraction, *reflection, *refraction, and *scattering. These include *glories, *haloes, *iridescence, *mirages, rainbows (*see feature*), and *scintillation. The study of such phenomena.

**atmospheric pressure** The average sea-level pressure is shown in the figures overleaf. *See also* PRESSURE.

**atmospheric pressure sensor (APS)** An advanced sensor under development to determine surface pressure from space. By monitoring several wavelengths, it measures the quantity of oxygen in the column between the satellite sensor and the surface. This may be directly related to the surface pressure. Initial airborne tests indicate an accuracy of about 1 hPa.

**atmospherics** Natural electrical discharges at radio wavelengths, audible as constantly present background static that fluctuates in strength. The discharges arise from *lightning, and their study (*sferics) may be used to locate currently active *thunderstorms.

**atmospheric tides** Oscillations in the atmosphere caused by the influence of the Sun and Moon, predominantly affecting pressure and the wind, and analogous to the more familiar oceanic tides (*see feature*). Although gravitational effects play a part, heating by the Sun is of greater importance. The predominant fluctuation is a semi-diurnal variation in pressure (*see* DIURNAL VARIATION), but other periods of 24, 6, and 8 hours are also observed. In the *troposphere, the tidal contribution to the wind is negligible, but it increases in the upper atmosphere and may reach 30 $m s^{-1}$ at around 100 km, where it may become the dominant component.

**ATOC** *See* ACOUSTIC THERMOMETRY OF OCEAN CLIMATE.

**A-Train** *See* AFTERNOON-TRAIN.

**attenuation** The reduction in the intensity of solar radiation (or other electromagnetic radiation) through *absorption and *scattering in the Earth's atmosphere. *See also* TURBIDITY.

**Aura** A NASA *polar-orbiting satellite, part of the Earth Observing System, the third to be launched (after the *Terra and *Aqua satellites). Its principal mission is to monitor atmospheric chemistry and dynamics, particularly the anticipated recovery in the *ozone layer and reduction in the *ozone holes. Its instruments are primarily designed to study the distribution and concentration of trace gases and *aerosols in the upper *troposphere and *stratosphere. It is currently last in the *Afternoon-Train satellite constellation. *See also* CALIPSO; CLOUDSAT; ORBITING CARBON OBSERVATORY; PARASOL.

**aureole** The inner portion of a *corona (1) that appears as a bluish-white disk around the Sun or Moon, with a brownish-red outer edge. When the cloud particles causing the *diffraction cover a wide range of sizes, rather than being uniform, the aureole may be the only part of the corona that is visible.

**aurora** [Latin: 'dawn'] A luminous phenomenon occurring in the upper atmosphere at altitudes of between approximately 100 and 1000 km, and generally visible from high geomagnetic latitudes, particularly in the *auroral zone. Infrequent major displays may be seen down as far as the geomagnetic equator.

The emission arises when energetic particles (protons and electrons) from the Sun enter the Earth's magnetosphere and are captured in the *magnetotail. When disturbances occur in the *solar wind or there is an energetic *solar flare, the particles are accelerated from remote regions of the tail towards the Earth, where they precipitate into the upper atmosphere, generally at high latitudes.

Primary or secondary collisions between the electrons and atmospheric *oxygen and *nitrogen excite the latter to higher energy levels, from which they decay with the emission of light. Faint displays appear almost colourless to most observers. In brighter displays, the dominant colour is often the characteristic 'auroral green': a 'forbidden' oxygen emission at 557.7 nm,

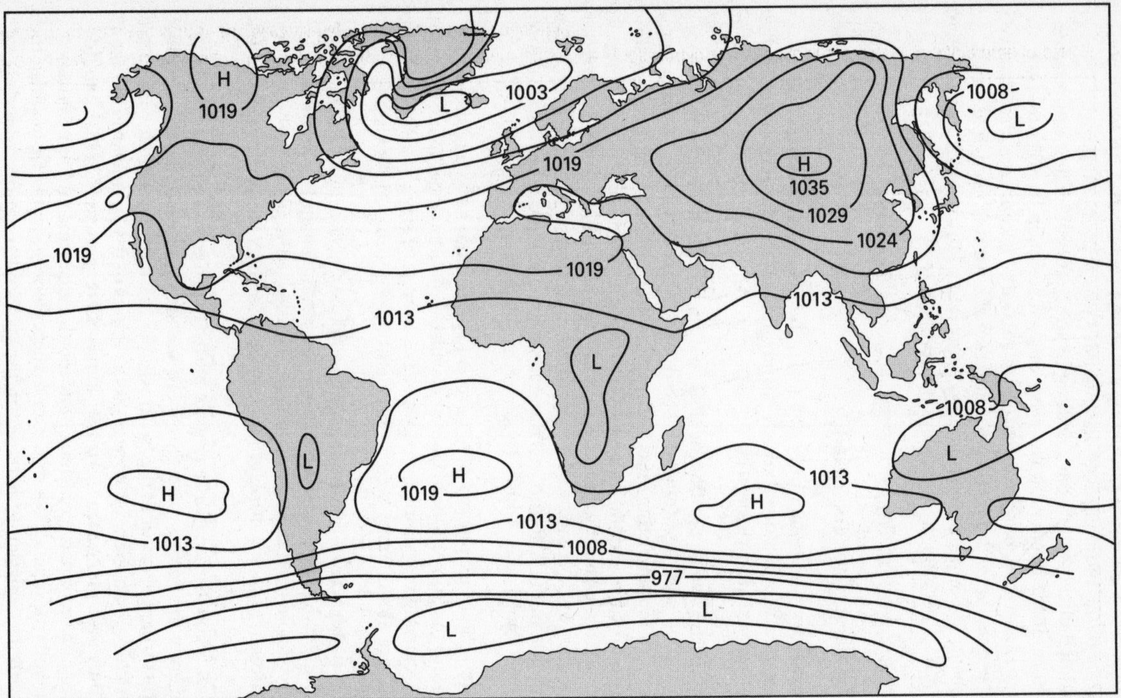


**Atmospheric pressure—January** Dominant features are the *Icelandic and *Aleutian lows, the *Siberian anticyclone, and the semi-permanent low-pressure zone over the southern oceans.

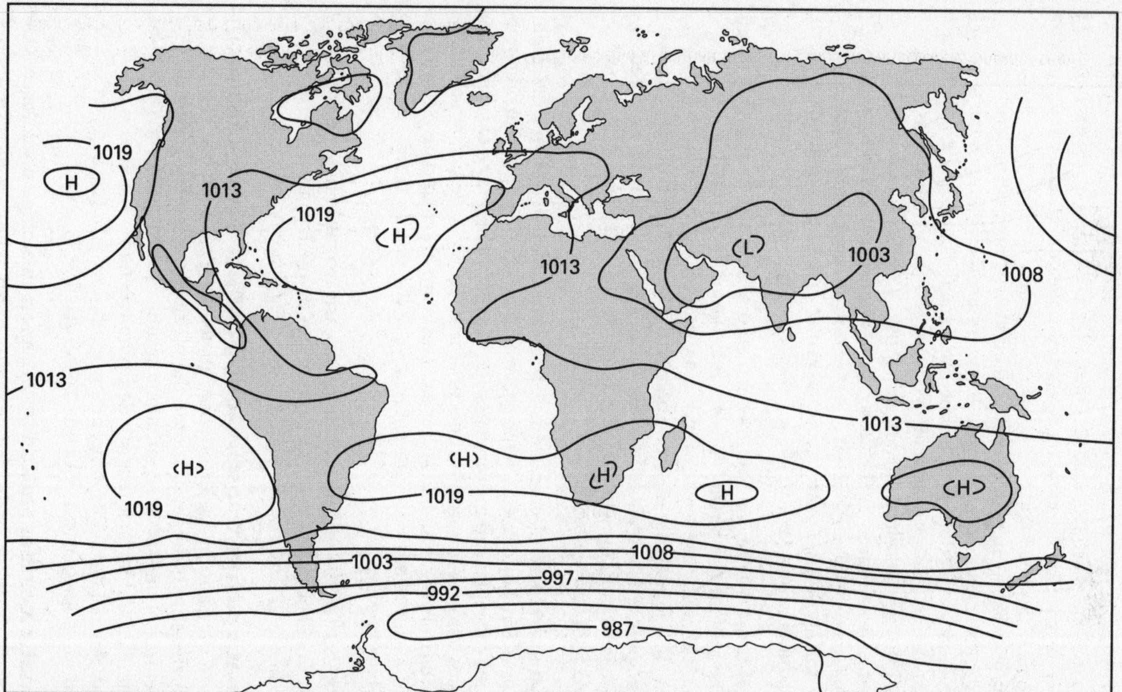


**Atmospheric pressure—June** Dominant features are the *Azores high, the *monsoon low over Asia, and the high-pressure area in the northern Pacific. The shift of the oceanic high-pressure zone is noticeable.

typically at altitudes of 90–150 km. Red oxygen emission (630.0 and 636.4 nm) is often seen at higher altitudes. Auroral *rays that reach 1000 km or more are sometimes in sunlight, when there is bluish-violet emission from ionized nitrogen molecules at 391.4 and 427.8 nm. When energetic electrons penetrate to low levels (*c.*100 km), neutral nitrogen molecules may produce red emission between 661.1 and 686.1 nm.

An auroral display may begin with a faint **glow**, an overall **veil**, or weak **patches** in the sky towards the pole. It commonly then assumes the form of an *arc, and may develop *rays. Strong displays may develop multiple *bands, and if the display moves overhead, the rayed structure may appear as a *corona (3). The intensity may vary slowly over a period of minutes as **pulsations**, more rapidly as **flickering**, or as **flaming**: surges of brightness that sweep upwards from the horizon. *See also* AURORAL SUBSTORM; GEOMAGNETIC STORM; POLAR CAP ABSORPTION; POLAR CUSP AURORA; STABLE AURORAL RED ARC.

**auroral oval** The region within which auroral activity occurs at any one time. The oval form is a reflection of the distortion of the Earth's *magnetosphere by the *solar wind, such that the long axis of each oval lies along the line joining the magnetic pole to the Sun, being farthest from the magnetic pole on the night-side of the Earth. The ovals may be regarded as relatively fixed in space while the Earth rotates beneath them. Not to be confused with *auroral zone.

**auroral substorm** A characteristic sequence of activity when a quiescent *aurora suddenly intensifies. In the **growth phase**, the *auroral oval brightens, and in the following **expansion phase** the activity spreads both eastwards and westwards. Growth towards the pole produces a wavy structure, which moves westwards from the midnight sector as a **westward-travelling surge**, while the morning sector of the auroral oval breaks up. The display fades away in the final **recovery phase**.

**auroral zone** The region, *c.*15–30° from each magnetic pole, with the greatest frequency of aurorae. The maximum frequency of aurorae at the *zenith lies *c.*22.5° from the magnetic poles. *See figure opposite.*

**automatic picture transmission (APT)** The continuous transmission of images of the Earth by *polar-orbiting *meteorological satellites. The *radiometer scan is perpendicular to the plane of the satellite's orbit, so that the image consists of a continuous swathe around the Earth. Data from two channels are normally transmitted simultaneously: visible and one infrared channel during daytime, and two infrared channels at night. *See also* HIGH-RESOLUTION PICTURE TRANSMISSION; WEFAX.

**automatic weather station (AWS)** Any meteorological *station at which all measurements are made by automatic sensors. The data may be stored locally until retrieved on a site visit, transmitted on interrogation or at fixed

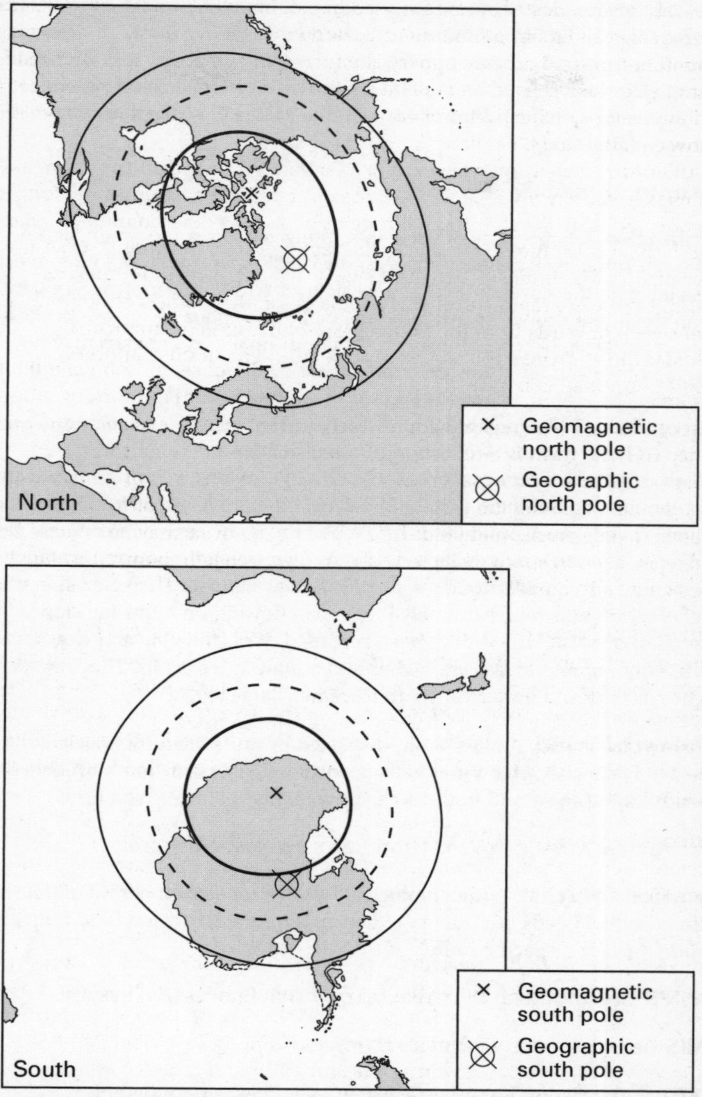


**Auroral zone** The auroral zones form rings, roughly centred on the geomagnetic poles, whose approximate positions are shown. The greatest annual frequency of overhead aurorae occurs around the dashed lines.

intervals, or broadcast continuously. Although initially introduced to obtain meteorological data from remote or hazardous locations (such as exposed mountain-tops and oceanic buoys), the increasing reliability and decreasing cost of electronic sensors mean that stations are now installed to monitor conditions for specific purposes (as with the roadside stations used by highway authorities).

**autumn** *See* SEASON.

**avalanche** A large mass of snow that plunges down a mountain slope. Although some avalanches occur as snow is falling and accumulating, most arise through the development of a weak layer within the snowpack, often when the initial snow crystals *sublime, with the vapour being redeposited as dense, but loose, ice crystals (known as **depth hoar**). Avalanches may be triggered by a sudden thaw, perhaps caused by the onset of *föhn conditions or rainfall, or else by vibrations caused by thunder, earth tremors, or human activity (including explosions and disturbance by skiers). Avalanches often transport large quantities of rock and other debris.

**Slab avalanches** occur when a large slab of a relatively strong surface layer initially moves as a single unit, before breaking up into individual icy blocks. Typical speeds are around $60\,km\,h^{-1}$. When the snow cover is extremely deep and loose, the snow may collapse under its own weight in **powder avalanches**. These may also be initiated by a smaller slab avalanche. They entrain a great deal of air and become extremely fluid, even flowing up opposing slopes. Speeds of over $300\,km\,h^{-1}$ have been recorded. **Wet (thaw) avalanches** mainly occur when temperatures are higher and a thaw is setting in. They have a higher density and may give rise to extremely large slides.

**avalanche wind** A blast of air, displaced by an *avalanche, particularly a powder avalanche, which may cause considerable destruction both ahead of the advancing snow and in the area surrounding the avalanche itself.

**AVHRR** *See* ADVANCED VERY HIGH RESOLUTION RADIOMETER.

**aviation forecast** A meteorological forecast for a given period of time and tailored to the needs of aviation, either for a particular region (and individual airports) or for a specific route.

**AWIPS** *See* ADVANCED WEATHER INTERACTIVE PROCESSING SYSTEM.

**AWS** *See* AUTOMATIC WEATHER STATION.

**AXBT** Abbreviation for airborne expendable *bathythermograph.

**azimuth** In the horizontal coordinate system, the angle, measured clockwise in degrees in the plane of the horizon, between the reference point, true north (0°), and the vertical circle passing through the object in question.

**Azores high (Azores anticyclone)** A semi-permanent region of relatively consistent high pressure in the northern hemisphere. It is a deep *subtropical anticyclone and *centre of action, extending throughout the *troposphere, that is generally located near 30° N in winter (i.e. south of the Azores) with a ridge that extends towards the Iberian peninsula. In summer, the pressure centre lies near 35° N and the ridge often extends over France into northern Germany. In both winter and summer the central pressure is around 1024 hPa. Conditions are more variable in winter and are associated with different states of the *North Atlantic Oscillation. *See also* BERMUDA ANTICYCLONE.

# B

**Babinet's point** *See* POLARIZATION OF THE SKY.

**back** Of the wind: to change in an anticlockwise direction (e.g. south-west to south to south-east), and the opposite of *veer. Note that this also applies to the southern hemisphere. The practice by certain meteorologists and writers (particularly in the United States) to use the term to suggest change in the opposite direction in the southern hemisphere (i.e. clockwise) is incorrect.

**back-bent occlusion (also known as a bent-back occlusion, especially in North America, or occasionally as a wrap-around occlusion)** An *occlusion where the motion of the front reverses, being carried westwards or southwards (in the northern hemisphere) by the flow behind the *cold front. This occurs either with the development of a *secondary depression near the *triple point or through motion of the main low-pressure centre eastwards along the front.

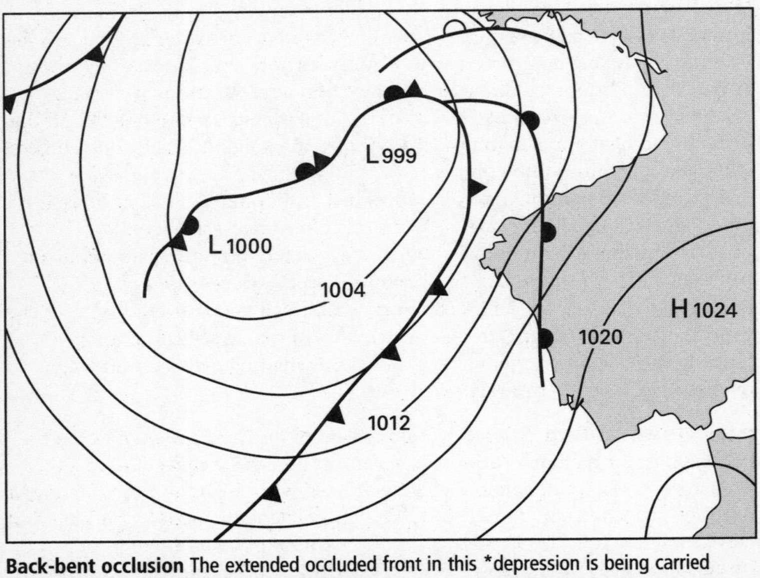


**Back-bent occlusion** The extended occluded front in this *depression is being carried southwards by the flow around the main low-pressure centre (999 hPa), and a secondary centre (1000 hPA) is beginning to develop.

**back-bent warm front** A *warm front that is detached from a *depression's *cold front and partially encircles the low-pressure core. During *cyclogenesis the original frontal zone may fracture near the centre of the depression into two separate zones with strong thermal gradients. The cold front advances eastwards, and the warm front develops westwards. The system appears, superficially, like the form more conventionally described as a *back-bent occlusion, but observations suggest that the revised model more satisfactorily accounts for features found in many depressions.

**back-building thunderstorm** A *thunderstorm in which new cells develop on the upwind rather than the more normal downwind side, i.e. to the west or south-west (in the northern hemisphere). Because of this behaviour, the storm may appear to be stationary, or even to develop 'backwards', that is, upwind.

**backdoor cold front** On the East Coast of the United States, a *cold front that moves from the north-east, rather than in the normal direction from the west or south-west.

**bad-i-sad-o-bist-roz** *See* SEISTAN.

**baguio** A local, Philippine name for a tropical cyclone (*see feature*), more generally known in the region as a *typhoon.

**ball lightning** A rare form of *lightning that appears in the form of a luminous sphere a few centimetres across that may persist for as long as a few minutes. It has been observed (mainly immediately after a nearby lightning strike) both outdoors and inside buildings and aircraft, and may remain relatively stationary or move (slowly or rapidly) in mid-air or along the ground or floor. No heat appears to be radiated from the ball, but it may burn objects with which it comes into contact.

Although ball lightning is often dismissed as an optical illusion, there is a great body of anecdotal evidence from reliable witnesses. Scientific observations have been hard to obtain and, despite numerous attempts, no truly satisfactory explanation has been proposed. A recent theory, yet unconfirmed, is that a discharge hitting suitable soil may turn silica into silicon vapour, which on cooling creates an *aerosol of minute silicon particles. If these particles clump together into a ball, this might glow as the silicon oxidizes. *See also* FULGURITE.

**balloon sounding** Securing observations of the atmosphere by use of a lighter-than-air balloon. Various forms are in use, the simplest being an uninstrumented *pilot balloon, usually inflated with *hydrogen, which may be used for the determination of cloud height. Tracking of a similar balloon that carries a radar reflector is used to determine upper winds. Ascents by *radiosonde provide measurements of pressure, temperature, and humidity, or (in an *ozonesonde) *ozone concentration. The largest balloons, inflated with *helium, are used for observations of the Sun or *cosmic rays. *See also* CONSTANT-LEVEL BALLOON; GLOBAL HORIZONTAL SOUNDING TECHNIQUE; ROCKOON; ULTRA-LONG DURATION BALLOON.

**band** **1.** A specific form of *aurora that consists of a folded, ribbon-like structure, and frequently associated with auroral *rays. A band often develops from an auroral *arc as the display intensifies, and multiple bands are commonly seen. Often known as an **auroral curtain.**

**2.** A form of *noctilucent cloud (classified as Type II), consisting of long streaks of cloud of a generally fibrous structure. A common form, often observed simultaneously with *billows (2).

**banner cloud** A cloud that is restricted to the area immediately downwind of an abrupt, isolated mountain peak, and which resembles a banner or flag. Two mechanisms are possibly involved, the principal one being the presence of a *lee eddy, which raises air behind the peak above the *condensation level. A reduction in pressure behind the obstacle would also facilitate condensation, similar to the way *condensation trails form in the central, low-pressure core of aircraft wing-tip vortices. The most famous banner cloud is that associated with the Matterhorn.

**bar** A c.g.s. unit of atmospheric *pressure equal to $10^6$ dynes $cm^{-2}$, and generally encountered in the form of its submultiple the millibar ($1\,mb = 10^{-3}$ bar). For scientific work, it has now been superseded by the SI unit the *pascal (Pa), and its convenient multiple the hectopascal ($1\,hPa = 10^2\,Pa = 1\,mb$). One bar is defined as being equal to the pressure of 750.062 mm of mercury at 0 °C under a gravitational acceleration of $9.80665\,m\,s^{-2}$ (i.e. one *standard gravity).

**barat** A squally west or north-west wind that occurs between December and February on the northern coast of the Philippine island of Celebes.

**barb** *See* WIND ARROW.

**barber** A wind that is carrying *sleet, *snow, or spray, when the air temperature is close to freezing. Named for the *precipitation's tendency to freeze on to the beard and hair.

**baroclinic** Applied to an atmosphere or *air mass in which a temperature (and thus density) gradient exists at a specific *isobaric surface, i.e. at a surface of constant pressure. Baroclinic zones therefore exist across *fronts, where pressures are equal but temperatures are not. *See also* BAROTROPIC.

**baroclinicity (baroclinity)** A measure of the degree to which a fluid such as the atmosphere is stratified, namely the extent to which *isobaric (constant-pressure) surfaces intersect constant-density surfaces.

**barogram** The continuous record of pressure variations produced by a *barograph, normally in the form of a paper chart.

**barograph** An instrument that provides a continuous record of atmospheric *pressure, most commonly in the form of an *aneroid barometer with a capsule (or stack of capsules) that acts to move a pen across a suitably driven paper chart.

**barometer** An instrument for determining atmospheric *pressure. There are four general types: mercurial, aneroid, hypsometric, and solid-state (piezoresistive or capacitance). The original mercurial design in which the weight of the column of air is balanced by a column of mercury is still used for *synoptic or *climatological purposes, the **Kew-pattern** and **Fortin** types being widely encountered. *Aneroid barometers, including the *precision aneroid type, make use of one or more evacuated capsules that deform with changes in pressure.

The hypsometric barometer (or *hypsometer) determines pressure by measuring the water vapour temperature (*boiling point). It was particularly useful for high-altitude work, where other forms were less accurate. Increasing use is now made of robust solid-state sensors, either the **piezoresistive** type, in which electrical resistance changes in response to pressure, or the **capacitance** form, in which flexure of a silicon diaphragm changes the device's capacitance. Versions of the latter exist for the complete pressure range, from surface measurements to use in *radiosondes, and this form has many other advantages, not least its accuracy and repeatability.

**barometric tendency** *See* PRESSURE TENDENCY.

**barothermograph** An instrument for simultaneously recording pressure and temperature on one chart.

**barotropic** Of an atmosphere or *air mass in which the surfaces of constant pressure and constant density (or constant temperature) do not intersect. Such theoretically ideal conditions (known as zero *baroclinicity) are a useful starting point for models and calculations. *Compare* BAROCLINIC.

**barotropic leaf** *See* CLOUD LEAF.

**barrage cloud** A band of thick *orographic cloud that arises when a moist airstream rises over a mountain barrier. This generally produces heavy, long-lasting *precipitation on the windward slopes, and may create *föhn conditions to leeward.

**barrier jet** A narrow band of strong winds resulting from an *orographic obstruction such as that presented by a range of mountains. Such a jet often arises when a surge of polar air encounters a mountain barrier. Unlike a *jet stream it is located near the surface, not near the *tropopause.

**bathythermograph** An oceanographic instrument for obtaining a record of temperature versus depth. A version dropped from aircraft, the **airborne**

**expendable bathythermograph** (AXBT), determines temperatures down to 300m.

**bead(ed) lightning (chain lightning, pearl-necklace lightning)** A very rare form of *lightning in which variations in brightness along the discharge channel produce a transient appearance similar to beads on a string.

**Beaufort, Rear-Admiral Sir Francis** (b. Navan, Ireland, 7 May 1774; d. Greenwich, 17 December 1857) British naval officer and hydrographer, who in 1806, as Commander, published a method of measuring the wind at sea, initially based on the effects upon a frigate (such as the specific sails that could be carried). The *Beaufort scale was eventually adopted by the Royal Navy in 1838. Beaufort became head of the Admiralty's Hydrographic Office in 1829, and transformed the office into the world's foremost hydrographic organization. He also introduced a set of letters for reporting the state of past or present weather. *See* BEAUFORT LETTERS.

**Beaufort letters** A standardized system of code letters for the logging and reporting of past and current weather, as well as significant atmospheric phenomena. Introduced by Admiral *Beaufort, but subsequently greatly altered and extended in scope.

**Beaufort scale** A standard scale, running from force 0 to force 12, for the description of wind speed, first introduced by Admiral *Beaufort. The original scale applied to conditions at sea, and was described in terms of the effect upon a frigate of the period. It was subsequently modified for use on land, and all the descriptions were generalized to be universally applicable.

**The Beaufort scale (for use at sea)**

| FORCE | DESCRIPTION | SEA STATE | SPEED | |
|---|---|---|---|---|
| | | | KNOTS | $m\ s^{-1}$ |
| 0 | calm | like a mirror | <1 | 0.0–0.2 |
| 1 | light air | ripples, no foam | 1–3 | 0.3–1.5 |
| 2 | light breeze | small wavelets, smooth crests | 4–6 | 1.6–3.3 |
| 3 | gentle breeze | large wavelets, some crests break, a few white horses | 7–10 | 3.4–5.4 |
| 4 | moderate breeze | small waves, frequent white horses | 11–16 | 5.5–7.9 |
| 5 | fresh breeze | moderate, fairly long waves, many white horses, some spray | 17–21 | 8.0–10.7 |
| 6 | strong breeze | some large waves, extensive white foaming crests, some spray | 22–27 | 10.8–13.8 |
| 7 | near gale | sea heaping up, streaks of foam blowing in the wind | 28–33 | 13.9–17.1 |

| | | | | |
|---|---|---|---|---|
| 8 | gale | fairly long and high waves, crests breaking into spindrift, foam in prominent streaks | 34–40 | 17.2–20.7 |
| 9 | strong gale | high waves, dense foam in wind, wave-crests topple and roll over, spray interferes with visibility | 41–47 | 20.8–24.4 |
| 10 | storm | very high waves with overhanging crests, dense blowing foam, sea appears white, heavy tumbling sea, poor visibility | 48–55 | 24.5–28.4 |
| 11 | violent storm | exceptionally high waves may hide small ships, sea covered in long, white patches of foam, waves blown into froth, poor visibility | 56–63 | 28.5–32.6 |
| 12 | hurricane | air filled with foam and spray, visibility extremely bad | ≥64 | ≥32.7 |

## The Beaufort scale (adapted for use on land)

| FORCE | DESCRIPTION | EVENTS ON LAND | SPEED | |
|---|---|---|---|---|
| | | | $km\,h^{-1}$ | $m\,s^{-1}$ |
| 0 | calm | smoke rises vertically | <1 | 0.0–0.2 |
| 1 | light air | direction of wind shown by smoke but not by wind vane | 1–5 | 0.3–1.5 |
| 2 | light breeze | wind felt on face, leaves rustle, wind vane turns to wind | 6–11 | 1.6–3.3 |
| 3 | gentle breeze | leaves and small twigs in motion, wind spreads small flags | 12–19 | 3.4–5.4 |
| 4 | moderate breeze | wind raises dust and loose paper, small branches move | 20–29 | 5.5–7.9 |
| 5 | fresh breeze | small leafy trees start to sway, wavelets with crests on inland waters | 30–39 | 8.0–10.7 |
| 6 | strong breeze | large branches in motion, whistling in telephone wires, difficult to use umbrellas | 40–50 | 10.8–13.8 |
| 7 | near gale | whole trees in motion, difficult to walk against wind | 51–61 | 13.9–17.1 |
| 8 | gale | twigs break from trees, difficult to walk | 62–74 | 17.2–20.7 |
| 9 | strong gale | slight structural damage to buildings; chimney pots, tiles, and aerials removed | 75–87 | 20.8–24.4 |
| 10 | storm | trees uprooted, considerable damage to buildings | 88–10 | 24.5–28.4 |

**The Beaufort scale (adapted for use on land) Cont.**

| FORCE | DESCRIPTION | EVENTS ON LAND | SPEED | |
|---|---|---|---|---|
| | | | $km h^{-1}$ | $m s^{-1}$ |
| 11 | violent storm | widespread damage to all types of building | 102–117 | 28.5–32.6 |
| 12 | hurricane | widespread destruction, only specially constructed buildings survive | ≥118 | ≥32.7 |

**Benguela Current** The fast, cold oceanic current that forms part of the South Atlantic *gyre and flows north along the west coast of southern Africa. Because of the *Ekman effect, *upwelling occurs from depths of 100–300 m. As the Benguela Current nears the Equator, it turns westward and becomes the *South Equatorial Current.

**Bergeron, Tor Harald Percival** (b. Godstone, England, 15 August 1891; d. Stockholm, 13 June 1977) Swedish meteorologist, particularly noted for his work in *cloud physics. He was a student of and collaborated with Vilhelm *Bjerknes. He introduced the concept of *air masses in 1928, studied frontal theory and the influence of upper-atmospheric processes on climate, and proposed the theory of *precipitation later known as the *Bergeron process.

**Bergeron process (Bergeron-Findeisen process, ice-crystal theory)** The process by which *precipitation is initiated in a *mixed cloud with a temperature below freezing. Because the equilibrium *vapour pressure of water vapour with respect to ice is less than that with respect to liquid water, ice crystals grow at the expense of *supercooled water droplets. Eventually the crystals become sufficiently heavy to begin to fall, and may subsequently be modified by other processes such as *accretion, melting, *coalescence, and *evaporation. The theory was introduced by *Bergeron in 1933 and subsequently extended by W. Findeisen.

**berg wind** A hot, dry *föhn wind originating over the high South African plateau. It is particularly common during the winter season, and especially affects the western coastal region.

**bergy bit** The remnant of a large *iceberg, some of which remains visible above the sea's surface. *Compare* GROWLER.

**Bermuda anticyclone (Bermuda high)** A semi-permanent region of high pressure (a *subtropical anticyclone) located over the western North Atlantic Ocean at approximately 30° N. It is a westward extension of the *Azores high.

**billows** **1.** Approximately parallel bands of cloud that lie perpendicular to the wind direction, with relatively narrow spaces between them. *Compare* CORRUGATIONS. **2.** A form of *noctilucent cloud (classified as Type III) consisting of approximately parallel, regularly spaced streaks of cloud. Sometimes observed in combination with *bands (2), and often exhibiting relatively rapid changes over a period of several minutes or tens of minutes.

**bioclimatology** The study of the effects of *climate on living organisms.

**biosphere** The portion of the Earth that is inhabited by living organisms of any kind. Apart from the *hydrosphere, the surface of the land and the immediate subsoil, the lower atmosphere (primarily the *troposphere), and large parts of the *cryosphere, it also includes subsurface rock environments in which bacteria exist.

**bise** A cold north or north-easterly wind, blowing from the mountains, that affects western Switzerland and eastern France. Although normally relatively dry, in winter it may bring heavy snow or hail.

**Bishop's ring** A pale disk of light surrounding the Sun, with a faint reddish-brown outer zone having inner and outer boundaries of about 10° and 20° when the Sun is highest. The radii increase at lower Sun elevations. Other colours (notably a bluish tint) have been reported but may arise through observers' different colour perceptions. First reported after the eruption of Krakatau in 1883, and observed after certain subsequent volcanic eruptions, the phenomenon is thought to be caused by *diffraction of sunlight by fine dust or *sulphur dioxide particles in the upper atmosphere.

**Bjerknes, Jacob Aall Bonnevie** (b. Stockholm, 2 November 1897; d. Los Angeles, 7 July 1975) Norwegian-American meteorologist and son of Vilhelm *Bjerknes. Collaborated with his father on the theory of frontal systems, and later made considerable advances in understanding the structure and dynamics of *depressions and other atmospheric systems. After moving to the United States in 1939, he specialized in the study of the general atmospheric circulation, identifying and naming the *Walker Circulation. He was one of the first to use photographs from research rockets for the analysis of weather systems and forecasting.

**Bjerknes, Vilhelm Friman Koren** (b. Christiania, Norway, 14 March 1862; d. Oslo, 9 April 1951) Norwegian meteorologist and physicist. After early work that was later of great significance in the development of radio, Bjerknes discovered mathematical tools applicable to hydrodynamics and thermodynamics. He applied these to large-scale motions in the atmosphere and oceans. During World War I, he set up a network of meteorological stations in Norway, and data from these stations, together with his theoretical studies, led to the concept of *fronts and *air masses. The founder of the so-called **Bergen** or **Norwegian school** of meteorologists. *See also* BERGERON.

**black-body radiation** The radiation emitted by a hypothetical black body, i.e. one that is capable of both absorbing all radiation incident upon it and radiating the maximum possible intensity of radiation for a given temperature.

**black-bulb thermometer** A mercurial, maximum thermometer with a blackened bulb enclosed in an evacuated outer bulb, once used for estimating solar radiation. Because of its low accuracy, it has been superseded by various forms of *pyranometer. Also known as a **solar radiation thermometer**.

**black frost** A frost that occurs without the formation of *hoar-frost crystals (i.e. when the atmosphere is dry) and which produces extensive damage and blackening of plants (hence the name). Without a covering of insulating hoar-frost (and the release of *latent heat that accompanies the freezing of water), plants are particularly susceptible to damage.

**black ice** **1.** A popular name for *glaze, often used in connection with its occurrence on roads, where it may form a significant hazard to traffic. **2.** Thin ice on fresh or salt water that is transparent and thus appears dark.

**black snow** Snow contaminated with soot, particularly that released by the extensive burning of coal, often at a considerable distance from the site of snowfall. Prevalent in Scandinavia during the 19th and early 20th centuries as a result of coal-burning in Great Britain.

**blizzard** A very cold, strong wind (force 7 on the *Beaufort scale or greater) that is laden with large amounts of *blowing snow raised from the surface. In the United Kingdom (and in popular usage in North America) the term is used for any heavy fall of snow accompanied by high winds.

**blocking** An interruption of the normal westerly *zonal flow, where there is a significant interruption to the normal eastward progression of *anticyclones, *depressions, and individual *ridges and *troughs, usually for several days, but occasionally for much longer periods. A significant *meridional flow develops, initially at upper levels, with an upper-level ridge diverting the *jet stream to create an *omega high, and eventually resulting in the formation of *cut-off highs (*anticyclonic circulation) poleward of the original flow, and *cut-off lows (*cyclonic circulation) on the equatorial side. The whole **block** generally extends slowly westwards. *See also* BLOCKING ANTICYCLONE; ZONAL INDEX.

**blocking anticyclone (blocking high)** A deep *anticyclone that extends throughout the depth of the *troposphere and tends to obstruct the normal passage of *depressions from west to east. The situation is less persistent than with large-scale *blocking (*See figure opposite.*) *See* OMEGA HIGH.

**blocking region** A region in which *blocking exists or frequently occurs. In the northern hemisphere, blocking regions are found in the eastern portions of the North Atlantic and North Pacific. Western Europe is particularly affected when blocking occurs over Scandinavia.

**blowing dust/sand** A term used in *synoptic meteorology to indicate that dust or sand is being transported by the wind, at heights greater than 1.5 m, but that visibility is greater than 1000 m. *See* DRIFTING DUST/SAND; DUST STORM.

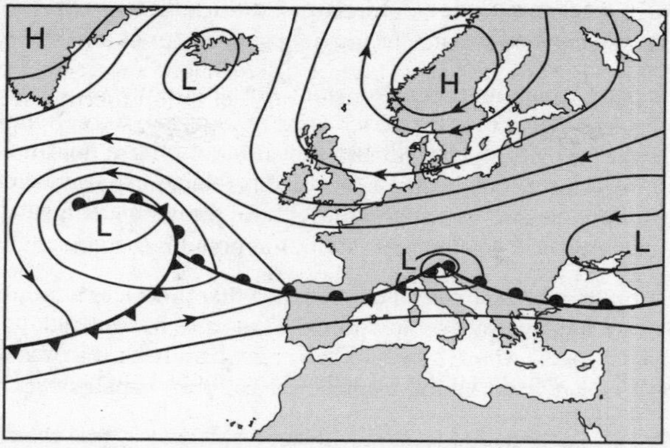


**Blocking anticyclone** A typical blocking anticyclone over Scandinavia during winter that brings an easterly flow of very cold air over western Europe. The arrows indicate the *geostrophic flow, rather than that at the surface.

**blowing snow** Loose snow that is being transported by the wind and has been raised to a height of 2 m or more (*compare* DRIFTING SNOW) in sufficient quantities to restrict horizontal visibility.

**blue jet** A faint pulse of blue light that arises from the top of a *thunderstorm and propagates upwards in the form of a narrow cone to altitudes of 40–50 km (at about $100 \text{km s}^{-1}$). Blue jets therefore extend approximately from bottom to top of the *stratosphere. They were first observed scientifically in the early 1990s, but the mechanism by which they are formed is currently unknown. They are much rarer than the higher *sprites and *elves, to which they are apparently unrelated. They are not detectable from the ground, because of strong *scattering of their blue light by the lower atmosphere. *See also* GIGANTIC JET.

**blue moon (blue sun)** The rare phenomenon when the Moon or Sun appears blue, caused by preferential *scattering of long-wavelength light by particles in the atmosphere so that only blue remains. Particles of the requisite size (0.35–0.65 μm) appear to be created most prolifically by forest fires, although wind-blown loess and *volcanic dust sometimes create the same effect. A strong green colour has also been reported on similar occasions.

**blue sky** The colour of the sky arises from *Rayleigh scattering of incident radiation from the Sun by gaseous molecules (primarily nitrogen and oxygen) in the atmosphere. Scattering is strongly wavelength dependent: the short wavelengths of blue and violet light are strongly deflected, whereas the longer-wavelength red, orange, and yellow are largely unaffected. Blue light reaching an observer has been subject to multiple scattering and is therefore classed as **diffuse radiation**. In the absence of larger solid particles or water droplets, the blue is intense, but when they are present, *Mie scattering of longer wavelengths introduces white light, causing a paler tint of blue.

**boiling point** The temperature at which the saturation *vapour pressure of a liquid becomes equal to the external atmospheric *pressure. Bubbles (the gaseous phase) form in the liquid, and the addition of heat does not cause a rise in temperature. Because the boiling point is dependent on pressure, it is normally quoted for one *standard atmospheric pressure (1013.25 hPa). For water it decreases by 3.3 deg C per 1000 m increase in altitude.

**bolide** A brilliant *meteor (*see* FIREBALL (2)) that produces a terminal explosion, and may cause a sonic boom that reaches the ground.

**bolometer** An instrument for measuring the intensity of incident radiation. *See also* SPECTROBOLOMETER.

**bolster eddy** An *eddy that sometimes forms to *windward of a steep rise, such as a vertical cliff, and which gives rise to a surface wind blowing away from the obstruction.

**bolt from the blue** A *lightning stroke that appears to come from a clear sky. In reality, such strokes originate in a storm system, but may strike some kilometres away from the main storm cell or cells, occasionally well outside the area covered by cloud. Tornadic *low-precipitation supercell systems are particularly prone to produce lightning strokes that propagate from the overhanging *anvil through the clear (i.e. precipitation-free) air beneath it to the ground.

**bomb (maritime bomb)** An *extratropical cyclone that deepens extremely rapidly, with the central pressure dropping by an average of at least 1 hPa per hour for 24 hours. Bombs occur where there is an *upper-level trough and great temperature contrast between a pool of warm oceanic water and its surroundings. Such conditions often exist in the west-central North Atlantic and on the eastern seaboard of Australia, where they are also known as **east coast cyclones** or **east coast lows**. Bombs obtain much of their energy from the release of *latent heat and thus bear some similarity to tropical cyclones (*see feature*).

**bora** A cold, dry downslope wind (often with violent *squalls) that affects the Dalmatian coast when there is low pressure over the Mediterranean and high pressure over Central Europe and the Balkans. Strongest in winter, it may be accompanied by heavy cloud, rain, or snow when a *depression lies over the Adriatic. The term is sometimes applied to similar *fall winds in other parts of the world.

**bore** A large body of water that advances up an estuary or river in the form of an abrupt wave or waves. Water level behind the bore may be far higher than in front. Known technically as a **hydraulic jump**, a bore is normally caused by the interaction of a rising (flood) tide (*see feature*) with the flow of water in a river, but may also be created by a *tsunami. The shape of the estuary and the width and depth of the river are crucial elements in determining whether a bore occurs. Rivers with notable bores are the Quiantong Jiang (Fu-ch'un Jiang) in China (with a recorded rise of as much as 7.5m at spring tides), the Para (part of the Amazon system; 4m), and the Severn (1.5m). A similar hydraulic jump in the atmosphere is responsible for the *morning glory, and a lesser form has recently been recognized as being quite widespread and has been named an *undular bore.

**boreal climate** A climatic regime characterized by snowy winters and warm summers, exhibiting a wide annual range of temperatures. It is found over large areas of the continental interiors of Asia, Europe, and North America. *See* Köppen classification.

**bottom water** The water found at the greatest depths in an oceanic water column, and distinguished by its physical properties such as *salinity and oxygen content. Depending on the topography of the ocean floor, this water may take part in the general circulation (such as the *Great Ocean Conveyor Belt), or may be relatively isolated from other parts of the oceans.

**Bouguer's halo (Ulloa's circle)** A rare optical phenomenon, consisting of a white circle or arc, centred on the *antisolar point, and with a reported radius of approximately 33.5°. It is not known whether this is a true *halo effect (i.e. through refraction or reflection by ice crystals) or is a misidentified *fogbow. Jean Bouguer and Juan de Ulloa were both members of La Condamine's expedition to Peru in 1735, during which they observed the effect.

**boundary current** A distinct current on either the western or eastern side of an oceanic basin. Western boundary currents are relatively narrow and rapid. They include the *Agulhas, *Brazil, *Gulf Stream, and *Kuro Shio Currents. Their poleward motion contributes anticyclonic *vorticity that reinforces the effects of the prevailing winds, resulting in their being considerably stronger and faster than the corresponding eastern boundary currents. The latter tend to be shallow, broader, and slower and because of the *Ekman effect are deflected away from the coast, producing upwelling of cold subsurface water. They include the *Benguela, *California, *Canary, and *Peru Currents.

**boundary layer** The thin layer of fluid (i.e. gas or liquid) in contact with a surface where adhesion causes a different form of flow than that in the freely moving fluid. In meteorology, a distinction is drawn between the *planetary boundary layer and the *surface boundary layer.

**bounded weak echo region (BWER)** A localized region within a *thunderstorm which shows low radar reflectivity, also known as a **vault**. Invisible except to radar, the BWER corresponds to the region with the most intense *updraught. Although *precipitation exists within the rising air, the velocity is so high that the particles have insufficient time to grow to large sizes, thus producing a weak radar echo.

**bow echo** A radar echo in the form of an arc, often some 100km in length. Damaging straight-line winds (*see* DERECHO) sometimes occur along the central region of such an echo, and the *cyclonic circulation at the northern end (in the northern hemisphere) may promote the development of tornadoes (*see feature*).

**Brandes, Heinrich Wilhelm** (b. Groden, Germany, 27 July 1777; d. Leipzig, 17 May 1834) German astronomer and meteorologist, who is generally credited with being the first to produce weather charts depicting the distribution of pressure with *isobars, and with recognizing the inflow of air to low-pressure centres. However, although he is said to have drawn maps for every day of the year 1783, none has apparently survived. With Benzenberg, he was the first to determine the height of *meteors, and later studied meteorological optics.

**Brazil Current** A warm, saline oceanic current that forms part of the South Atlantic *gyre. It is an extension of the South Atlantic Equatorial Current and flows south along the coast of Brazil. It later merges with the *Falkland Current to become the *South Atlantic Current.

**breakaway depression** A *warm-front wave or *warm occlusion depression that moves away from the parent depression in a generally eastward direction.

**breeze** **1.** In general usage, any light wind. **2.** On the *Beaufort scale, winds of *force 2–7.

**Brewster's point** *See* POLARIZATION OF THE SKY.

**brickfielder** A strong, hot, dry, and dusty wind in southern Australia (particularly New South Wales) which precedes the passage of a *depression.

**bright band (also known as the melting band)** In *radar meteorology, a narrow band of strongly enhanced echoes produced just below the level where ice crystals are beginning to melt (the *melting level). The brightness arises from the high reflectivity of ice or snow particles that are surrounded by a film of liquid water.

**brightness temperature** The apparent temperature of an object (such as a portion of the Earth's surface or cloud cover), based on the assumption that the object in question radiates as a *black body, i.e. that its brightness may be related to its temperature. In infrared images of the Earth, the radiation detected by a *radiometer may be used directly as an indicator of temperature, or converted to a greyscale value to give an image in which hot surfaces (such as daytime deserts and tropical oceans) are generally shown as black, and cold cloud tops as white.

**Brocken spectre** The apparently greatly magnified shadow of an observer cast on to the surrounding mist. The effect is an illusion. Depth perception is altered by the mist, causing the shadow to appear more distant and to be interpreted as larger than normally expected. Named after the Brocken, a peak in the Harz Mountains in Germany, with which the effect is associated.

**broken cloud** Cloud that covers 60–90 per cent of the sky. *See also* OVERCAST.

**brush discharge** *See* CORONA DISCHARGE.

**bubble high** A *mesoscale area of high pressure (i.e. one 80–250 km across), usually associated with the cold *downdraught from one or more thunderstorms. A *gust front may separate the bubble high from the surrounding warmer air, and this may develop into a *squall line if the bubble is overrun by unstable air.

**Buchan spell** A period of the year during which it was suggested that particularly cold or warm conditions might be expected in Britain. Nine such periods were proposed by Alexander Buchan in 1867, but subsequent statistical analysis has shown that there is little foundation in fact for the proposal.

**bull's-eye squall** A violent *squall that occurs in otherwise fine weather off southern Africa. In its initial stages the circulation of *updraughts and *downdraughts as a *vortex is largely invisible except for a small cap of cloud with an odd-looking structure, from which the squall derives its name.

**buoyancy** The vertical force acting upon a system (an object or air parcel) as a result of a difference in density between the system and its surroundings. The force is defined as being positive when acting upwards, and negative downwards. In meteorology, the density difference usually arises because of a temperature difference between an air parcel and the surrounding environment, and is the source of the motion that occurs in *convection.

**buran** A strong north-easterly wind in Siberia and Central Asia, particularly the extreme *blizzard conditions that occur in winter. Over the *tundra it is also known as *purga.

**burga** A severe north-easterly wind in Alaska that brings *snow or *ice pellets. *Compare* BURAN.

**butterfly effect** A term commonly interpreted to mean that small disturbances in the atmosphere may become amplified to give rise to large (or even catastrophic) effects, as might be suggested by *chaos theory. It is derived from the title of a paper, '*Predictability: Does the Flap of a Butterfly's Wings in Brazil Set off a Tornado in Texas?*', given in 1972 by Edward Lorenz, whose theme was whether the atmosphere is unstable to small perturbations. Lorenz concluded, not that small changes would necessarily lead to large effects, but that the question is currently unanswerable and will probably remain so for the foreseeable future. Errors in weather forecasting are inevitable and come from inadequate data coverage, incomplete knowledge of the underlying physics, and the fact that approximations cannot be avoided in formulating the equations used for human or computer predictions. The term—formally known as 'sensitive dependence on initial conditioning'—should be used to suggest that there are limits to predictability, not that such an effect exists.

**Buys Ballot, Christopher Hendrik Diederik** (b. Kloetinge, the Netherlands, 10 October 1817; d. Utrecht, 3 February 1890) Dutch meteorologist, best known for *Buys Ballot's law, who founded the Netherlands Meteorological Institute.

**Buys Ballot's law** A rule, first expressed on an empirical basis by *Buys Ballot in 1857: If, in the northern hemisphere, one faces downwind, low pressure is on the left, and high on the right. The opposite applies in the southern hemisphere. This law, describing the flow of an idealized *geostrophic wind along the *isobars, had earlier been derived theoretically by *Ferrel. For the *surface wind, the low-pressure centre will normally be 10–50° further forward, rather than directly to the left, the amount depending on the nature of the surface and the resulting *friction.

**BWER** *See* BOUNDED WEAK ECHO REGION.

# C

**California Current** The cold ocean current (a branch of the *Aleutian Current) that flows south along the West Coast of North America. Because of the *Ekman effect it tends to be deflected to the west away from the coast, causing *upwelling of cold water from intermediate depths. This is particularly strong between March and July.

**Calipso** The Cloud-Aerosol Lidar and Infrared Pathfinder Satellite Observation satellite is a joint project by NASA and the French Centre National d'Études Spatiales (CNES) to study the role that clouds (particularly thin clouds) and *aerosols play in determining weather, climate, and air quality. It has an active *lidar system, and passive infrared and visible imaging systems. The satellite is in a polar orbit and is part of the *Afternoon-Train constellation. Its observations are designed to complement those made by *Cloudsat (launched by the same rocket on 26 April 2006), and the two satellites are positioned so that the interval between them is no more than 15 seconds, thus providing essentially simultaneous observations of the same region. *See also* AQUA; AURA; ORBITING CARBON OBSERVATORY; PARASOL.

**calm** An absence (or near-absence) of motion of the air. Force 0 on the *Beaufort scale is reported when smoke rises vertically and water appears as flat as glass. The wind speed must not exceed 1 knot.

**calvus (cal)** [Latin: 'bald'] A *cloud species. A *cumulonimbus cloud in which the tops of certain towers have started to lose their hard *cumuliform outline, but *cirriform characteristics are not yet readily apparent. Some striations may be evident. This stage indicates that *glaciation has begun in the top of the cloud.

**camanchaca** *See* GARÚA.

**Campbell–Stokes recorder** A form of *sunshine recorder in which a glass sphere focuses sunlight on to a card (changed daily), leaving a burnt trace that may be measured to determine the duration of sunshine.

**Canary Current** The portion of the North Atlantic *gyre that flows south along the eastern side of the ocean, carrying cool water towards the Equator. It then turns west to become part of the *North Equatorial Current.

**canicula** *See* VERANILLO.

**cap cloud** A *stratiform *orographic cloud that remains essentially stationary either above or actually enshrouding an isolated hill or mountain. The term is also occasionally used for *pileus.

**CAPE** *See* CONVECTIVE AVAILABLE POTENTIAL ENERGY.

**capillatus (cap)** [Latin: 'hairy'] A *cloud species. A *cumulonimbus in which the tops of the cloud have developed into *cirrus, with a distinct fibrous or striated structure. The cirrus may be in the form of *virga, or appear as a massive plume or flattened anvil (*incus).

**CAPPI** *See* CONSTANT ALTITUDE PLAN POSITION INDICATOR.

**capping inversion** *See* INVERSION.

**carbon capture and storage (CCS)** *See* CARBON SEQUESTRATION.

**carbon credit** *See feature* CARBON EMISSION TRADING.

**carbon cycle** The set of processes by which carbon is exchanged between the various global reservoirs: sedimentary rocks, the *atmosphere, *hydrosphere, and *biosphere. Carbon in the atmosphere is predominantly in the form of *carbon dioxide, and its increasing level (largely as a result of the burning of fossil fuels) is the source of global warming and an important factor in climatic change (*see features*). (Other contributions come from *carbon monoxide and *methane.) The carbon dioxide content of the oceans is *c.*60 times that of the atmosphere, and is largely determined by water temperature and acidity. There is a general release in tropical oceans and a sink in polar seas, with a turnover time of *c.*5 years.

The largest reservoir consists of sedimentary rocks, such as limestone, chalk, marble, and dolomite, and contains *c.*600 times as much carbon as the atmosphere and oceans combined.

**carbon dating** The determination of the age of organic material from the amount of radioactive $^{14}C$ (carbon-14) present. All living organisms incorporate the three carbon *isotopes into their cells, but assimilation ceases when the organism dies, after which the $^{14}C$ decays to $^{14}N$, changing the ratios between the isotopes. From the known original ratios (particularly the $^{14}C : {}^{12}C$ ratio) and half-life of $^{14}C$ (*c.*5 500 years) an estimate of the age may be obtained.

Reasonably accurate dates may be obtained back to about 40000–50000 BP. Dating is subject to errors, however, partly because of uncertainties about the abundance of $^{14}C$ during past epochs. Carbon-14 is produced from atmospheric nitrogen by the action of *cosmic rays, whose intensity fluctuates inversely with the strength of the *solar wind. The latter itself depends on solar activity, which varies semi-regularly in the *solar cycle, and also on longer time-scales. During the *Maunder Minimum, for example, solar activity was low, and the cosmic-ray flux and $^{14}C$ production were high. For the recent past, dates obtained by *dendrochronology provide calibration data, reducing errors of dating.

## Carbon emission trading (carbon trading, emission trading)

The trading of permits (**carbon credits**) to emit *carbon dioxide and other *greenhouse gases. A provision of the *Kyoto Protocol was that nations should be compensated for retaining or increasing the size of their forests in particular, and (by extension), to other methods of *carbon sequestration. Because trees remove carbon dioxide from the atmosphere and incorporate carbon into their tissues, nations that increase their forested area are awarded credits that may be set against their individual targets for the reduction in carbon dioxide emissions. In addition, countries whose emissions are below the established levels are able to trade credits with countries that are unable to meet their greenhouse-gas reduction targets.

Although trading has already commenced, there are still serious scientific reservations about the extent to which forests sequestrate carbon dioxide, and for how long. Other proposed methods, such as the 'seeding' of tropical waters with iron to encourage the growth of phytoplankton (photosynthetic plankton), have even greater scientific uncertainties. Only geological and oceanic sequestration are currently apparently free from such problems. Despite this, carbon trading is generally viewed as a valid method of encouraging a reduction in carbon-dioxide emissions or, at least, an initial step in this process.

At present the only mandatory trading scheme is that instituted by the European Union, the European Carbon Exchange (ECX). Other, non-mandatory trading schemes include the Chicago Climate Exchange (CCX) and the Asia Carbon Exchange (ACX). In April 2007, the United Nations Climate Change Secretariat launched a global carbon exchange, the International Transaction Log (ITL), to allow developing nations to sell pollution reduction credits—called Certified Emission Reductions (CER)—to rich nations, which may then count these credits towards their greenhouse-gas emissions targets under the Kyoto Protocol, from 2008–12.

**carbon dioxide ($CO_2$)** A colourless, odourless gas which is the fourth most abundant component of the *air in the *homosphere, amounting to about 0.035 per cent by volume. It is released by volcanoes, part of the source material being carbonate rocks, and produced by metabolic processes in animals and by the combustion of hydrocarbons. It is broken down by *photochemical processes in the upper atmosphere and by photosynthesis in plants. The most important *greenhouse gas, it is the major factor in global warming (*see feature*). It is readily soluble in water, and the oceans act as the principal reservoir (containing approximately 99 per cent of terrestrial carbon dioxide), from which it is deposited as carbonate rocks such as limestone. Its solubility is strongly dependent upon temperature, leading to considerable variation in oceanic concentration at different times and places. Its role in global warming was first suggested by Svante *Arrhenius in 1896.

**carbon-isotope determination** The measurement of the ratio of the carbon isotopes $^{12}C$ and $^{13}C$ in sedimentary deposits to determine the extent of plant life at the time. Plants metabolize $^{12}C$ in preference to $^{13}C$, so that when plants are abundant, the atmosphere is enhanced in $^{13}C$. Residues in ancient deposits contain resistant organic molecules known as biomarkers. Determination of the residual isotope ratios in such biomarkers shows that there were dramatic changes in the abundance of land and oceanic plants at the times of the mass-extinctions at the end of the Permian and Triassic, possibly caused by the abrupt, catastrophic release of *hydrogen sulphide from the oceans.

**carbon management** The concept that the impact of burning fossil fuels on global warming (*see feature*) might be minimized by separating carbon emissions (primarily *carbon dioxide) as they are produced. The carbon dioxide could then be removed from the atmosphere by *carbon sequestration, or even used as a feedstock (together with hydrogen) for the production of methanol, a 'clean' fuel.

**carbon monoxide (CO)** A colourless, odourless, toxic gas which may be created by the incomplete combustion of carbon or carbon compounds. It may be used as a fuel, but is more commonly encountered as a pollutant. As such, its lifetime in the atmosphere is about 60 days. In a vehicle's catalytic converter, platinum particles assist its oxidation to carbon dioxide.

**carbon sequestration (carbon capture and storage, CCS)** The storage of the *greenhouse gas *carbon dioxide in natural sites, thus removing it from the atmosphere and reducing global warming (*see feature*). Carbon may be sequestered in plants (especially trees), but other potentially useful reservoirs are depleted oil and gas wells, underground geological formations, and deep in the ocean. (Provided the gas is injected below the oceanic *thermocline, it takes many hundreds of years to mix with the surface waters.) *See also* CARBON MANAGEMENT.

**carbon tetrachloride ($CCl_4$)** Tetrachloromethane, a heavy, colourless liquid, formerly extensively used as a solvent. Because of its significant contribution to the destruction of the *ozone layer, its use was discontinued under the *Montreal Protocol in 1996.

**carbon trading** *See feature* CARBON EMISSION TRADING.

**castellanus (castellatus, cas)** A *cloud species in which vertical *cumuliform turrets arise from a lower line or layer of cloud. Many of the cloud turrets are higher than they are wide, giving a clear indication of *instability at that level. The species may occur with *cirrus, *cirrostratus, *altocumulus, and *stratocumulus.

Strictly speaking, castellatus [Latin: 'turreted'] is the correct term. Castellanus [from Latin *castellum*: 'castle'] appears to have been included accidentally in a standard terminology issued by the *World Meteorological Organization, perhaps through a dictionary error. It has become the most commonly used form.

**CAT** *See* CLEAR-AIR TURBULENCE.

**catchment area** The area that collects *precipitation and feeds into a single stream, river, or water-supply system.

**cat's paws** A small area of ripples on a water surface which indicates the presence of a light puff of wind.

**CCB** *See* COLD CONVEYOR BELT.

**CCS** *See* CARBON SEQUESTRATION.

**ceiling** **1.** The maximum height that a *thermal, aircraft, or balloon may attain. **2.** In aviation, and more generally in the United States, the height of the lowest cloud layer that covers more than 50 per cent of the sky, i.e. classed as *broken cloud, *overcast, or obscured, rather than thin or partial cover.

**ceilometer** *See* CLOUD-BASE RECORDER.

**celestial equator** The projection of the Earth's Equator on to the *celestial sphere.

**celestial sphere** The imaginary sphere, centred on the observer, on which all celestial bodies such as the Sun, Moon, and stars appear to be located. The positions of certain atmospheric *optical phenomena are often referred to points on the celestial sphere, such as the *antisolar point.

**cellular convection** *Convection that is organized into a pattern of polygonal individual cells, normally approximately hexagonal in shape. Such convection is commonly seen in satellite images of *layer clouds, with cloud occurring where the air is rising, and clear sky where it is descending. There are two basic forms: **closed cells**, where the centre of each cell is occupied by cloud, and **open cells**, where the centre is clear and the cloud around the edges forms a reticulated pattern.

**Celsius, Anders** (b. Uppsala, Sweden, 27 November 1701; d. Uppsala, 25 April 1744) Swedish astronomer and physicist, who made extensive studies of the *aurora and linked its activity with variations in the Earth's magnetic field. His thermometer, with 100 graduations between the boiling and *freezing points of water, subsequently became the basis of the *Celsius temperature scale that is now used for all scientific measurements.

**Celsius scale** The most commonly used temperature scale, based on the *freezing or ice point of water (0 °C) and its *boiling point (100 °C). The scale is derived (by inversion) from one proposed by Anders *Celsius in 1742, in which the freezing point was designated as 100°, and the boiling point as 0°. For many physical calculations and experimental purposes, it is often replaced by the *Kelvin temperature scale, with an equivalent unit, but which is defined in terms of the *triple point of pure water.

**centigrade** A term commonly, but incorrectly, used as a synonym for the *Celsius scale, or a degree Celsius. Although formerly used as the name for the temperature scale, it was discontinued by international agreement in 1948. It may, in fact, be used to describe any scale that has one hundred divisions or units.

**Central England Temperature (CET)** A series of average temperatures for rural sites in central England, consisting of monthly figures from 1659, and daily ones from 1772 to date. Originally compiled by Professor Gordon Manley in 1973, it is the longest homogeneous meteorological record in the world and is now regularly kept up to date by the UK Meteorological Office.

**centre of action** A semi-permanent pressure centre that regularly appears in one particular area on charts of mean sea-level pressure. Centres of action are created by the overall general circulation (*see feature*) of the atmosphere and have significant effects upon the weather in surrounding regions. The *Aleutian low, the *Azores high, and the *Icelandic low are examples.

**centrifugal force** An apparent but fictitious force, directed radially outwards from the centre of rotation, that seemingly acts on a body moving in a circular, elliptical, or curved path. The magnitude of this apparent force is equal to the *centripetal force but opposite in direction. In reality, the forces acting on any such body are the centripetal and inertial forces. The latter, arising from the body's intrinsic mass, is directed along a line that at any particular instant is tangent to the actual path.

Although non-existent, centrifugal force is often a useful fiction that simplifies the description of rotating systems.

**centripetal force** The force, acting radially towards the centre, that constrains a body to move in a circular, elliptical, or otherwise curved path. For a planet or comet orbiting the Sun, or a satellite orbiting a planet, the centripetal force is the gravitational acceleration between the two bodies. For the flow of air around a low-pressure centre, the centripetal force is created by the *pressure gradient, and for *geostrophic wind is equal to the Coriolis acceleration (*see* CORIOLIS EFFECT).

**CERES** The Clouds and the Earth's Radiant Energy System instrument initially carried on the *Tropical Rainfall Measurement Mission (1997) and currently active on the *Terra satellite. A broad-band scanning *radiometer designed to monitor the radiation flux throughout the depth of the atmosphere. It primarily provides data about the influence of clouds on the radiation budget. (Approximately 8 per cent of incoming radiation cannot be accounted for in current atmospheric models.) The instrument is scheduled to be carried by five separate satellite missions to provide a 15-year database.

**CET** *See* CENTRAL ENGLAND TEMPERATURE.

**CFC** *See* CHLOROFLUOROCARBON.

**CGCM** *See* GENERAL CIRCULATION MODEL.

**chaff** Extremely fine ribbons or wires of metal, metallized glass fibre, or plastic that act as passive radar reflectors. Clouds of chaff, released by sounding rockets, may be tracked to monitor upper-atmosphere winds. It is also used for radar calibration purposes.

**chain lightning** *See* BEAD LIGHTNING.

**chaos theory** A field of mathematics that examines the behaviour of mechanical or physical systems (such as the weather) that are found to behave unpredictably over time, despite being apparently simple and governed by well-known physical laws. In such systems a small alteration to the initial conditions may produce very different end results. Despite the unpredictability, however, many systems have quasi-periodic behaviour that may be modelled accurately. *See also* BUTTERFLY EFFECT.

**Chapman, Sydney** (b. Eccles, England, 29 January 1888; d. Boulder, Colorado, 16 June 1970) British mathematician and geophysicist, noted for his work on the kinetic theory of gases (which he later applied to the study of the upper atmosphere) and *geomagnetism. He carried out extensive and fundamental investigations of the *aurora, the Earth's geomagnetic field, and magnetic storms.

**Charney, Jule Gregory** (b. San Francisco, 1 January 1917; d. Boston, 16 June 1981) American meteorologist noted for his major contribution to deriving the equations describing atmospheric motion, and advancing understanding of the general circulation (*see feature*). He was a pioneer in the use of these models for *numerical weather prediction, including the use of digital computers.

**chemocline** An oceanic interface, separating oxygenated surface waters from anoxic, deep water saturated with *hydrogen sulphide ($H_2S$), produced by anaerobic organisms. Green and purple $H_2S$-consuming bacteria are concentrated at the chemocline. Anoxic conditions are found in the lower layers of the Black Sea (for example).

**chemosphere** A term sometimes used for the region of the atmosphere at about 20–40 km (i.e. within the *stratosphere) in which extensive *photochemistry occurs.

**chili** A local name in Morocco, Algeria, and Tunisia for a hot, dry wind similar to the *sirocco.

**chinook** [Salish: 'snow-eater'] A strong, warm, and exceptionally dry *föhn wind that descends the eastern slope of the Rockies, and may bring an abrupt temperature rise, the disappearance of lying snow, and a risk of desiccation and consequent fire hazard. Often heralded by a line of low cloud (the **chinook arch**) parallel to the Rockies. A chinook was responsible for the most rapid temperature rise ever recorded: 27 deg C in 2 minutes at Spearfish, South Dakota, on 23 January 1943. Chinooks are often extremely violent, with one gusting to 220 $km\,h^{-1}$ at Boulder, Colorado in 1982. The term is occasionally applied to a warm, moist wind from the sea that affects the coasts of Washington and Oregon.

**chlorofluorocarbon (CFC)** Any of various organic compounds primarily consisting of carbon, chlorine, and fluorine, such as trichlorofluoromethane, $Cl_3CF$ (CFC-11), and dichlorodifluoromethane, $Cl_2CF_2$ (CFC-12). Once used extensively as refrigerants, aerosol propellants, foaming agents, and solvents, they have been found to be major components responsible for the destruction of atmospheric *ozone, the creation of the *ozone holes, and (in certain cases) contributing to global warming (*see feature*). CFC-12, for example, is *c.*12 000 times as effective at absorbing solar infrared radiation as *carbon dioxide. Their withdrawal from use worldwide is governed by the *Montreal Protocol.

**cierzo** A dry, cold, north-westerly wind that flows down the Ebro Valley in Spain during the winter months.

**circulation, atmospheric** *See feature* GENERAL CIRCULATION.

**circumhorizontal arc** An occasionally spectacular *halo phenomenon that consists of a brilliantly coloured arc lying parallel to the horizon and sometimes extending for as much as 120°. It arises from refraction through ice-crystal faces with a 90° included angle, when light enters through a vertical face and exits from a horizontal one (*compare* CIRCUMZENITHAL ARC). It may be observed only when the solar elevation exceeds 58°, and is thus most readily visible from lower latitudes. If the 46° halo is present the arc normally lies slightly below its lowest point, becoming tangential at a solar elevation of 68°.

**circumzenithal arc** A *halo phenomenon, consisting of a short arc, showing brilliant spectral colours, centred on the *zenith, and normally symmetrical about the line joining the Sun and zenith. It arises from refraction through ice-crystal faces with a 90° included angle, with the light entering a horizontal face and leaving through a vertical one. If the 46° halo is present, the arc may appear slightly above, or tangential to it at its highest point, depending on solar elevation.

**cirriform** Clouds that consist solely of ice crystals and thus have a generally fibrous structure: *cirrus, *cirrostratus, and *cirrocumulus. The term is also applied to the cirrus shields or plumes of *cumulonimbus *incus.

**cirrocumulus (Cc)** [Latin: 'wisp' + 'heap'] One of the main *cloud genera: a high-level cloud (*see* ÉTAGE) that occurs as a layer or patch of more or less separate cloudlets in the form of heaps, rolls, or pancakes (Plate 5). The cloud elements have an apparent width of less than 1° (by which they are distinguished from the lower *altocumulus). They are white, without shading. Cirrocumulus occurs in the *cloud species *castellatus, *floccus, *lenticularis, and *stratiformis, and the *cloud varieties *lacunosus and *undulatus.

**cirrostratus (Cs)** [Latin: 'wisp' + 'layer'] One of the main *cloud genera: a high-level cloud (*see* ÉTAGE) that occurs as a thin sheet, occasionally in the form of a thin featureless white veil, but normally showing some fibrous structure (Plate 6). Cirrostratus consists of ice crystals and is the cloud genus most prone to exhibit *halo phenomena. It may occur as the *cloud species *fibratus and *nebulosus, and the *cloud varieties *duplicatus and *undulatus.

**cirrus (Ci)** [Latin: 'curl', 'tuft', 'wisp'] One of the main *cloud genera: a high-level cloud (*see* ÉTAGE) that occurs as white or mainly white patches or bands, generally with a fibrous appearance (Plate 4). Cirrus clouds consist of ice particles, and although they do not exhibit many *halo phenomena (commonly seen in *cirrostratus), may show other optical effects, especially a bright form of *parhelion. Cirrus may occur as various *cloud species (*castellatus, *fibratus, *floccus, *spissatus, and *uncinus), *cloud varieties (*duplicatus, *intortus, *radiatus, and *vertebratus), and as *praecipitatio and *virga. Dense cirrus, particularly the spissatus variety, may appear dark grey when seen against the light.

**clathrate** A solid chemical compound, in which molecules of one substance are physically trapped within a lattice formed by a second substance. One such clathrate is *methane hydrate, which is found mainly on and beneath the sea-floor.

**clear-air turbulence (CAT)** A form of *turbulence that is distinct from the eddies caused by friction with the ground, and from the turbulent motion created by *convection in clouds such as *cumulus and *cumulonimbus. It arises in regions of marked *wind shear, particularly at the boundaries of *jet streams, but may also be found in strong *lee waves (**mountain/terrain-induced turbulence**), and disturbances that propagate from the centres of extremely vigorous convection such as *supercell thunderstorms. Despite its name, it is frequently associated with thin *cirrus cloud.

Clear-air turbulence is not normally a hazard to the structural integrity of aircraft in the upper troposphere or lower stratosphere—although objects on board may be tossed around and passengers injured—but is a significant danger when it occurs at low level, especially near airports. Airborne detection systems are under development, using *lidar that senses the motion of *aerosols a few kilometres ahead of the aircraft. It is also believed that sound waves are produced by turbulence and that these may be detectable at longer ranges. Experiments are under way to test Doppler wind *profilers and lidar systems for the detection of low-level turbulence near airports.

**clear ice** *See* GLAZE.

**clear slot** A region of higher cloud base and clearer skies to the south-west or west (in the northern hemisphere) of a *wall cloud beneath a *supercell thunderstorm. Normally considered to be a sign of a *rear-flank downdraught.

**climate** The description of the variability of weather conditions prevailing in a particular region or latitude zone over a specific period of time, as derived from statistical information about various meteorological elements. Because of climatic change (*see feature*), the precise interval to which the data apply should always be specified. Less rigorously, and in general popular usage, climate may be regarded as the average values encountered for the various elements, together with information about the extreme ranges that have been observed, and their probability of recurrence. Factors that are generally taken into account are temperature, *precipitation, wind speed, *humidity, and duration of sunshine, together with details of the frequency of *fog, *frost, *hail, *snow, *thunder, etc.

The climate of any particular location is primarily governed by latitude, position relative to *centres of action, *continentality or *oceanicity, altitude, and local topography. A very broad division may be made into *continental climate and *maritime climate, but there are various detailed classification schemes, with the *Köppen classification being widely used.

**climate classification** *See* KÖPPEN CLASSIFICATION.

**climate modelling** The study of climate systems by the use of computer programs, such as the various forms of *general circulation model. Because of the short period for which instrumental records are available (a few hundred years), the creation of suitable models involves the study of past climates (*palaeoclimatology) as well as present systems. The modelling is complex, because it must include the effects of various long-term cycles, such as the *carbon cycle and the *nitrogen cycle, which are not considered in *numerical weather prediction models.

**Climate Variability and Predictability Programme (CLIVAR)** An international programme coordinating studies of natural climate variability on scales of 1–100 years and estimation of climatic change (*see feature*) through the increase in *greenhouse gases. Four main fields are covered: El Niño Southern Oscillation (*see feature*); variability of *monsoon regimes; variability of the *thermohaline circulation; and global climate and sea-level change.

**climatic optimum** A period during which temperatures are consistently higher than the long-term average, and which may display other signs of climatic change (*see feature*). One such period was the **Holocene climatic optimum** extending from about 8 000 to 4 000 BP. Between 7 000 and 5 500 BP mid-latitude temperatures in the northern hemisphere were about 2 deg C higher than the average for the 20th century. The summer *monsoon was stronger and there was much greater rainfall over northern Africa, particularly over the Sahara. The treeline was some 200–300 km farther north in Siberia and northern Canada. Although a deterioration began about 5 500 BP, a marked

## Climatic change (climate change)

A significant change in the *climate of a region as shown by a secular trend in individual parameters such as *temperature and *rainfall as averaged over a decade or longer. Over geological time there have been several periods when the Earth has experienced major *ice ages, consisting of alternating *glacial and *interglacial phases, and there are indications of even more extreme *glaciation (*see* SNOWBALL EARTH), as well as periods when the Earth has been essentially free from ice. Major factors have included evolution of the Earth's atmosphere (*see* ATMOSPHERE (1)); changes in the areas and distribution of oceans and continents (*see* PLATE TECTONICS), and orogeny (mountain building); and volcanic activity (particularly episodes of widespread volcanism, such as that found in the Deccan Traps in India, the Columbia River Basin flood basalts, and the Siberian volcanic province). In addition, there are astronomical factors, such as changes in the *solar constant and the variations in *insolation suggested by Milankovich theory (*see feature*). Fluctuations in the concentration of *greenhouse gases, particularly *carbon dioxide, have caused episodes of global warming (*see feature*) and cooling, and it is believed that variations in the heat storage capacity of the oceans have also occurred.

The immediate cause of drastic temperature changes appears to be major alterations in the carbon dioxide concentration, with the Milankovich fluctuations acting as the principal *forcing mechanisms for cyclic variations. Much current research is devoted to an evaluation of changes brought about by human activity, particularly the burning of fossil fuels.

change around 4 000 BP may have contributed to the decline of various ancient civilizations, notably the fall of the Egyptian Old Kingdom around 2200 BC.

The **medieval climatic optimum** lasted from about AD 1000 to 1350, when summer temperatures in north-western Europe appear to have been slightly warmer than those in the 20th century. Unlike the Holocene optimum, the effect was regional rather than global in extent.

**climatic zone** A region of the Earth's surface in which the *climate is essentially consistent. The simplest classification divides the surface on the basis of latitude and *insolation, with five basic zones: polar, temperate, subtropical, tropical, and equatorial. More sophisticated classification schemes take into account differences in the climatic regime resulting from *continentality and *oceanicity, and allow for factors such as yearly *rainfall patterns. *See* KÖPPEN CLASSIFICATION.

**climatological station** An observing *station from which climatological reports are submitted. There are various categories of station, depending on instrumentation and frequency of reporting, but observations of the main meteorological elements will be made at least once a day (at 09:00 GMT). Certain stations provide tabulations of hourly data derived from automatic instruments (such as *barographs, *thermographs, etc.), whereas others may report just a single element (e.g. *rainfall stations). *See also* SYNOPTIC STATION.

**climatology** The study of *climate in all its aspects and of the phenomena associated with it. *See also* PALAEOCLIMATOLOGY; PHYSICAL CLIMATOLOGY; SYNOPTIC CLIMATOLOGY.

**CLIVAR** *See* CLIMATE VARIABILITY AND PREDICTABILITY PROGRAMME.

**cloud** A visible accumulation of minute water droplets or ice crystals (or both) suspended in the atmosphere, created by the condensation or freezing of *water vapour, and with its base above the ground. When in contact with the surface, the same collection of cloud particles would be known as *fog, *ice fog, or *mist. Most clouds occur within the *troposphere, with the exception of some *cirrus in the lower *stratosphere, and the rare *nacreous, *noctilucent, *polar mesospheric, and *polar stratospheric clouds at greater altitudes.

Most clouds arise when vertical motion of the air, caused by *convection, forced uplift over high ground (giving *orographic cloud), or major motions within a *depression, raises water vapour above the *condensation level. Other processes, such as *turbulence and reduced air pressure, may occasionally be involved. Cloud droplets range 1–100 μm in diameter, with typical concentrations of 100–800 million per $m^3$. They are frequently *supercooled and some may remain liquid as low as −40 °C.

The varied forms exhibited by clouds (which reflect their origin and history) are described in standardized ways by the *cloud classification scheme.

**cloud amount** The portion of sky covered by cloud, either of a specific type (**partial cloud amount**) or of any type (**total cloud amount**). The cover is estimated to the nearest *okta (eighth) on a scale of 0 to 8, where 0 is clear sky, and 8 is completely cloud covered. In North America, a scale of tenths is used instead of oktas. *See also* CLOUDINESS; CLOUDY DAY; STATE OF SKY.

**cloud base** The lowest level at which cloud water droplets or ice crystals are observed, as distinct from clear air or *haze. It therefore frequently corresponds to the *condensation level. Its height is reported as height above actual ground level, except in forecasting, when it is frequently given as a *pressure altitude.

**cloud-base recorder** An instrument for determining and recording the altitude of the cloud base. Modern designs are a type of *lidar, where a pulsed laser beam is directed vertically and the return pulses, reflected by the cloud base, are picked up by a suitable detector.

**cloud bow** *See* FOGBOW.

**cloudburst** A popular term for an extremely heavy *shower that may be accompanied by *hail or *lightning, and which gives rise to a high rate of rainfall in a short space of time.

**cloud ceiling** In aviation, the height of the lowest layer of cloud that is more than 4 *oktas in extent, measured from the height of the airfield.

**cloud classification** A method of dividing clouds into various categories. Two formal systems exist, together with less rigid descriptions based upon form or composition.

**1.** A division, based upon an original classification proposed by *Howard, into *cloud genera (ten basic cloud forms), subdivided when necessary into *cloud species (describing structure or shape), and *cloud variety (denoting transparency and the arrangement of cloud elements). Additional terms, categorized as *accessory clouds and *supplementary cloud features, define particular types of cloud or morphological forms that accompany certain specific genera or species. Two additional terms, *genitus and *mutatus, are used to describe the growth or alteration of one cloud type into another. The terminology, and the standard two- and three-letter abbreviations, are defined by the *World Meteorological Organization in its *International Cloud Atlas*.

**2.** A triple division of clouds into high, medium, or low, based upon the range of altitude (or *étage) in which a *cloud genus normally occurs. They are *cirrus, *cirrocumulus, and *cirrostratus (high); *altocumulus, *altostratus, and *nimbostratus (middle); *cumulus, *cumulonimbus, *stratocumulus, and *stratus (low). Note that certain cloud types often occur (or extend) into another étage: altostratus into high; nimbostratus into high or low; cumulus and cumulonimbus into both middle and high.

To describe the state of the sky (rather than cloud type) for *synoptic purposes, the three étages are coded as $C_H$, $C_M$, and $C_L$, and each is subdivided into nine states. Examples are $C_H4$, cirrus progressively invading the sky; $C_M6$, altocumulus from the spreading out of cumulus; and $C_L9$, cumulonimbus with anvil.

**3.** A description of clouds either by their form—heap (*cumuliform) or layer (*stratiform) clouds—or by their composition: water-droplet, ice-crystal, or mixed clouds.

**cloud genus (pl. genera)** One of the ten fundamental cloud types: *cirrus (Ci), *cirrocumulus (Cc), *cirrostratus (Cs), *altocumulus (Ac), *altostratus (As), *nimbostratus (Ns), *stratocumulus (Sc), *stratus (St), *cumulus (Cu), and *cumulonimbus (Cb). *See* CLOUD CLASSIFICATION.

**cloud head** The smooth bulge of *cirrus cloud (usually semi circular in outline) that forms on the polar side of a vigorously developing depression. It marks the position of part of the *cold conveyor belt.

**cloudiness** The amount of sky covered by cloud, i.e. the total *cloud amount, irrespective of cloud type.

**cloud leaf** *See* LEAF CLOUD.

**cloud particles** Primarily water droplets, generally in the range *c.*1–100 μm, with a tendency for larger droplets to be found in *cumuliform rather than *stratiform clouds. *Ice crystals of various physical forms are also present in certain *cirriform and *mixed cloud types.

**cloud physics** The branch of study concerned with the processes involved in the formation and evolution of all forms of *clouds (including *fog) and *precipitation, as well as *atmospheric electricity.

**Cloudsat** A NASA *polar-orbiting satellite, launched (together with *Calipso) on 26 April 2006, and now part of the *Afternoon-Train constellation. It is specifically designed to monitor clouds with a millimetre-wavelength radar system, and thus provide data about a poorly understood aspect of global climate models. The short wavelength enables it to detect much smaller liquid and frozen cloud particles than previous radars. It flies in close formation with Calipso so that the two satellites obtain essentially simultaneous data of the same region of the atmosphere. *See also* AQUA; AURA; ORBITING CARBON OBSERVATORY; PARASOL.

**cloud searchlight** An instrument for projecting a beam of light vertically above the station to illuminate the base of the clouds. By triangulation, using an alidade at a known distance from the searchlight, the height of the *cloud base may then be determined. Now being replaced by automatic *cloud-base recorders.

**cloud seeding** The technique of adding suitable particles to clouds in an attempt to cause them to dissipate, modify their structure, or alter the intensity of associated phenomena, such as wind speeds or hail. The process generally involves the initiation of *precipitation through the use of *dry ice to lower the temperature locally to below −40 °C, causing *glaciation; *silver iodide to provide *freezing nuclei; and either a fine spray of water or hygroscopic salt crystals to promote *coalescence.

Limited success has been obtained in dispersing *stratus cloud and in initiating rainfall. Despite many claims, however, no quantifiable results have been produced in attempts to minimize damaging *hail.

**cloud species** Part of the system of *cloud classification, consisting of a subdivision of *cloud genera according to structure or shape. Cloud species are: *fibratus, *spissatus, *uncinus, *castellatus, *floccus, *stratiformis, *nebulosus, *lenticularis, *fractus, *humilis, *mediocris, *congestus, *calvus, and *capillatus. The standard abbreviations consist of three letters. To describe a specific cloud, none, one, or more of the terms may be applied to an individual cloud genus. *See also* CLOUD VARIETY.

**cloud street** A line of cumulus clouds that extends downwind parallel to the wind direction (Plate 7). Multiple sources of *thermals and a constant wind speed often give rise to a series of cloud streets. The effect may be accentuated when *convection is restricted by an overlying *inversion, when the spacing between the streets is about two or three times the depth of the convective layer.

**cloud-to-cloud discharge** A *lightning discharge that occurs between parts of the same cloud, or between different *thunderstorm cells, rather than between the cloud and the ground.

**cloud variety** Part of the system of *cloud classification, which subdivides the *cloud genera according to transparency and the arrangement of cloud elements. Cloud varieties are: *duplicatus, *intortus, *lacunosus, *opacus, *perlucidus, *radiatus, *translucidus, *undulatus, and *vertebratus. The standard abbreviations consist of two letters. To describe a specific cloud, none, one, or more of the terms may be applied to an individual cloud genus. *See also* CLOUD SPECIES.

**cloudy day** A day on which the average amount of *cloudiness at the standard times of observation is greater than 6 *oktas. In North American usage, it is a day when overall cloudiness has exceeded seven-tenths.

**CMA** Calcium magnesium acetate, a chemical used to disperse ice by lowering its *freezing point. Unlike salt, CMA also has the property of causing snow to remain as separate crystals that do not bond together, and are thus easily removed. Although far less damaging to the environment and structures than salt, CMA is much more expensive.

**coalescence (collision-coalescence process)** A term used in *cloud physics for the process by which a liquid *precipitation droplet grows through collision with another liquid droplet of whatever size (*compare* ACCRETION). The process occurs naturally in clouds because the largest droplets have a higher rate of fall (or slower rate of rise) and thus tend to overtake (or be overtaken by) smaller droplets. Any electrical charge on the droplets, or the overall electrical field that exists in most clouds, assists in the process. The **Langmuir chain-reaction theory** suggests that droplets that are of such a size to have a high collection efficiency eventually grow so large that they break up into smaller droplets that are returned to higher levels of the cloud by the updraughts, and are subjected to the same process. This appears to account for the sudden release of heavy rain following *cloud seeding.

**col** The point on a *weather map that lies between two *depressions and two *anticyclones, arranged alternately. The pressure distribution may be envisaged as the intersection of a *trough with a *ridge, in a saddle-shaped region. When the two high-pressure regions are closer together than the low-pressure areas, the situation is described as an **anticyclonic col**, and when the low-pressure regions are closer, a **cyclonic col**. The slightest change in the pressure pattern may cause a substantial alteration in the position of a col, which therefore tends to be a transitory feature. *See figure overleaf.*

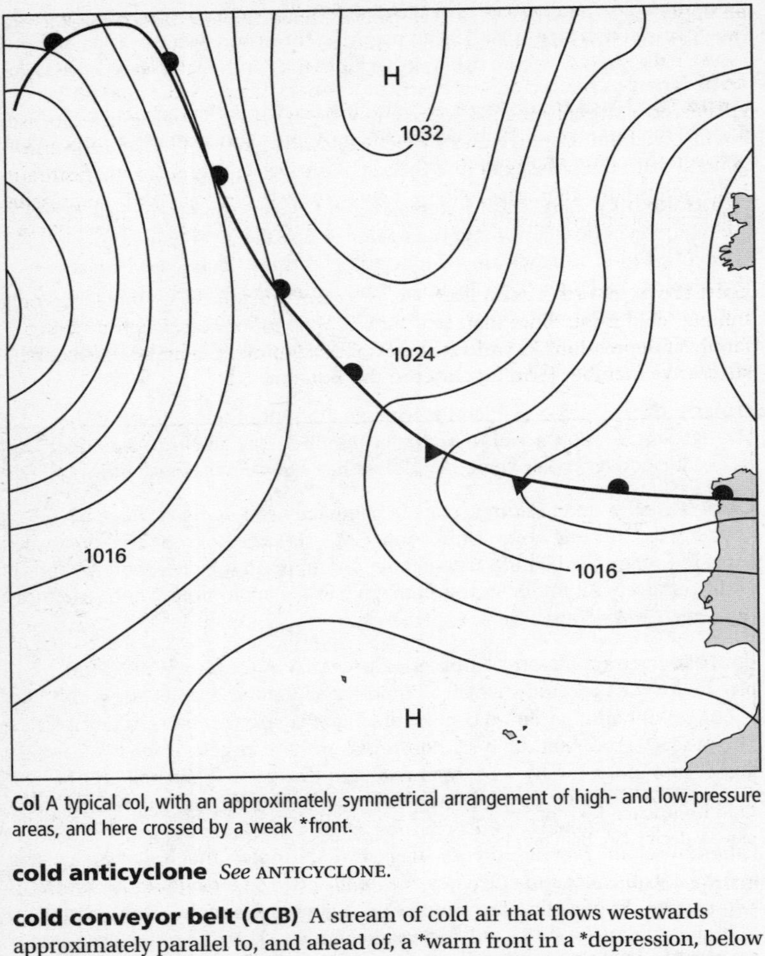


**Col** A typical col, with an approximately symmetrical arrangement of high- and low-pressure areas, and here crossed by a weak *front.

**cold anticyclone** *See* ANTICYCLONE.

**cold conveyor belt (CCB)** A stream of cold air that flows westwards approximately parallel to, and ahead of, a *warm front in a *depression, below the encroaching warm air. It generally rises away from the surface to the middle levels near the tip of the *warm sector, where it encounters the *air mass behind the *cold front. It then tends to split with the poleward flow creating the distinctive *cloud head of the depression. The other, equatorward flow causes a descent of cold, dry air from the upper *troposphere and lower *stratosphere, producing the *dry intrusion behind the cold front. *See also* SPLIT FRONT; WARM CONVEYOR BELT.

**cold dome** **1.** A mass of cold air, with an overall shape like that of an inverted saucer, that forms over areas such as Antarctica or Greenland. **2.** On an upper-air *contour chart, a region surrounded by closed isobars, which is thus isolated from the main mass of cold air.

**cold front** A boundary between two *air masses, where the cold air is advancing, and replacing the warm air at the surface. (When three air masses are involved the *front is more correctly described as a cold *occlusion.) In general, with the passage of a cold front, the wind will *veer (in the northern hemisphere), the temperature and *dew point will decrease, and the pressure will rise. Cold fronts have a typical slope of 1:50 to 1:75, roughly twice that of a *warm front.

**cold-front wave** A common form of *secondary depression that forms on a trailing *cold front. More than one such depression may arise, producing a family of depressions in various stages of development, generally with each successive member forming closer to the Equator.

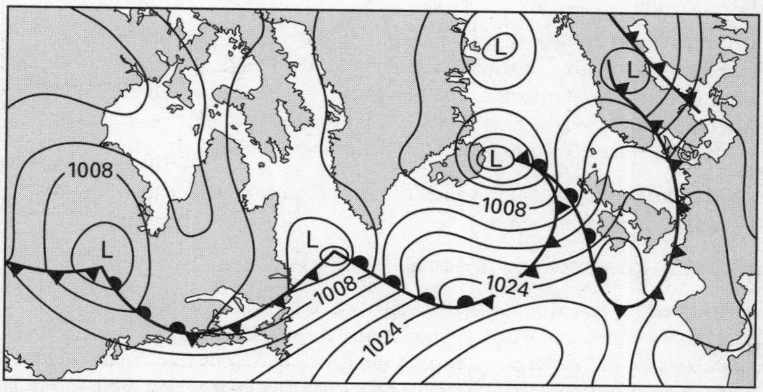


**Cold-front wave** Two *depressions are developing on the *cold front that trails behind the primary depression. Both may be expected to deepen as they move eastwards.

**cold-occlusion depression** A *secondary depression that forms at the *triple point where warm and cold *fronts combine to create a cold *occlusion.

**cold pole (pole of cold) 1.** The point in each hemisphere where the lowest temperature has been measured. In the northern hemisphere this is in Siberia, either Verkhoyansk (67° 21′ N, 133° 15′ E) or Oymyakon (63° 30′ N, 142° 44′ E), with temperatures of –68 °C on 5 and 7 February 1892 and 6 February 1933, respectively. (Oymyakon unofficially recorded –72 °C at a later date.) In the southern hemisphere, the Vostok scientific station (78° 27′ S, 106° 52′ E) recorded –89.2 °C on 21 July 1983.

**2.** In an alternative definition, the point in each hemisphere with the lowest mean annual temperature. In the southern hemisphere this is the Pole of Inaccessibility at approximately 82° S, 58° E, where a mean annual temperature of −58 °C has been recorded. The situation in the northern hemisphere is more debatable, but the pole undoubtedly lies in Siberia, probably close to Oymyakon.

**cold pool** An area of relatively cold air, surrounded by closed *isotherms and isolated from the main body of cold air closer to the pole. At the surface, such a cold pool is a region of relative *stability, whereas one at middle levels tends to have low stability.

**cold rain** An informal term for rain that originates through the initial deposition of water vapour on to a *freezing nucleus in the *Bergeron process, and subsequent melting, rather than by *coalescence. *See also* WARM RAIN.

**cold sector** The region of a *depression where cold air is in contact with the surface. It usually amounts to at least 50 per cent of the area in a newly formed system, and increases in size to 100 per cent as the system develops and then dissipates.

**Cold Tongue Index (CT)** A measure of the *anomaly in *sea-surface temperature in the central and eastern equatorial Pacific relative to the annual cycle. It is averaged over the area between 5° N and 5° S, and between the International Date Line and 90° W. It is anti-correlated with the *Southern Oscillation Index. *See also feature* EL NIÑO SOUTHERN OSCILLATION.

**cold trough** A *trough in which the temperature is colder than in immediately adjacent areas.

**collision-coalescence process** *See* COALESCENCE.

**coloured rain** Rain that leaves a deposit of fine dust on the ground, cars, and other objects, and which is often noticeably pink or red. The dust normally originates in arid regions, and under the right conditions may be carried for considerable distances before being swept out by precipitation falling from a higher layer of cloud. Saharan dust frequently affects Western Europe in this way.

**comfort zone** The range of *effective temperature in which most persons work with maximum efficiency. This range is generally taken to be 19–24 °C, with the optimum temperature being 21 °C.

**comma cloud** An area of cloud on satellite imagery that has the shape of a comma. On a *synoptic scale, this feature is often seen in the relatively early stages of the development of a *depression. Also frequently associated with *mesoscale convective clouds. *See also* DRY SLOT; LEAF CLOUD.

**concrete minimum temperature** The temperature recorded by an accredited standard minimum thermometer, exposed to the open air, with its bulb just in contact with a concrete slab that is essentially level with the ground surface. *See also* GRASS MINIMUM TEMPERATURE.

**condensation** The process by which a substance such as water vapour changes from a gaseous phase into a liquid; the opposite of *evaporation. The process is accompanied by the release of *latent heat. In the formation of cloud droplets from water vapour a *condensation nucleus is also required.

**condensation level** The level at which a parcel of air reaches the *dew point. For a rising parcel of air that previously cooled by expansion at the *dry adiabatic lapse rate, it will subsequently cool at the *saturated (or *pseudo-) adiabatic lapse rate. When a mass of air has been cooled by mixing, the **mixing condensation level** is the lowest level at which condensation occurs.

**condensation nucleus** A *nucleus that promotes the *condensation of water vapour into a water droplet. Typically, a suitable nucleus has a radius greater than 0.1 μm, which thus excludes the smallest (and most numerous) nuclei.

**condensation point** *See* DEW POINT.

**condensation trail (contrail)** A trail of water droplets or ice crystals produced by condensation or freezing of the water vapour present in the exhaust from aircraft engines. Also known as a **vapour trail**. Glaciated (frozen) trails may be extremely persistent, depending on the precise conditions at altitude. *See also* DISTRAIL.

**conditional instability** The condition in which a parcel of air becomes unstable if *condensation occurs (perhaps through uplift in a *lee wave). Such circumstances apply when the *environmental lapse rate lies between the *dry adiabatic lapse rate and *saturated adiabatic lapse rate. Also known as **potential instability**. *See* STABILITY.

**conduction** The transfer of heat from points of higher temperature to those of lower temperature through the collision of atoms or molecules that possess a high *kinetic energy (energy of motion) with those of lesser energy. Conduction is of particular importance in heat transport in solids, and to a lesser extent in fluids, where *convection plays a greater role. It is of least significance in gases, which are generally poor conductors. *See also* CONVECTION; RADIATION.

**confluence** A reduction in the separation of adjacent *streamlines in the flow of any fluid such as air or water. It is the opposite of *diffluence. Confluence may be regarded as the reduction in area covered by a particular airstream, and is generally accompanied by an increase in wind speed downstream. *Compare* CONVERGENCE.

**congestus** A *cloud species, commonly known as **towering cumulus**. Cumulus cloud of considerable vertical extent, growing vigorously, with a firm cauliflower-like head that shows no sign of *glaciation (which would cause the cloud to be classed as *cumulonimbus). Cumulus congestus may produce *precipitation through the *coalescence process and their strong *convection may sometimes give rise to *funnel clouds, *landspouts, or *waterspouts.

**constant altitude plan position indicator** A composite display of radar data, consisting of information obtained at several elevations of the radar beam. The data are combined to produce an image of conditions at a specific altitude.

**constant-level balloon** An instrumented balloon that is designed to remain at a specific level in the atmosphere for a long time, generally achieved by overpressure within the balloon envelope. The level may be either one of constant pressure or of constant atmospheric density. *See also* EOLE; GLOBAL HORIZONTAL SOUNDING TECHNIQUE; ULTRA-LONG DURATION BALLOON.

**constant-pressure chart** *See* ISOBARIC CHART.

**constant-pressure surface** *See* ISOBARIC SURFACE.

**Contessa del Vento** [Italian: 'countess of the wind'] *Lenticular clouds (often occurring as a *pile d'assiettes) that form above Mount Etna in Sicily.

**continental air** An air mass whose *source region is over a continent or very large land mass. Such air masses are normally very dry, but are rapidly modified when they move across oceanic areas. *See* ARCTIC AIR; CONTINENTAL POLAR AIR; CONTINENTAL TROPICAL AIR.

**continental Arctic air** *See* ARCTIC AIR.

**continental climate** A climatic regime that occurs in the interior of continental land masses. Such a climate tends to have low *precipitation and low *humidity and to exhibit a very large diurnal temperature range. At higher latitudes, there may also be a large range of temperatures during the year, with extremely low temperatures in winter, and high ones in summer.

**continental drift** The theory that the continents have slowly changed their relative positions throughout geological time. First proposed in modern form by *Wegener, and actively promoted by the British geologist Arthur Holmes and the South African Alexander du Toit, the theory was not generally accepted until a suitable transport mechanism (sea-floor spreading) was found. Subsequently modified and refined into the modern theory of *plate tectonics.

**continentality** The degree to which a specific region of the Earth is subject to the influence of a neighbouring land mass, and thus the opposite of *oceanicity. In *climatology it is a numerical index, based upon the average annual temperature range. *See also* CONTINENTAL CLIMATE.

**continental polar air (cP)** Cold, dry, and generally stable air that originates in the continental *anticyclones in the northern hemisphere, primarily the Canadian and *Siberian anticyclones. These are particularly strong in winter, but weaken in summer, when the *source regions shrink to a relatively narrow belt in the northern part of the continents. There are no source regions in the southern hemisphere. Strong cooling at the surface leads to stable air, which combines with *subsidence to produce a deep *inversion, typically up to 850 hPa. When continental polar air flows out over open water, it gains in temperature and humidity, and becomes unstable *maritime polar air. *See also* ARCTIC AIR; MARITIME ARCTIC AIR.

**continental tropical air (cT)** Warm dry air that originates over a continental land mass. Major semi-permanent *source regions in the northern hemisphere are the *subtropical anticyclones over North Africa and Arabia, together with parts of Central Asia, where there are light winds and upper tropospheric subsidence. An additional source region occurs over the southwestern United States and Mexico. In summer, the Eurasian source region expands to become a wide belt running from the western Mediterranean almost to the Sea of Japan. In the southern hemisphere, only central Australia is a permanent source. In winter, the only other significant source lies in southeastern Africa, but this shrinks in summer, when *maritime tropical air invades the continent. A small summertime source region exists over Argentina. The cT air that originates in the subtropical anticyclones in summer is extremely hot and dry, but often unstable at low levels.

**contour** In *synoptic meteorology, a line on an *isobaric chart that represents a constant height on the corresponding *isobaric surface.

**contrail** *See* CONDENSATION TRAIL.

**convection** The transport and mixing of heat and other properties of a fluid through mass motion. A primary means of energy transfer, together with *conduction and *radiation. In meteorology, convection is of great significance in producing circulation on all scales, including the general circulation (*see feature*) and cloud formation. It is generally taken to imply vertical motion, being produced by differences in *buoyancy, arising from variations in density; horizontal transport being described as *advection.

**convective available potential energy (CAPE)** The maximum amount of energy available to a rising parcel of air, arising from its higher temperature relative to its environment.

**convective cloud** Any form of cloud that arises directly as a result of *convection. Although *stratiform clouds often break up into *altocumulus, *cirrocumulus, or *stratocumulus as a result of shallow convection occurring within the layer, the term is normally reserved for the types that make convection readily visible, namely the *cumulus species *mediocris and *congestus, and *cumulonimbus.

**convective rain** Rain that occurs as a result of the active convection within a *cumulonimbus (or *shower) cloud, particularly when this forms part of a larger, organized system, such as a *multicell storm, a *supercell thunderstorm, or a *squall line. Convective rain is typically more intense and of shorter duration than *cyclonic rain or *orographic rain.

**Convention on Long-Range Transboundary Air Pollution** An international agreement, formalized in 1979, setting out a framework for the development of programmes for reducing the sources of long-range air pollution. *See also* GOTHENBURG PROTOCOL.

**convergence** **1.** A net horizontal inflow of a fluid (such as air or water) into a region: the opposite of *divergence. Convergence is always accompanied by a compensatory vertical motion. At some intermediate level, the rate of inflow falls to zero, and beyond that point divergence begins. In the *atmosphere (such as in a *depression), the compensating flow is upwards to higher levels (with divergence above the system), whereas in the oceans the vertical motion is a *downwelling. **2.** The zone where convergence occurs, such as the (atmospheric) *Intertropical Convergence Zone or the (oceanic) *Antarctic Convergence. *See also* CONFLUENCE.

**conveyor belt** **1.** The principal flow of air in a system such as a *depression, or associated with a specific *front. *See* COLD CONVEYOR BELT; WARM CONVEYOR BELT. **2.** A term sometimes used for major oceanic circulation systems, such as the *Great Ocean Conveyor Belt.

**cordonazo** In Mexico, the term for a tropical cyclone (*see feature*) that travels north-east in the eastern Pacific and affects the coasts of Mexico and other Central American states. The term is also applied to the hurricane-force winds that accompany such tropical cyclones.

**Coriolis effect** The apparent tendency for an object or a parcel of air to deviate from a straight path when viewed in a rotating frame of reference (such as the Earth). In meteorology, the horizontal component of this **Coriolis acceleration** (or **Coriolis force**) is of primary importance. It is zero at the Equator, and becomes a maximum at the poles. It acts towards the right in the northern hemisphere, and towards the left in the southern. Although it affects all moving objects, it is particularly significant with regard to winds, ocean currents, and tidal streams.

**corner region** The area at the surface where the flow of air into a tornado (*see feature*) changes from horizontal to vertical.

**corona** [Latin: 'crown'] **1.** An optical phenomenon consisting of one or more sets of coloured rings that surround the Sun or Moon. The rings are produced by *diffraction by water droplets in thin *stratiform cloud. A full corona shows an inner *aureole, and an outer set (or sets) of rings ranging in colour from violet on the inside to red on the outside. Uniform drop size gives the purest colours, but when there is a wide range of sizes only the aureole may be visible. The radius of a corona is inversely proportional to the droplet size.

**2.** An auroral form (*see* AURORA) in which *rays appear to converge high overhead (in fact at the magnetic zenith).

**3.** The extremely hot ($c.2 \times 10^6$ K) outer atmosphere of the Sun, above the visible *photosphere, which is normally seen only at total solar eclipses, when it shines with a characteristic pearly light. It is the source of the *solar wind, and associated with *solar flares and events known as **coronal mass ejections**, which cause major changes in the Sun's magnetic field and may affect the Earth as a *geomagnetic storm.

**corona discharge** The discharge of electricity (also known as a **brush discharge**) from the tips of sharp objects, such as lightning conductors, ships' masts, and aircraft propellers and wing-tips. Corona discharges are often accompanied by a fizzing or crackling sound, and a luminous glow. *See* ST ELMO'S FIRE.

**coronal hole** A region in the solar *corona (3) in which the Sun's magnetic field lines are not closed loops, but extend out into interplanetary space. Solar *plasma is able to escape along these lines, forming high-speed streams within the *solar wind, which may interact with the Earth's magnetic field to give an *aurora or a *geomagnetic storm.

**corposant** *See* ST ELMO'S FIRE.

**corrugations** Fine undulations in a cloud that lie parallel to the wind direction. Generally observed superimposed upon larger *billows. *See also* STRIATION.

**cosmic rays** Highly energetic atomic and subatomic particles that travel at almost the speed of light and enter the Earth's atmosphere from space. Approximately 90 per cent consist of hydrogen nuclei (protons) and most of the remainder are helium nuclei, although minute quantities of the nuclei of almost all known elements have also been detected. Energies range from $c.10^8$ eV to over $3 \times 10^{20}$ eV. Low- and medium-energy cosmic rays are believed to be produced by supernova explosions, but the origin of the most energetic particles remains a mystery. They may be new particles predicted by supersymmetry theories and emitted by distant quasars, involve the decay of superheavy particles ($10^{12}$ proton masses) formed in the early universe, or be created by collisions between cosmic strings.

The cosmic-ray flux is inversely related to the strength of the *solar wind. Collisions with atmospheric nuclei produce large showers of secondary particles that are detected at the surface. Cosmic rays occasionally interact with nitrogen to produce $^{14}C$, measurement of which is the basis of *carbon dating. They may be responsible for the formation of ion clusters that clump together to create stable *aerosol particles.

**counterglow** **1.** The coloured band or bands visible at the edge of the *shadow of the Earth. Although normally red only is present, under certain circumstances yellow and green tints may also appear. All are caused by the *refraction and selective *absorption of sunlight by the atmosphere. Also known as the **countertwilight**, **antitwilight**, or **antitwilight arch**. **2.** An alternative name for the *gegenschein.

**counter sun** *See* ANTHELION.

**countertwilight** *See* COUNTERGLOW.

**coupled general circulation model** *See* GENERAL CIRCULATION MODEL.

**crachin** Persistent *drizzle and extensive low cloud (*stratus, *fog, or *mist) along the coast of southern China in the early part of the year. Normally associated with low pressure over China and an *anticyclone over the South China Sea, which causes dry *continental air to be replaced by a moist, *tropical maritime *air mass.

**crepuscular rays** Alternating light and dark rays that, because of perspective, appear to diverge from the Sun's position, and arise when clouds (or occasionally distant hills or mountains) cast shadows on to the atmosphere. They are most noticeable when considerable *scattering is present. When the Sun is low, towards twilight (from which the rays derive their name), the light rays often appear reddish, whereas the dark shadows have a green tinge. Such rays may appear to converge at the *antisolar point to form *anticrepuscular rays.

With a high Sun, light rays may be seen penetrating breaks in *stratocumulus clouds (for example) and this form is known by many popular terms, including 'the Sun drawing water', 'Apollo's backstays', and 'Jacob's ladder' (Plate 11).

**critical point (critical temperature)** The maximum temperature at which gaseous and liquid phases of a substance may exist in equilibrium. Above the critical point a substance will be in a gaseous state, but below it the substance may be liquefied by application of a sufficiently high pressure.

**Cromwell Current (also known as the Equatorial Undercurrent)** A subsurface current with a very large volume (approximately equal to that of the *Gulf Stream) that flows eastwards in the Pacific Ocean, between about 150° E and 95° W, and close to the Equator. It arises from the *Ekman effect, which acts opposite to the normal westward surface flow, and is aided by the gradient towards the east as a result of the sloping water surface. During the *El Niño phase of an El Niño Southern Oscillation event (*see feature*) it is replaced by the *Equatorial Counter Current.

**Cryosat** A European Space Agency satellite, designed specifically to monitor long-term changes in the *cryosphere. The initial satellite was lost through a launcher failure in October 2005, but a replacement satellite, Cryosat-2, is scheduled for launch in March 2009.

**cryosphere** The portion of the Earth's environment in which water is in the frozen state. The cryosphere thus consists of *ice sheets, *glaciers, *permafrost regions, *sea ice, and those areas covered in seasonal snow or ice. It contains *c.*70 per cent of all fresh water. The term is sometimes restricted to areas that are permanently frozen.

**cum sole** [Latin: 'with the Sun'] In the same direction as the Sun, i.e. clockwise in the northern hemisphere, anticlockwise in the southern. This is a useful term for describing the curvature of the path of a tidal current (for example), particularly when the forces involved give rise to opposite curvature in the two hemispheres (as in the *Coriolis effect).

**cumuliform** Clouds that occur as specific heaps or masses (i.e. as a result of *convection): *cumulus, *cumulonimbus, *stratocumulus, *altocumulus, and *cirrocumulus. The term is often confined to the first two of these types, with the last three being regarded as having some *stratiform characteristics.

**cumulonimbus (Cb)** [Latin: 'heap' + 'raincloud'] One of the *cloud genera. A large, towering cloud of great vertical extent, with a dark, ragged base from which rain, hail, or snow is falling, often in the form of *virga (Plate 8). The cloud usually shows signs of vigorous convection, and the tops (which appear brilliantly white when illuminated by the Sun) are *glaciated and may appear smooth (Plate 8), fibrous, or striated, or be flattened into an anvil (*incus: Plate 9). Individual cumulonimbus cells pass through three stages: growing (lasting *c.*20 minutes), mature (*c.*20 minutes), and decaying (30–120 minutes). Associated *cloud species and *accessory clouds are *calvus and *capillatus; *pannus, *pileus, and *velum. Cumulonimbus are *shower clouds, and may occur as *thunderstorms, *multicell storms, and *supercell thunderstorms. *Supplementary cloud features sometimes visible include *arcus, *mamma, and *tuba. *See also* SHOWER.

**cumulus (Cu)** [Latin: 'heap'] One of the *cloud genera. A cloud in the form of a relatively well-defined individual mound or heap, occasionally ragged, but generally with a flat, darkish base and rounded, white upper regions, often actively growing vertically (Plate 7). Cumulus clouds originate from *convection in the form of *thermals, but *orographic effects often contribute to their growth. Cumulus occurs in four *cloud species: *fractus, *humilis, *mediocris, and *congestus.

**cup anemometer** An *anemometer consisting of a vertical shaft that is rotated through the action of the wind on three or four conical or hemispherical cups at the end of cross bars. The shaft's rotation may either operate a counter once per revolution (in what is known as a **cup-contact anemometer**), or generate an electrical current. The former requires conversion to obtain the wind speed, whereas the latter may give a direct reading.

**current, ocean** A permanent or semi-permanent major flow of water across one of the oceans of the world. Unlike most tidal streams, where the direction and velocity vary continuously, ocean currents flow in a single general direction and tend to show seasonal fluctuations in volume and rate. Their motion arises from several causes, including *wind stress (particularly from the dominant *westerlies) and density differences, such as those in a *thermohaline circulation. The displacement of water may be balanced by a counter current, flowing either at the surface or at depth.

The main ocean currents are a major factor in determining *climate, because they transport large quantities of heat from equatorial regions to higher latitudes and, conversely, feed cool water to lower latitudes. They are of great significance in *climate modelling, and major climatic changes (*see feature*), such as the *Little Ice Age, are believed to have been initiated by alterations in current patterns. *See also* BOTTOM WATER; GREAT OCEAN CONVEYOR BELT; GYRE; INDIVIDUAL CURRENTS.

**curvature** A derived quantity that acts as a measure of the change in direction of an airstream, and is defined as the wind speed divided by the radius of curvature. By convention, *cyclonic circulation has positive curvature.

**CUTLASS (Co-operative UK Twin Located Auroral Sounding System)** A pair of backscatter radar systems for studying the interactions between the *solar wind, the *magnetosphere, and the *ionosphere. The radars are located at Pykkvibaer in Iceland and Hankasalmi in Finland and survey a large segment of the northern *auroral oval.

**cut-off high** A warm *anticyclone that lies at high latitudes and which is isolated from the main *subtropical anticyclones. The situation is normally associated with *blocking. A cut-off high usually begins as a *ridge in the upper-air westerly flow, then becomes a closed circulation and subsequently extends down to the surface. *See also* WARM POOL.

**cut-off low (closed low)** A *depression that lies at a much lower latitude than the normal track of depressions and which is separated from *polar air by the zonal westerlies. The feature usually begins as a *trough in the upper-air flow, which becomes a closed circulation and then extends down to the surface. *See also* COLD POOL.

**cyclogenesis** The development or strengthening of *cyclonic circulation, such as the initial formation of a *trough or *depression, or the intensification of circulation in an existing system. The opposite of *cyclolysis. Although, in a depression, cyclogenesis often occurs at the same time as *deepening, the latter is not essential for the process.

**cyclolysis** The weakening of *cyclonic circulation, and the opposite of *cyclogenesis. Although in a *depression it is often accompanied by *filling, the latter is not an essential part of the process.

**cyclone** **1.** The generic term for a system in which, on a *synoptic chart, a central low-pressure region is surrounded by a set of closed *isobars (approximately circular, elliptical, or oval in shape). There is *cyclonic circulation with the winds flowing inwards. Also known as a *depression or **low**. Outside the tropics, such a system is often known as an **extratropical cyclone** or, when differing air masses are involved, a **frontal cyclone**. **2.** The local name for a tropical cyclone (*see feature*) in the Indian and western South Pacific Oceans.

**cyclonic circulation (cyclonic curvature)** Airflow around the local vertical that is in the same sense as the Earth's rotation, i.e. anticlockwise in the northern hemisphere, and clockwise in the southern.

**cyclonic rain** Rain that results from the vertical motion in a *synoptic-scale system, such as a *depression, or at a *front. *See also* CONVECTIVE RAIN; OROGRAPHIC PRECIPITATION.

**cyclostrophic wind** A wind with a marked curved flow in a system, such as a tropical cyclone (*see feature*), where the Coriolis acceleration (*see* CORIOLIS EFFECT) is weak compared with the *centripetal acceleration.

# D

**Dansgaard–Oeschger event** A period of extremely rapid temperature change (increases of 10 deg C or more) that occurred during the last *ice age. Strictly speaking, our knowledge of these events is confined to Greenland, where 23 have been identified by the *oxygen-isotope method from deep ice cores, and where they are correlated with snow deposition rates. The ice cores show, however, that the events also coincide with large changes in methane concentrations and dust deposition rates, which probably relate to conditions in tropical wetlands and in eastern Asia, respectively. These imply major climatic changes throughout the northern hemisphere. *See also* HEINRICH EVENT.

**DART** *See* DEEP-OCEAN ASSESSMENT AND REPORTING OF TSUNAMIS.

**dart leader** A streamer that follows the same path as the first *return stroke in a *lightning discharge, but in the reverse direction. It is distinguished from a *stepped leader by having no branches. When multiple strokes occur along the same channel, dart leaders precede each return stroke.

**data collection platform (DCP)** A term rather loosely used either for a complete *automatic weather station or similar hydrological station, or for the actual electronic equipment, including the radio transmitter by which measurements are returned to the operating centre.

**dayglow** A form of *airglow, similar to the better-known *nightglow, but stronger and with more emission lines. It is difficult to study because of the dominance of the far more intense scattered daylight, but observations are made by *polar-orbiting satellites. There are oxygen bands in the infrared, bands from various nitrogen species ($N_2$, $N_2^+$, NO) in the ultraviolet, and oxygen, nitrogen, and hydrogen emissions at far-ultraviolet wavelengths.

**daylight recorder** The intensity of daylight is measured by an instrument commonly known as an **illuminometer**. Daylight is received by a photocell or similar device after having passed through a diffuser. The output is proportional to the intensity of the illumination.

**DCP** *See* DATA COLLECTION PLATFORM.

**debris cloud** The visible cloud of dust or debris that occurs at the point where a *landspout or tornado (*see feature*) touches the surface. It may sometimes indicate the presence of a tornado, even if the vortex itself is not visible, and often extends many metres into the air.

**declination** The angular distance of a celestial body north (+) or south (–) of the *celestial equator, measured along a great circle passing through the object and the north and south celestial poles.

**declination, magnetic (sometimes known as magnetic variation)** The deviation of a compass needle from true north, measured in degrees, defined as positive towards the east and negative towards the west. The amount of deviation varies with geographical location, and also, because of secular changes in the orientation of the Earth's magnetic field, alters with the passage of time.

**deepening** A decrease in the central pressure of a low-pressure system (*depression, tropical cyclone (*see feature*), etc.), normally associated with an intensification of the accompanying winds. A high, consistent rate of deepening for 24 hours occurs with a *bomb. The converse is *filling.

**Deep-Ocean Assessment and Reporting of Tsunamis (DART)** A system for detecting *tsunamis deployed by the *National Oceanic and Atmospheric Administration. Ocean-floor pressure sensors relay data acoustically to anchored surface buoys, which forward the data via satellite links to tsunami centres. The system consists of six units (being increased to ten) around the northern Pacific basin.

**deflation** The removal of loose surface material, such as snow, dust, or sand, by the action of the wind.

**degree-day** A measure of the accumulated deviation of the mean temperature over a period from a specific standard value, assessed as one degree-day per degree above or below the standard for one day. Originally applied to the growth of plants, which shows a direct relationship to the cumulative temperature above 5 °C.

**degrees of frost** An obsolete term for the number of degrees (Fahrenheit) that the air temperature falls below the freezing point. Rarely, if ever, used with degrees Celsius.

**dendrochronology** The dating of events or objects by the study of *tree rings (commonly known as **tree-ring dating**). The pattern of the varying widths of tree rings from a particular climatic region may be matched across various timber samples. From certain known dates, a series of chronologies may be derived far longer than the lifetime of individual trees. For European oak and North American bristlecone pine, in particular, such chronologies extend back to approximately 10000 BP.

**dendroclimatology** The determination of past *climate from the examination of the variations in width or density of *tree rings, including the 'false rings' caused within annual rings by various forms of stress. Because of the number of variables involved in trees' response to climatic factors, such study is more complex than simple dating. *See* DENDROCHRONOLOGY.

**deposit gauge** An instrument used in studies of *pollution and designed to trap both *aerosols and solid particles in the atmosphere for subsequent analysis.

**deposition** The process by which *water vapour changes phase directly into a solid without passing through an intermediate liquid phase; the opposite of *sublimation. Examples are the formation of *frost and the growth of *ice crystals at low temperatures. The water vapour releases *latent heat into the system as it changes phase.

**depression** An area of low pressure, characterized by closed isobars (approximately circular, elliptical, or oval in shape), accompanied by *cyclonic circulation with winds that spiral in towards the centre, leading to *convergence. Also known as a **cyclone** or **low**. The term 'depression' is most commonly applied to those systems outside the tropics (**extratropical cyclones**) which are associated with different *air masses and *fronts (**frontal cyclones**), but non-frontal forms are *orographic depressions, polar-air depressions (*see* POLAR MESOCYCLONE), and *thermal lows. *Tropical depressions are weak early or late stages in the evolution of (non-frontal) tropical cyclones (*see feature*).

A typical, middle-latitude depression goes through three major stages during its lifetime. The first is a wave on the *Polar Front known as a *wave depression. This subsequently develops a distinct *cold front, *warm front, and *warm sector, becoming a *warm-sector depression, during which time it normally deepens and its motion accelerates. In the final stage, the *occluded depression, the cold front overtakes the warm front, creating an *occlusion, and the systems begins to fill and decay. At the mature stage, depressions have

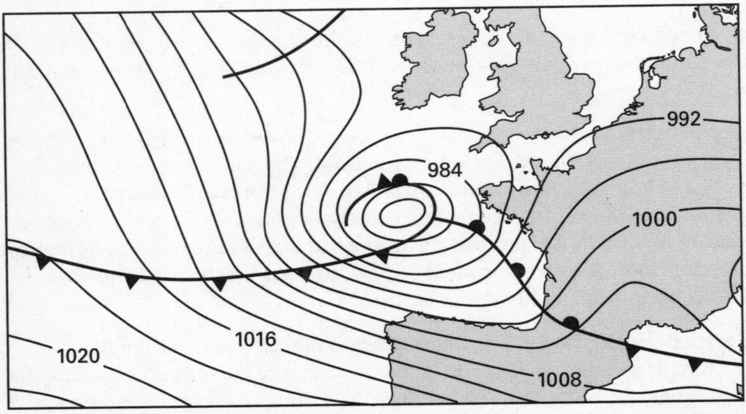


**Depression** A typical middle-latitude depression, with distinct *cold, *occluded, and *warm fronts.

a typical diameter of 2000–4000 km. Their motion tends to be from west to east with the overall *zonal flow, at speeds of up to 50 knots (*c.*90 km $h^{-1}$), but individual depressions may move in any direction, and large systems often remain stationary for several days.

Depressions are extremely important features of the general circulation (*see feature*), and are the principal means for the interchange of air and the transfer of heat from subtropical regions to higher latitudes.

**derecho** [Spanish: 'straight'] A damaging straight-line wind (i.e. one not associated with rotation as in a tornado—*see feature*), produced by organized *downdraughts such as those found in a *squall line or *mesoscale convective system. The areas affected may be hundreds of kilometres long and up to 150 km in width.

**descuernacabras** [Spanish: 'dehorns goats'] A bitterly cold, and often gusty, northerly wind, occurring around Christmas in the mountainous regions of Spain. *See also* MATACABRAS.

**desert** A region that has an *arid climate and is able to support very little (or no) flora or fauna, and where the species that do exist are adapted for an extreme climate. Although normally envisaged as the hot, sand and rock deserts such as the Sahara and the Arabian Desert, the term also applies to other regions where the extremely low *precipitation is caused by the local topography, such as the Gobi and Atacama Deserts, and also to certain high-altitude regions, such as the Andean Altiplano and the central plateau of Antarctica.

**desertification** The process by which a semi-arid region is transformed into a *desert, often by the growth of existing *arid regions. The alteration may occur as a result of global or regional climatic change (*see feature*) or variation, or through human intervention such as deforestation or overgrazing by domestic livestock. The decrease in ground cover and erosion also result in changes in *albedo, which may also contribute to a reduction in *precipitation, and thus intensify the process of degradation.

**desiccation** Permanent decrease in the amount of water available in a region, primarily through a reduction in *precipitation as a result of climatic change (*see feature*).

**deuterium (D)** An *isotope of *hydrogen ($^2H$), with a nucleus that consists of one proton and one neutron. On Earth, it is present in natural sources of hydrogen at an abundance of approximately $1.5 \times 10^{-2}$ per cent, and is also present in water as the oxides HDO and $D_2O$ (heavy water). Its chemical properties are essentially identical to those of hydrogen, but reactions generally occur more slowly on account of its greater atomic mass.

**development** The intensification of either *anticyclonic or *cyclonic circulation.

**devil** The name used for various forms of *whirl and usually descriptive of the type of material being raised from the surface, such as *dust devil, snow devil, steam devil, and water devil. Some dust devils may involve strong convection, but other devils may arise from weak convection or be vortices produced by the funnelling effects of the local terrain.

**dew** Water droplets that have condensed on to grass, plants, and other objects near the ground surface, as a result of radiative cooling overnight. The temperature has declined to the *dew point, but not below freezing, when water vapour would be directly deposited as ice crystals in the form of *frost. Droplets that subsequently freeze are known as *silver frost. Dew should not be confused with *guttation drops.

**dewbow** A coloured bow, analogous to a rainbow (*see feature*), observed on a dew-covered surface. The optical causes are as the same as those in a rainbow, but because the dewdrops are located on a more or less horizontal plane, a dewbow generally appears as an ellipse or hyperbola. Similar bows have been reported from water droplets lying on the surface film of a pond.

**dew point** The temperature at which a parcel of air (at constant pressure and *humidity mixing ratio) reaches saturation, and at which condensation forms as cloud droplets, *dew, *mist, or *fog. The dew point may be measured directly by a dew-point *hygrometer, or indirectly from wet- and dry-bulb thermometer readings. Also known as the **condensation point.**

**dew pond** An artificial pond created by farmers to provide water for livestock on hills or other places with no natural water source. It originally consisted of a shallow excavation lined with impermeable clay. Nowadays a plastic membrane is normally used. Dew ponds conserve natural rainfall; the idea that dew is the source of the water is incorrect.

**diabatic** A process in which heat enters or leaves a system (*compare* ADIABATIC). Important meteorological examples of such processes are *condensation, *evaporation, *freezing, and *sublimation, as well as the absorption or emission of radiation. Because of the significance of adiabatic processes in meteorology, the (grammatically incorrect) term 'non-adiabatic' is often used instead.

**diablo** A hot, drying wind from the interior that affects the San Francisco area of California and which is similar in nature to the *Santa Ana that occurs farther south around Los Angeles.

**diamond dust (occasionally known as frost in the air, frost mist, snow mist)** Extremely fine *precipitation in the form of tiny, unbranched ice crystals which occurs at temperatures below –30 °C. The crystals are so fine that they do not affect visibility and are seen only by their glitter in sunlight or artificial light. They produce *halo phenomena, including some rare effects that are suspected not to occur with larger crystals. Diamond dust forms a very significant fraction of the annual accumulation of ice and snow over the high Antarctic plateau.

**diffluence** An increase in the separation of the *streamlines in a flow such as an airstream. It is the opposite of *confluence and may be considered as an increase in the area covered by the flow. It is generally accompanied by a decrease in wind speed downwind, and a decrease in pressure. Diffluence in a *jet stream, with the corresponding reduction in pressure at height, is a major factor governing the formation of *depressions. *See also* CONVERGENCE; DIVERGENCE.

**diffraction** The dispersion or deviation of any form of waves as they encounter the edge of a physical barrier or pass through an aperture. The resulting different path-lengths generate interference patterns. In meteorology, the most common effects of diffraction are found in various optical phenomena such as *Bishop's ring, *corona (1), *fogbow, *glory, *iridescence, and the supernumerary bows associated with rainbows (*see feature*).

**diffuse front** A *front that is poorly defined, where the two *air masses have similar characteristics, such as *temperature and *humidity, and across which there is little change in wind direction.

**diffuse radiation** Radiation that reaches a surface or detector from all directions. Generally applied to the portion of solar radiation that reaches the surface after *scattering by the *aerosols and molecules in the atmosphere.

**diffusion** The process by which two gases or fluids, originally separate, become mixed through molecular motion. It is a far less efficient and a slower process than turbulent mixing. In the atmosphere, it becomes of importance primarily in the upper atmosphere. *See* HETEROSPHERE; TURBULENCE.

**Dines, William Henry** (b. London, 5 August 1855; d. Benson, England, 24 December 1927) British meteorologist who invented various instruments to measure atmospheric properties, notably the pressure-tube *anemometer and a lightweight *meteorograph. He carried out extensive research and analysis of conditions in the upper *troposphere and *stratosphere and made an important contribution to understanding the dynamical properties of *cyclones and *anticyclones.

**direct circulation** A closed, vertical circulation in which warm air rises and cold air sinks, or where the *potential energy of adjacent warm and cold air masses is converted into *kinetic energy. In the general circulation (*see feature*), the cells between the Equator and the *subtropical anticyclones are the principal examples. Both *land- and *sea-breezes are smaller-scale examples.

**discontinuity** A point at which the value of a specific atmospheric property undergoes an extreme, abrupt change, rather than varying relatively smoothly, as is normally the case. Typical discontinuities are found at major *fronts, in hydraulic jumps (such as *bores and the *morning glory), and shockwaves in general.

**disdrometer** A device for determining the sizes and size distribution of raindrops.

**dissipation trail (distrail)** A clear lane in cloud produced by an aircraft, and thus the converse of a *condensation trail. Three mechanisms may be involved: (a) although the engine exhaust adds water vapour, its heating effect may be sufficient to evaporate pre-existing cloud droplets; (b) the aircraft's wake vortices may entrain sufficient dry air from above the cloud to lower the *relative humidity below the *dew point; (c) the exhaust may add sufficient *nuclei to initiate *glaciation giving rise to a clear lane in a manner similar to the formation of *fallstreak holes.

**dissociation** The breakdown of molecules into smaller molecules or individual atoms, normally through the absorption of *ultraviolet radiation. In the atmosphere, the most significant dissociation is that of oxygen molecules into atomic oxygen, which occurs at altitudes above 80 km.

**distrail** *See* DISSIPATION TRAIL.

**disturbance** A term sometimes used for *depression or *trough, especially in descriptions of the general situation, when the more specific terms may be inappropriate.

**disturbance line** An organized system of thundery *showers and *squalls, several hundred kilometres long, that travels westwards across West Africa. Disturbance lines originate in summer when south-westerly *monsoon air is overrun by dry air from the Sahara, particularly when there is *divergence in the upper *troposphere to the north of the easterly tropical *jet stream.

**diurnal variation** A fluctuation in an atmospheric element over the course of a day, obtained by averaging observations over a sufficient length of time to remove all non-periodic variations. Temperature and *relative humidity show distinct diurnal variations. The former reaches a maximum about 2 hours after local noon in winter, 3 hours after noon in summer, and a minimum around sunrise. The variation in humidity is approximately opposite in phase.

Pressure shows a semi-diurnal variation arising from *atmospheric tides, with maxima at around 10:00 and 22:00 local time, and minima at 04:00 and 16:00. The *amplitude is greatest at the Equator, where it reaches about ±2 hPa, and decreases polewards, being about ±0.7 hPa at latitude 50°. At low latitudes the pressure variation is clearly visible on *barograph traces.

**divergence** A net horizontal outflow of fluid, such as air or water, from a region, and the opposite of *convergence. Vertical motion is required to compensate for the outflow. In the atmosphere, large-scale divergence is encountered at upper levels above *depressions and tropical cyclones (*see feature*), often marked by an extensive *cirrus shield. In the oceans, divergence caused by the *Ekman effect on currents, for example, leads to *upwelling on the western side of continents (as with the *Benguela, *California, *Canary, and *Peru Currents). *See also* DIFFLUENCE.

**D-layer** *See* IONOSPHERE.

**Dobson spectrophotometer** An instrument used for measuring the abundance of atmospheric *ozone. By monitoring the strength of absorption at two specific wavelengths in the *ultraviolet, it is possible to determine the amount of ozone in the vertical column above the instrument.

**Dobson unit (DU)** A measure of atmospheric *ozone: 1 Dobson unit is equivalent to 0.01 mm of ozone. For the purposes of measurement, it is assumed that all the ozone, throughout the atmospheric column, exists as a single layer at 0 °C and a pressure of one *standard atmospheric pressure (1013.25 hPa). The range normally recorded is between 200 and 500 DU, but values of less than 100 DU have been measured for the Antarctic *ozone hole at its greatest extent. Recently, the abundance of certain other gases, such as *sulphur dioxide, has also been quoted in Dobson units, using similar criteria.

**doctor** A name used in various parts of the world for a wind that is invigorating or brings relief from uncomfortably hot (or hot and humid) conditions. Specifically applied to the *harmattan on the Guinea coast of north-west Africa, various *sea-breezes (especially the very strong **Cape Doctor** around Cape Town in South Africa and the **Freemantle Doctor** in Western Australia), and to certain *fall winds, as in Jamaica.

**doldrums** The nautical term for the *equatorial trough, the zone of light and variable winds that lies within approximately 5–10° of the Equator, where the *trade winds converge at the *Intertropical Convergence Zone.

**Doppler effect** The difference between the frequency of electromagnetic radiation or sound emitted by a source and that received by an observer, caused by the relative motion of the two. When the distance between the two is decreasing, the received frequency increases. Electromagnetic radiation is shifted towards the blue end of the spectrum, and sound is raised in pitch. Conversely, when the distance is increasing, electromagnetic radiation is shifted towards the red, and sound decreases in pitch. Because the change in frequency is proportional to the relative velocity, the effect may be used to determine the radial motion and velocity of a distant source. In the case of a radar system, the 'source' is the object reflecting the radar signal.

**Doppler On Wheels (DOW)** A mobile radar system consisting of a *Doppler radar mounted on a vehicle, and designed to be transported and set up rapidly to record details of tornadic *supercell systems and tornadoes (*see feature*).

**Doppler radar** A radar system that is able to determine the motion, towards or away from the radar aerial, of the object reflecting the radio waves. This makes use of the *Doppler effect, in which the frequency of the returned signal is increased when motion is towards the radar, and decreased when motion is away from it. Meteorological radars (such as the *NEXRAD system) determine both the distribution and velocity of *precipitation particles and may therefore be used to estimate the severity of *thunderstorms or the likelihood of tornado development (*see feature*).

**Douglas sea and swell scale** A two-digit scale (proposed by Captain Douglas, Hydrographer to the Navy in the 1920s) for reporting the height of waves as observed at sea. It allows for the distinction between the *sea* (waves generated locally by the wind) and *swell.

**DOW** *See* DOPPLER ON WHEELS.

**downburst** An extremely strong *downdraught that occurs beneath a vigorous *cumulonimbus and affects an area more than 4km across (*compare* MICROBURST). It fans out horizontally on reaching the ground, and then may cause considerable damage, which tends to show a starburst pattern. Depending on conditions, vigorous downbursts may give rise to major *gust fronts. Large downbursts are sometimes known as **macrobursts**.

**downdraught** A relatively confined, descending current of air within a *cumulonimbus cloud or other storm system. Generally initiated by the drag exerted by the severe *precipitation in a *shower cloud or *thunderstorm, downdraughts are colder than the surrounding air, because they warm at the *saturated adiabatic lapse rate, which is less than the *environmental lapse rate. The air may also be cooled by evaporation of *precipitation, thus becoming denser and either starting the downdraught or increasing the rate of descent. Strong downdraughts fan out on reaching the ground, and may extend tens of kilometres ahead of the system, as well as producing violent *squalls. *See also* UPDRAUGHT.

**downshear** In the direction of maximum vertical *wind shear, normally in relation to the main convective circulation (i.e. *updraught). The *anvils of *cumulonimbus clouds are a maximum extent on the downshear side. *See also* UPSHEAR.

**downwelling** A subsidence of oceanic water from the surface, usually as a result of cooling or an increase in salinity, both of which result in an increase in density. There are two principal regions: the North Atlantic and the Weddell Sea. In the former, the principal driving mechanism is evaporative cooling, and in the latter the release of brine by the formation of *sea ice in winter. *See also* GREAT OCEAN CONVEYOR BELT; THERMOHALINE CIRCULATION.

**draa** A giant sand dune or ridge (sometimes tens of kilometres long), occurring in *ergs, on which smaller dunes are superimposed. The largest such dunes are known to reach heights of *c.*360 m in the Namib Desert; *c.*460 m in the Sahara; and are reputed to attain 500 m in the Badan Jaran Desert in China.

**drag-force anemometer** An anemometer that determines wind speed from the pressure exerted on a cylinder or sphere. Although it has certain disadvantages, both speed and direction may be measured with one device, and this is of value in severe environments.

**drainage area** *See* CATCHMENT AREA.

**drainage gauge** *See* LYSIMETER.

**drainage wind** *See* KATABATIC WIND.

**drift** **1.** Material that has been transported and formed into a pile by the wind, such as a snowdrift. **2.** The velocity of an ocean *current, often specified in *knots. **3.** An ocean current, e.g. *North Atlantic Drift. **4.** Loose rocky material that has been transported by *glaciers or *ice sheets and deposited from the ice or meltwater.

**drifting dust/sand** A term used in *synoptic meteorology for material that has been raised by the wind, but which does not obscure visibility at eye level, and is thus less than 1.5 m above the ground (*compare* BLOWING DUST/SAND). A similar term applies to snow.

**drizzle** Liquid *precipitation in the form of water droplets 200–500 μm in diameter, generally arising from *stratus cloud. The droplets form by *coalescence of cloud particles, assisted by the weak *turbulence within the cloud. Provided the lowest layer is sufficiently humid, the droplets in drizzle are sufficiently heavy to fall to the surface, unlike *fog droplets.

**drop** A particle of water, generally taken to mean one that is sufficiently large to fall as *precipitation (either in the form of *rain or *drizzle), with a diameter of 200 μm or more, as distinct from a *fog (*c.*100–200 μm) or *cloud droplet (*c.*1–100 μm). *See also* RAINDROP.

**dropsonde** An instrument package (similar to a *radiosonde) that is designed to be deployed from a manned or unmanned aircraft where meteorological data are required, such as in areas within a tropical cyclone (*see feature*). The descent is controlled by a parachute. Two versions are commonly employed, the first reporting *pressure, *temperature, and *humidity (PTU) readings, and the second having the additional capacity for position-finding from *GPS or *LORAN signals, from which wind vectors may be obtained. The latter form is sometimes referred to as a *dropwindsonde.

**dropwindsonde** A version of *dropsonde that has intrinsic wind-finding capability through the reception of *GPS or *LORAN radio-location signals. Some versions do not carry any sensors for pressure, temperature, or humidity. Like *rawinsonde, the term is not used consistently.

**drosometer** An instrument for the measurement of the amount of *dew that is deposited on an artificial surface. The method may involve weighing (which is used in recording devices), or direct comparison with standardized photographs obtained under known conditions.

**drought** An extended period with abnormally low rainfall that is long enough to cause specific effects upon crops, vegetation, or water supply in the area concerned. Because of the great variation in moisture requirements, both between different crops and also with regard to domestic and industrial water supply, no duration or percentage is universally applicable, but a drought is generally held to prevail when an area experiences at least three weeks with less than one-third of normal precipitation for the time of year. Various drought indices exist based upon the balance between supply (*precipitation and soil moisture) and demand (from *evapotranspiration and lake, river, and groundwater recharge), together with shorter-term crop moisture indices based upon rainfall, temperature, and the specific requirements of individual crops.

**droxtal** [from 'drop' + 'crystal'] A rarely used term for a small particle of ice, 10–20 μm in diameter, that forms directly from *supercooled water at temperatures below –30 °C. Such tiny particles are largely responsible for *diamond dust and *ice fog.

**dry adiabatic lapse rate** The rate of change of temperature with altitude of a dry (unsaturated) air parcel under *adiabatic conditions, and which is numerically equal to 9.767 deg C $km^{-1}$. *See* LAPSE RATE; PSEUDO-ADIABATIC LAPSE RATE; SATURATED ADIABATIC LAPSE RATE.

**dry air** **1.** In physical meteorology, air that contains no water vapour whatsoever. **2.** In *synoptic meteorology, air with a low *relative humidity, often defined as below 60 per cent.

**dry- and wet-bulb hygrometer** *See* PSYCHROMETER.

**dry ice** The common name for solid *carbon dioxide, which *sublimes at –78.5 °C. It may be used for *cloud seeding, in which it is released into cloud containing *supercooled water. It creates localized cooling, causing the water to freeze into ice crystals, which then fall out as *precipitation.

**dry intrusion (dry wedge, dry slot, occasionally known as the surge region)** An area associated with the *cold front of a *depression, where the equatorward branch of the *cold conveyor belt causes cold, dry air from the upper *troposphere and lower *stratosphere to descend towards the surface. When the eastwards motion of the *warm conveyor belt and the cold air are approximately equal, the dry intrusion reaches the surface and produces a clearly defined cloud-free zone behind the cold front. This initially appears as a notch in the rear of a *leaf cloud, and develops into a distinct clear sector at the *comma-cloud stage. The area is not only conspicuously clear of cloud and sharply defined on water-vapour satellite images, but has gusty winds because the air originates near the *jet stream, which normally crosses the system at this position. The dry intrusion is associated with the *sting jet found in *explosive cyclogenesis. When the dry intrusion overruns the warm conveyor belt it creates a *split front, with a distinct pattern of cloud cover and precipitation that differs considerably from the 'classic' cold front first described by Vilhelm *Bjerknes.

**dry line (dryline)** The boundary between warm, humid air and hot, dry air. It is therefore distinct from a *front, but has inherent instability, because humid air has a lower density than dry air (at the same temperature and pressure). The term is particularly used in North America for the boundary between humid air from the Gulf of Mexico on the east and dry desert air on the west. Normally lying approximately north–south over the Midwestern states, the dry line often advances eastwards in the day and retreats westward at night. Severe *thunderstorms and tornado activity (*see feature*) frequently develop along the dry line or immediately to the east of it.

**dry season** In tropical and subtropical regions (in particular), a period that recurs every year in which *precipitation is low or completely absent. In *Mediterranean climates, the dry season occurs in summer, but in the tropics it is generally in the 'winter' period, i.e. the period when the Sun is on the opposite side of the Equator.

**dry slot** **1.** A distinct cloud-free area visible on satellite images of developing *depressions (*see* DRY INTRUSION). **2.** A similar region that develops behind certain major *thunderstorms and *supercell thunderstorms, and visible on radar images, where dry air descends from middle levels.

**dry spell** An imprecisely defined period during which *precipitation is lower than normal for the time of year. The criteria for determining a dry period differ depending on the specific application.

**ducting** *See* ANOMALOUS PROPAGATION.

**duplicatus (du)** A *cloud variety. Either *stratiform or *cumuliform cloud that occurs as more than one layer at different heights. Most frequently applies to *altocumulus, *altostratus, *cirrus, *cirrostratus, and *stratocumulus.

**dust** Solid particles that may be suspended in the atmosphere and carried for long distances, and as *volcanic dust, dust and sand raised from the surface, material from *meteors (and *micrometeorites), and smoke particles. They give rise to *scattering and when sufficiently dense are the cause of *haze. At lesser concentrations, scattering and *diffraction, which are strongly dependent on particle size, may produce specific optical effects, including *Bishop's ring, a *blue moon or sun, and strong *sunrise/sunset colours. Certain particles also act as *condensation nuclei.

**dust counter** An instrument for determining the number of dust particles in a known volume of air, also sometimes known as a **konimeter**. In one form, the volume may be expanded when the individual dust particles function as *condensation nuclei, giving rise to water droplets that may be more easily counted.

**dust devil** A rapidly rotating column of air carrying dust and debris into the air, sometimes for as much as 500 m. Dust devils arise in hot, dry environments when the layer of air closest to the ground has a *superadiabatic lapse rate, and often occur ahead of an advancing *dust storm. Air converging on a particularly hot spot on the ground starts the rotation, which may be in either direction, and is often initiated by local differences in friction with the ground. Also known as a **dust whirl** or (more rarely) **sand pillar** and by various local names, such as 'cock-eyed Bobs' in Western Australia. *See also* WHIRL.

**dust haze** A form of *haze where the particles have been raised into the atmosphere by a recent *dust storm or *sandstorm. Saharan dust is often transported widely across southern Europe and the Atlantic, to create a summertime haze in the eastern United States.

**dust storm** Fine dust raised from the surface by strong and often highly turbulent winds, usually after a period of *drought. Dust storms are relatively infrequent and mainly occur over arable land. They may be differentiated from the more common *sandstorms that occur in desert areas. Visibility must be reduced to less than 1000 m for the term to be used in synoptic reports. *See* BLOWING DUST.

**Dust Veil Index** An index, originally developed by Hubert Lamb in 1970, that provides an approximate measure of the influence of volcanic dust on climate. Subsequently replaced by the *Volcanic Explosivity Index.

**dust whirl** *See* DUST DEVIL.

**dynamical meteorology** The study of both the causes and the intensity of the motion of air throughout the atmosphere.

# E

**earthquake light** A luminous phenomenon preceding or coinciding with an earthquake, frequently mentioned in anecdotal accounts, but of which few scientific observations have been made. Various glows, streamers, and luminous spheres have been reported. Numerous speculative theories as to the origin of the light exist, including the escape of gas from the ground (*see* *IGNIS FATUUS) and *corona discharges (similar to *St Elmo's fire). Other suggestions are that the light may be the phenomenon known as triboluminescence, where the fracture of crystals produces a momentary glow, or caused by the piezoelectric effect, in which charges separate in response to mechanical pressure.

**earthshine** The faint illumination of the dark portion of the Moon that is particularly notable during the crescent phases and is also known as **earthlight**. It consists of backscattered sunlight that has been previously reflected from the Earth's surface and atmosphere, and is particularly strong when the relevant region of the Earth has extensive cloud cover. Because of *Rayleigh scattering, the blue component is stronger than in the original sunlight.

**east coast cyclone (east coast low)** *See* BOMB.

**easterlies** A belt of winds that blow from a generally easterly direction, in particular the polar easterlies, and the *equatorial easterlies or *trade winds.

**easterly** Any wind that blows from the east.

**easterly wave** *See* TROPICAL WAVE.

**East Wind Drift** *See* ANTARCTIC COASTAL CURRENT.

**eclipse wind** A localized wind that occurs during a solar eclipse, which is of doubtful scientific validity, despite many anecdotal accounts. It appears to be a largely subjective phenomenon, although there is some evidence to suggest that short-lived *fall winds or a temporary change in the strength of a prevailing *sea-breeze may sometimes occur.

**ecliptic** The Sun's apparent path against the *celestial sphere, which is, in fact, the projection of the Earth's orbit in space. Because the Earth's axis is not perpendicular to the plane of its orbit, the ecliptic cuts the *celestial equator at two points, the vernal (spring) and autumnal *equinoxes. It reaches its greatest elevation above and below the celestial equator at the summer and winter *solstices, respectively.

**ECMWF** *See* EUROPEAN CENTRE FOR MEDIUM-RANGE WEATHER FORECASTS.

**ecoclimatology** The study of living organisms and their relationship to *climate.

**eddy** A whirl that occurs within a fluid flow, where a discrete packet of air or liquid follows an approximately circular motion different from, but within, the overall flow. Eddies occur at all scales from the largely horizontal eddies in the form of *depressions and tropical cyclones (*see feature*) to microscopic eddies created by airflow over grass or leaves. They occur in regions of *wind shear and are directly involved in *turbulence and *mixing.

**effective height** The estimated height, assuming open ground free from obstructions, that would experience wind speeds equivalent to those recorded by a specific *anemometer. The latter may well be affected by local obstructions and eddies, which may so affect the instrument that different effective heights apply for different wind directions. *See also* EXPOSURE.

**effective temperature** The *temperature at which still, saturated air would produce the same sensation of comfort (or discomfort) in a sedentary worker wearing normal indoor clothing as that created by the environmental air under actual conditions of temperature, *humidity, and air motion.

**EISCAT (European Incoherent Scatter)** An international radar system that studies the interaction between the *solar wind and the Earth's *magnetosphere and *ionosphere. The main radar transmitters are located near Tromsø in Norway and Longyearbyen on Svalbard (Spitzbergen). Receiving sites are near Kiruna in Sweden, and Sodankylä in Finland. The EISCAT Scientific Association also cooperates with other installations, such as *CUTLASS, to carry out experiments in high-latitude, ionospheric *plasma physics.

**Ekman effect** The change in direction of the ocean currents with depth. The flow is deflected increasingly towards the right (in the northern hemisphere) and decreases in velocity as the effect of *wind stress lessens with greater depth. Away from the Equator, where the effect is virtually non-existent, the deep flow may be in the opposite direction to the surface current. The depth of the current reversal decreases towards the poles but is at around 50–100 m over large regions of the middle latitudes. The effect was first observed and qualitatively understood by Fridtjof Nansen, the Norwegian oceanographer and polar explorer, who asked Ekman to derive the mathematical proof. The term has since been extended to the change in direction and speed of the wind from the surface, where the flow is subject to the greatest friction, to the level at which it becomes *geostrophic. *See also* EKMAN SPIRAL.

**Ekman layer** The portion of the *planetary boundary layer that lies between the *surface boundary layer and some higher level at which the wind flow becomes *geostrophic. Within this layer, surface *friction is significant, and the wind vectors display an *Ekman spiral. Its depth is largely determined by the extent of *convection and *turbulence.

**Ekman spiral** The figure obtained by plotting either the vectors of oceanic current flow with increasing depth, or those of the wind with increasing altitude. In both cases, because of the *Ekman effect, the result is an equiangular spiral, which graphically indicates the changes in direction and velocity.

**E-layer** *See* IONOSPHERE.

**electricity, atmospheric** *See* ATMOSPHERIC ELECTRICITY.

**electric storm** An obsolescent term for *thunderstorm.

**electrojet** A strong and relatively narrow electrical current that flows in the lower *ionosphere above the Equator.

**electrometeor** An obsolescent term for any phenomenon involving atmospheric electricity: *lightning, *thunder, *St Elmo's fire, etc., and, by extension, phenomena such as the *airglow and *aurora.

**elevation** The height of the ground above mean sea level. *Compare* ALTITUDE (2).

**elf (pl. elves)** An extremely short-lived optical phenomenon, first observed scientifically in the early 1990s, that occurs high above certain vigorous *thunderstorms. Lasting approximately one millisecond, these transient flashes, invisible visually, occur at heights of 85–105km, often in conjunction with *sprites. They begin at a central point and expand (faster than the speed of light) as a torus, to diameters of several hundred kilometres. This agrees with the theory that elves occur when an electromagnetic pulse from a powerful lightning discharge reaches the *ionosphere, where it accelerates electrons that collide with nitrogen molecules, which then radiate the faint red light.

**El Niño** [Spanish: 'the (Christ) child'] A warm southward-flowing current that appears along the coast of Ecuador and Peru around Christmas. It was considered to bring a pleasant respite from the cold *Peru Current, and provide welcome rains to the barren coastal region. The term was subsequently extended to apply to the much more dramatic events that occur every few years when high sea-surface temperatures apply over a wide area of the central and eastern Pacific. Now viewed as the 'warm' phase of the overall El Niño Southern Oscillation (*see feature*). *See also* LA NIÑA.

## El Niño Southern Oscillation (ENSO)

The coupled interactions between the ocean and atmosphere that occur in the tropical Pacific, now known to link what were once thought to be the separate phenomena of *El Niño and the *Southern Oscillation. In the normal state, the easterly *trades are strong; there is a deep pool of warm water in the western Pacific (where the *thermocline is at a depth of *c.*150 m) with strong atmospheric convection above it. In the eastern Pacific, equatorial upwelling brings cold water to the surface, and the thermocline is at shallow depths.

In an El Niño event, the easterly trade winds weaken, and this decreases the upwelling of cold water in the eastern Pacific. The warm pool in the western Pacific migrates eastwards, leading to decreased convection and *precipitation in the west, and increased convection in the central and eastern Pacific. This 'warm' phase ceases when the oceanic *Rossby waves that were generated by the relaxation of the trade winds, and which have propagated westwards, are reflected by the Indonesian archipelago, and return as subsurface *Kelvin waves. These travel eastwards along the thermocline beneath the Equator, at 2–3 $m\,s^{-1}$ and with amplitudes of *c.*100 m. They bring cold water back towards the surface along the coast of South America, thus shutting down the cycle.

In the La Niña 'cold' phase, there are stronger easterly trades, accentuated upwelling off South America, leading to a much larger pool of cold water in the eastern Pacific, and a deep pool of warm water in the west, with greatly enhanced convection and precipitation. By contrast the eastern Pacific region is abnormally dry.

The interval between El Niño events is 2–7 years and their duration 12–18 months. Their strength varies considerably and it has recently been announced that a scale will be introduced to quantify the effects, similar to the *Fujita scale for tornadoes (*see feature*) and the *Saffir-Simpson scale for tropical cyclones (*see feature*). Warm (El Niño) phases will be categorized as W1–W5, and cold (La Niña) phases as C1–C5. Through *teleconnections, the various phases of the ENSO phenomenon affect conditions in many other regions of the world, especially those in the northern Pacific and Indian Oceans. It has recently been established that the frequency of Atlantic *hurricanes is also associated with the ENSO phase. More hurricanes occur during the La Niña phase than during warm El Niño conditions. *See also* Antarctic Circumpolar Wave; oscillation; Pacific Decadal Oscillation; Walker circulation.

**emissions credit (emissions trading)** A provision of the *Kyoto Protocol that rewards nations for retaining or increasing the size of their forests. Because trees remove *carbon dioxide from the atmosphere and incorporate carbon into their tissues, nations that increase their forested area are to be awarded credits that may be set against their individual targets for the reduction in carbon dioxide emissions. It has also been proposed that countries whose emissions are below the established levels should be able to trade credits with countries that are unable to meet their greenhouse-gas reduction targets.

**Enhanced Fujita Scale** *See* FUJITA SCALE.

**ensemble forecasting** A method of forecasting in which a computer simulation is run numerous times, each with slightly different initial conditions. (The *European Centre for Medium-Range Weather Forecasts currently calculates 51 versions for its global ensemble forecast.) If the forecasts remain similar 7–10 days ahead, it may be assumed that the weather is in a quasi-stable state and that predictions are likely to be reasonably accurate. If, however, the forecasts deviate after a few days, conditions are more likely to break down suddenly and unpredictably. Ensemble forecasting therefore provides a method of estimating the probability that particular forecasts will be correct.

**ENSO** *See feature* EL NIÑO SOUTHERN OSCILLATION.

**entrainment** The mixing of surrounding air into a *thermal or other flow of air, which inevitably results in the alteration of the latter's properties. Entrainment of cool air significantly limits the growth of *cumulus clouds.

**entrance region** The region upstream of a localized wind-speed maximum in a *jet stream, where the air is accelerating. This causes *convergence in the left half of the entrance region, a compensating *subsidence from stratospheric levels, and *divergence at the surface. The overall vertical circulation tends to promote *frontolysis (as seen in the weakening of the cold front in the rear of a *depression).

**entropy (*S*)** A measure of the energy in a system that is no longer available to provide work in any particular process. A consequence of the second law of thermodynamics, which in simplified form states that all systems tend to greater disorder.

**environmental lapse rate** The actual *lapse rate that prevails in the atmosphere at a particular time, and which is, for example, recorded by a *radiosonde ascent. The difference between this and the *dry adiabatic lapse rate or *saturated adiabatic lapse rate determines the *stability or *instability of a specific parcel of air at any particular altitude.

**environmental pressure** The actual pressure at a particular site, with no correction to sea level.

**Envisat (Environmental Satellite)** A European *polar-orbiting satellite launched in 2002 and designed to monitor various atmospheric and surface properties, in part to ensure continuity with the measurements made by the ERS-1 and ERS-2 *European Remote Sensing satellites. The highly sophisticated instruments include the AATSR (*Advanced Along-Track Scanning Radiometer), GOMOS (*Global Ozone Monitoring by Occultation of Stars), and SCHIAMACHY (*Scanning Imaging Absorption Spectrometer for Atmospheric Chartography). It will be superseded by the Sentinel-1 satellites (*see* SENTINEL), the first of which is scheduled for launch in 2011.

**EOLE** A Franco-American project that launched 480 *constant-level balloons in the southern hemisphere, 1971–2, to study the *zonal flow and *planetary waves, and also more specific atmospheric parameters.

**EPS** *See* EUMETSAT POLAR SYSTEM.

**equatorial air** The air in the *equatorial trough, which although originating as *tropical air in the *subtropical anticyclones tends to be somewhat cooler than in the immediate areas of subsidence. There are no distinct differences in the physical properties of equatorial and *maritime tropical air. Any slight changes in stability may cause major changes in convective activity and the resulting weather.

**Equatorial Counter Current** An oceanic current that flows from west to east, approximately along the Equator, in the Atlantic, Indian, and Pacific Oceans. There is an accumulation of water (and higher levels) on the western sides of the main oceanic basins, partly because the *Ekman effect is absent at low latitudes, and also because water is displaced northwards by the strong, permanent *subtropical anticyclones in the southern hemisphere. Water flows eastwards down the hydraulic gradient. The Pacific Equatorial Counter Current is particularly strong during an *El Niño event. *See also* CROMWELL CURRENT.

**Equatorial Current** *See* NORTH EQUATORIAL CURRENT; SOUTH EQUATORIAL CURRENT.

**equatorial easterlies** A term sometimes applied to the *trade winds during the summer season, when they are particularly deep and may extend to altitudes of 8–10 km.

**Equatorial Jet Stream** An easterly *jet stream that develops in the eastern hemisphere, at approximately 10° N, during the northern summer (particularly in the months of July and August). It lies at altitudes of 15–20 km, and extends from Asia to Africa, but is not observed over the western hemisphere. It arises because of the steep temperature gradient from north to south in the upper troposphere—the coldest air lying over the Equator. A weaker, broad, low-level easterly jet, at 4–5 km altitude, may form over Africa during the same period and at approximately the same latitude. Oscillations beneath this jet create westward-moving *tropical waves, *disturbance lines, and *thunderstorms.

**equatorial trough** The more or less continuous region of low pressure that lies between the *subtropical anticyclones and within 5–10° of the Equator. It shifts slightly north and south with the Sun during the course of the year, but with a lag of between 1 and 2 months. This zone is occupied by *equatorial air, and over the oceans is also known as the *doldrums. It is less pronounced (or absent) over land areas in the northern summer. Certain regions experience *divergence at the surface and consequent *subsidence, giving rise to an **equatorial dry zone**, the most distinctive being an area south of the Equator in the central Pacific Ocean.

**Equatorial Undercurrent** *See* CROMWELL CURRENT.

**equatorial westerlies** The westerly winds that sometimes develop within the *equatorial trough when the *Intertropical Convergence Zone is well north or south of the Equator. The north-easterly or south-easterly *trade winds cross the Equator and, because of the reversal of the *Coriolis effect, acquire a westerly component. The term is also applied to the westerlies that are present throughout most of the year in the eastern Indian Ocean.

**equilibrium line** The boundary between a glacier's *accumulation and *ablation zones, also known as the **firn line**. Also applied to *ice caps and *ice sheets.

**equinoctial gales** A term frequently (but erroneously) used with the implication that gale-force winds are prevalent, or strongest, around the time of the *equinoxes in March and September. In fact, for the British Isles, gales occur most often around the winter *solstice in December, and are least frequent in June.

**equinox** **1.** The two points at which the *ecliptic intersects the *celestial equator. **2.** The dates at which the Sun reaches those points, crossing the celestial equator from south to north on 20 or 21 March and from north to south on 22 or 23 September. (The variations of date arise from the calendar that we employ.) At the equinoxes (neglecting *refraction and certain other effects) the Sun rises due east and sets due west, and day and night have equal durations. *See also* SEASONS; SOLSTICE.

**equivalent head wind** In aviation, the uniform *head wind, always directed in the opposite direction to an aircraft's track, that results in the same average ground speed as that actually attained. An equivalent *tail wind may be similarly defined.

**equivalent potential temperature** The *equivalent temperature adjusted at a *dry adiabatic lapse rate to a pressure level of 1000 hPa (i.e. nominally to sea level).

**equivalent temperature** The temperature that an air parcel would assume if all the *water vapour present were removed by *condensation, under conditions of constant pressure, with the *latent heat released being used to warm the parcel.

**erg** [from Arabic *irj*: 'dune field'] One of the three main desert landforms, and which consists of shifting sand dunes, with no exposed rock surfaces. *See also* HAMMADA; REG.

**ERS** *See* EUROPEAN REMOTE SENSING SATELLITE.

**ERTS (Earth Resources Technology Satellite)** *See* LANDSAT.

**escape velocity** The minimum velocity that a molecule or atom of gas (or larger material object) must obtain to escape from the gravitational field of a particular planetary body. For the Earth, this velocity is 11.2 m $s^{-1}$. Atmospheric components escape from the *exosphere, where there is a long *mean free path between collisions.

**ESOP** *See* EUROPEAN SUBPOLAR OCEAN PROGRAMME.

**étage** [French: 'level'] One of three height zones (high, middle, and low) into which the *troposphere is divided for the purposes of *cloud classification. The height is taken as that of the cloud base. High: between *c.*3–8 km (*c.*10000–25000 ft) in polar regions and *c.*6–18 km (*c.*20000–60000 ft) in the tropics. Middle: between *c.*2–4 km (*c.*6500–13000 ft) in polar regions and *c.*2–8 km (*c.*6500–25000 ft) in the tropics. Low: from the surface to *c.*2 km (*c.*6500 ft) at all latitudes.

**etesian winds** Generally northerly winds (between north-west and north-east) that blow in the summer season (between May and September) over the Aegean Sea and eastern Mediterranean. They are known as *meltemi* in Turkey.

**Eumetsat** The European Organization for Meteorological Satellites, an intergovernmental organization that operates the *Meteosat satellites, originally developed and launched by the European Space Agency. The organization currently also operates the Meteosat Second Generation (MSG) *geostationary satellites and the first of the series of *MetOp *polar-orbiting satellites.

**Eumetsat Polar System (EPS)** A series of *polar-orbiting meteorological satellites (the *MetOp satellites) to replace the current series of *NOAA morning satellites with an improved service. *See also* INITIAL JOINT POLAR SYSTEM.

**European Centre for Medium-Range Weather Forecasts (ECMWF)** An international organization, at present supported by a consortium of 25 European states, set up in 1973 for the development of numerical methods of medium-range forecasting (up to 10 days ahead), and the preparation of such forecasts for its members. Based in Reading, UK, ECMWF also conducts research aimed at improving its forecasts, which are widely regarded as the most accurate currently attainable.

**European Remote Sensing Satellite (ERS)** A pair of satellites (ERS-1 and ERS-2) launched by the European Space Agency. Although primarily designed as Earth-resource satellites, their instrumentation also obtained meteorological and climatological data. ERS-2, in particular, surveyed atmospheric *ozone with the *Global Ozone Monitoring Experiment.

**European Subpolar Ocean Programme (ESOP)** A European Union research programme devoted to the study of the Greenland Sea region and in particular the role of oceanic convection in the *carbon cycle.

**evaporation** The process by which a liquid or solid changes into a gaseous state at a temperature less than the *boiling point; the opposite of *condensation. In meteorology, it is generally restricted to the change between water and water vapour, or ice and water vapour, the latter phase change also being known as *sublimation. The molecules that escape from the surface have the highest *kinetic energy. The removal of this *latent heat by the water vapour leads to a reduction in the average kinetic energy of the remaining water molecules, and thus to a reduction in temperature of the liquid.

**evaporation fog** *Fog that forms through the *evaporation of relatively warm water or raindrops into cooler air. Examples are *steam fog and *frontal fog, respectively.

**evaporimeter** An instrument for determining the rate at which water evaporates into the atmosphere.

**evapotranspiration** The overall loss of *water vapour from the ground, consisting of *evaporation directly from the soil plus *transpiration from plants.

**exit region** The region downstream of a localized wind-speed maximum in a *jet stream, where the air is decelerating. This causes *divergence in the left half of the exit region (in the northern hemisphere), compensating upward motion in the underlying *troposphere, and *convergence at the surface. The opposite effect occurs in the right half of the exit region, and the combined overall vertical circulation tends to promote *frontogenesis, i.e. the formation of frontal systems. *See also* ENTRANCE REGION.

**exosphere** The outermost region of a planetary atmosphere, in which atoms and molecules may occasionally reach *escape velocity and leak away into space as a result of the low density, high temperatures, and long *mean free paths between collisions. For the Earth, this region may be considered to begin between altitudes of *c.*200 and *c.*700 km, depending on *solar activity. *See* THERMOPAUSE.

**explosive cyclogenesis** *Cyclogenesis that occurs when a *depression deepens by more than 25 hPa in 24 hours, often as a result of enhanced uplift linked with upper-level *divergence at a *jet streak. *See also* BOMB.

**exposure** The precise location of a meteorological instrument, which has a direct effect on the accuracy, repeatability, and representative nature of the readings given by that instrument. The requirements for temperature measurements are largely met by standardized *thermometer screens, but the correct siting of *anemometers and *rain gauges (for example) often involves considerable problems.

**extratropical cyclone** A system occurring outside the *tropics that has closed *isobars around a central low-pressure area, and *cyclonic circulation. The term is most commonly applied to those systems (*depressions) that arise through the interaction of two different *air masses (otherwise known as **frontal cyclones**). The mechanisms that sustain the system differ from those found in tropical cyclones (*see feature*). *See also* BOMB; CYCLONE (1).

**extratropics** A term sometimes used for those regions of the Earth that are poleward of the *tropics or of the tropical easterly (*trade wind) zone.

**extreme** A term used in climatology for the highest or lowest value of a specific meteorological parameter for the period under consideration (generally a month, season, or year). When applied to the whole period for which records are available it may sometimes be described as an **absolute extreme**.

**eye** The calm central region of a tropical cyclone (*see feature*), where the storm's temperature is highest. Its diameter is generally 10–50 km, but may reach 70 km or more in the most powerful storms. Some of the air that has risen to high levels in the *eye wall sinks into the eye and warms adiabatically as it does so, contributing to the high temperature. Cloud cover is often non-existent or extremely broken.

**eye of the wind** A nautical expression indicating that one is facing, or the vessel is pointing, directly into the wind.

**eye wall** The sharply defined inner edge of a ring of *cumulonimbus clouds that surrounds the central *eye in a tropical cyclone (*see feature*). The most intense convective activity and *updraughts occur here, together with the greatest *precipitation, the highest sustained wind speeds, and extreme *thunderstorm activity.

# F

**Fahrenheit, Gabriel Daniel** (b. Danzig, 14 May 1686; d. The Hague, 16 September 1736) German-Dutch physicist who developed the first truly accurate mercury thermometer in 1714. He defined a temperature scale (subsequently named after him), taking as his zero point the lowest temperature obtainable with an ice–salt mixture. He initially took his second fixed point as body temperature, dividing the interval into 96 degrees. He subsequently redefined the scale to give the *freezing point of pure water as 32 °F, and *boiling point as exactly 212 °F.

He also discovered that boiling point varies with pressure, and that water may remain liquid below its nominal freezing point (*see* SUPERCOOLING).

**Fahrenheit scale** A temperature scale based on the *freezing or ice point of water (32 °F) and its *boiling point (212 °F). Although no longer widely employed in most countries, and long superseded for scientific work by the *Celsius and *Kelvin scales, it remains in common usage in the United States. For conversions, *see* APPENDIX B.

**fair-weather cumulus** A popular term for *cumulus of the *humilis or *mediocris species, used because these forms are normally accompanied by significant amounts of sunshine but do not produce any rain.

**Falkland Current** A branch of the *Antarctic Circumpolar Current which turns north around Cape Horn and flows parallel to the coast of Argentina to about 30° S. It then combines with the opposing *Brazil Current, becoming the eastward-flowing *South Atlantic Current.

**fallout** The process by which dry solid material is deposited from a dust cloud, such as that created by a volcanic eruption or a dust storm; alternatively, the material itself. In general, large particles will be deposited close to the source, whereas finer ones may be carried to considerable distances. Note that, strictly speaking, no *precipitation is involved. The term is, however, often specifically applied to radioactive material produced by a nuclear test or accident, when the deposition by precipitation is frequently extremely significant.

**fallstreak** *See* VIRGA.

**fallstreak hole** A clear, approximately circular or elliptical patch in thin *stratiform cloud (particularly *altocumulus or *cirrocumulus). If *glaciation is initiated over a relatively restricted area, the ice crystals fall out as *virga (fallstreaks), leaving a sharply defined hole in the sheet of cloud.

**fall wind** A strong downslope wind that remains cold, despite *adiabatic heating (and thus differs from a *föhn). It is a *katabatic wind that has its origins in a pool of cold air over the high ground. The most extreme examples are the violent winds that sweep down from the Antarctic ice sheet, reaching a maximum of around 320 km $h^{-1}$, and which may maintain hurricane-force winds for days or weeks on end. Fall winds are largely responsible for the record annual mean wind speed of 67 km $h^{-1}$ recorded at Cape Dennison, Commonwealth Bay, Antarctica.

**false cirrus** A popular name for cirrus *spissatus.

**fast ice** Any lake, river, or sea ice that is attached to the shore (including ice shelves), grounded in shallow water, or frozen to the bottom.

**fata bromosa** A form of *fata morgana, in which the flat and uniformly illuminated surface of the sea appears as a vertical or slightly overhanging wall of fog. Astigmatic imaging blurs any detail and also causes the 'fog' to appear brighter than its surroundings.

**fata morgana** [Italian: 'Morgan the fairy'] A form of superior *mirage in which the image of the actual surface appears in the form of a wall. The effect occurs when the temperature profile has an inflection, but is also relatively gentle. The atmosphere exhibits lensing properties but these are astigmatic, resulting in a redistribution of brightness within the image, often creating the effect of light and dark arches, and distant buildings. Named after Morgan, the half-sister of King Arthur in Celtic legend, who was able to create such illusions.

**FCCC** *See* FRAMEWORK CONVENTION ON CLIMATE CHANGE.

**feather** *See* WIND ARROW.

**feeder band (inflow band)** A band of low-level cloud that flows in towards the *updraught region of a *thunderstorm or *supercell thunderstorm. Curvature of feeder bands is one indication of an incipient tornado (*see feature*).

**Ferrel, William** (b. Fulton Co., Pennsylvania, 29 January 1817; d. Maywood, Kansas, 18 September 1891) American meteorologist who carried out fundamental theoretical investigations of the atmospheric circulation that would occur on a rotating Earth. His work on how air circulates around high- and low-pressure centres pre-dated *Buys Ballot's empirical discovery of the law to which the latter's name was subsequently given.

**Ferrel cell** A mid-latitude circulation cell, described by *Ferrel, and once thought to exist between a low-latitude *Hadley cell and a polar cell. It is now known that the form of circulation originally suggested does not occur, but the term is often used to describe the indirect, meridional circulation at middle latitudes. *See feature* GENERAL CIRCULATION.

**fetch** The distance over which a wind has free passage across a sea or ocean before encountering a coastline. The length of fetch is a major factor in determining the height of waves and swell (to which the term is also occasionally applied).

**fibratus (fib)** [Latin: 'fibrous'] A *cloud species. Straight or slightly irregular, long streaks of cloud, without any obvious tufts (*generating heads) or hooks. Generally applied to thin sheets or detached patches of *cirrostratus or *cirrus.

**filling** An increase in the central pressure of a low-pressure system (*depression, tropical cyclone (*see feature*), etc.), which may normally be taken as an indication that the system is weakening. The converse is *deepening.

**fine-mesh model** *See* GRID.

**fireball** **1.** An alternative name for *ball lightning. **2.** A *meteor whose luminosity is greater than even the brightest planet (Venus), and thus exceeds magnitude −4. Those fireballs that end in a terminal explosion are sometimes known as *bolides. Because bright fireballs may result in *meteorite falls, their timing and track (and any sonic boom) should always be reported.

**firn** **1.** Snow that has become granular as a result of various processes (melting and refreezing, or *sublimation and *deposition) and is in the process of becoming compacted and consolidated into glacier ice. Also known as **névé**. **2.** Snow that has persisted through one summer.

**firn line** *See* EQUILIBRIUM LINE.

**firn wind** *See* GLACIER WIND.

**Fitzroy, Admiral Robert** (b. Bury St. Edmunds, England, 5 July 1805; d. Norwood, England, 30 April 1865) British naval officer, hydrographer, and meteorologist. Became head of the Meteorological Department of the Board of Trade in 1854. Advocated obtaining simultaneous (*synoptic) observations, and this was begun on a limited basis in 1857. He introduced the system of storm-warning signals that remained in use for over a century, and suggested that warnings should be issued by telegraph, but was anticipated in this by *Le Verrier. An exchange of data with the French was started in April 1860, and Fitzroy issued the first regular daily weather report on 3 September 1860.

**flanking line** A line of *cumulus or cumulus *congestus clouds that (in the northern hemisphere) lies to the south or south-west of a *supercell thunderstorm. The clouds increase in height towards the main storm, with which they progressively merge. They indicate the position of the *gust front or pseudo cold front: the boundary between the *rear-flank downdraught and the inflow region. The term is often used for the similar line of clouds found with some *multicell storms.

**flann** A violent gust on the island of Foula (to the west of Shetland), particularly notable for being greatly intensified by the topography of the high land at the centre of the island.

**flash flood** Sudden, intense flooding, normally strongly localized and associated with extreme *shower or *thunderstorm activity, when high rates of *precipitation occur in a short period of time. Flash floods may occur under many different conditions, but usually arise when rain falls over a small catchment basin, or one, such as an urban area, where the surface is unable to absorb large amounts of water. The terrain may channel the extreme amount of run-off to produce a flood peak that reaches its maximum in just a few minutes or hours.

**flashy catchment** A *catchment area that, because of geographic, topographic, and geological factors, shows an almost immediate response to intense rainfall, resulting in a *flash flood.

**F-layer** *See* IONOSPHERE.

**flight level** The designated level at which aircraft may fly, and which is a surface of constant pressure, relative to a specific zero point, defined as being 1035.25 hPa. Normally given to the nearest 1000 feet in units of 100 feet, in accordance with the *ICAO standard atmosphere.

**floccus (flo)** [Latin: 'tuft'] A *cloud species. *Cirrus, *cirrocumulus, or *altocumulus clouds in the form of small individual tufts, more or less rounded above but ragged below, frequently with trailing *virga. Occasionally called **shred cloud**, but this term should not be confused with *pannus.

**Florida Current** The warm-water current that flows northwards from the Straits of Florida to Cape Hatteras. It then leaves the continental slope as it turns towards the north-east and combines with the *Antilles Current to become the *Gulf Stream.

**fluoroform** The common name for trifluoromethane, $CF_3H$ (a *hydrofluorocarbon, HFC-23), which has recently been found to be a potent *greenhouse gas, with an estimated atmospheric lifetime of 260 years. It is a waste by-product of the manufacture of the hydrofluorocarbon HFC-22, which is itself being used as a replacement for certain *ozone-destroying *chlorofluorocarbons.

**foehn** *See* FÖHN.

**fog** A visible suspension of water droplets in the atmosphere near the surface, and defined by international agreement as reducing visibility to less than 1 km (*compare* MIST). There is no physical distinction between fog and cloud, other than the fact that the base of clouds is above the surface of the ground.

Fogs arise when the air temperature and *dew point become essentially identical (and in the presence of sufficient *condensation nuclei). Cooling of the air may be by *advection over a cold surface, by *radiation, or by uplift, giving rise, respectively, to *advection fog, *radiation fog, or *upslope fog. Fog may also form through the *evaporation of relatively warm raindrops or water into cooler air, as in the cases of *frontal fog, *saturation fog, and *steam fog.

**fogbow** A white arc, with a radius of approximately 42°, centred on the *antisolar point. The inner edge is often bluish and the outer reddish. The mechanism is that of the rainbow (*see feature*), but the water-droplet diameter is so small (below about 50 μm) that *diffraction broadens the bands of colour until they overlap and appear white. When seen from within cloud or fog, the bow may be a partial or full circle, but when seen from outside (as from an aircraft flying above cloud), it may appear as a hyperbola or ellipse, depending on solar elevation. Also known as a **cloud bow**.

**fog drip** Liquid water that arises from the collision of *fog droplets with trees or other objects. Also known as **fog precipitation**. The principle is used (often employing plastic mesh) to collect water in certain arid or water-poor regions, such as in the Atacama desert of Chile.

**fog point** The air temperature at which *fog forms.

**fog precipitation** *See* FOG DRIP.

**föhn (foehn)** A warm and often extremely dry wind that descends in the lee of a mountain barrier. Föhn winds occur under stable conditions that are conducive to the formation of large-amplitude *lee waves. Air from very high levels (well above the mountain-tops) may be brought down to the surface and undergo considerable *adiabatic heating. Föhn winds (such as the *chinook) are noted for their rapid temperature rise, their desiccating effect, and the rapid disappearance of snow cover. Although originally applied to winds in the Alpine region, the term is now used for all similar winds.

**föhn wall** An elongated, often dense and persistent *wave cloud that lies above, or enshrouds, a mountain ridge, and is produced by *föhn conditions. When plentiful moisture is present, air rising on the windward side undergoes condensation into a continuous line of cloud, before plunging down to produce a föhn to leeward. When the humidity is lower, isolated *lenticular clouds may appear above just the highest peaks. Föhn walls are sometimes known by local names, such as the **chinook arch**, and the **helm**, which, respectively, often accompany a *chinook, and the *helm wind of Cumbria.

**force** **1.** In physics, an influence that acts to set a body in motion or induce an elastic strain within it. **2.** In meteorology, a subdivision of the *Beaufort scale.

**forcing** **1.** Any factor that causes air that is not initially warmer or cooler than its surroundings to rise or sink. Some examples are *convergence, *divergence, *frontal lifting, and *orographic lifting. **2.** The effect of any factor on one or more other physical processes that results in positive feedback. For example, it appears that the astronomical effects proposed by the Milankovich theory (*see feature*) act as forcing for changes in the *carbon cycle, with consequences for climatic change (*see feature*).

**forecast** A prediction of the expected weather conditions for a certain period of time ahead. Depending on their intended use, forecasts may be prepared for a particular area, place, or route between two places. Forecasts may also be classified according to the length of time covered: short-term forecasts, 24 hours to 3 days; *medium-range forecasts, 3–10 days; *long-range forecasts, >10 days. Short-term forecasts are the province of *synoptic meteorology, and may be carried out by the standard methods of *analysis, although the majority of modern forecasts (for all three types) are now prepared using *numerical weather prediction. For medium- and long-range forecasts, the *ensemble forecasting technique is increasingly used, and *analogue methods have become less common. *See also* CHAOS THEORY; EUROPEAN CENTRE FOR MEDIUM-RANGE WEATHER FORECASTS; HINDCASTING; NOWCASTING; WEATHER ROUTING.

**forensic meteorology** The application of meteorological knowledge to a wide range of legal problems. A forensic meteorologist may provide meteorological information to assist police with preliminary enquiries (for example), or appear as an expert witness at a coroner's enquiry, or in judicial courts in both criminal and civil cases.

**fork(ed) lightning** A popular term for cloud-to-ground *lightning in which several branching channels are visible in addition to the main discharge channel (*compare* STREAK LIGHTNING). Often erroneously thought to be fundamentally different from *sheet lightning.

**FORTE (Fast On-orbit Recording of Transient Events)** A US military satellite designed to detect electromagnetic radiation from clandestine nuclear tests that has provided some fundamental information about *lightning and *trans-ionospheric pulse pairs. FORTE's results indicate that there is more inter-cloud lightning over the oceans than previously thought, and that lightning is a significant source of *ozone in the upper *troposphere.

**Fourier, Baron de (Jean Baptiste), Joseph** (b. Auxerre, 21 March 1768; d. Paris, 16 May 1830) French mathematician, who was particularly interested in the physical applications of mathematics. He was the first to show that the increasing concentration of atmospheric *carbon dioxide would lead to global warming (*see feature*). *See also* ARRHENIUS, SVANTE AUGUST.

**fractocumulus** An obsolescent, unofficial term for *cumulus *fractus.

**fractostratus** An obsolescent, unofficial term for *stratus *fractus.

**fractus (fra)** [Latin: 'broken'] A *cloud species: *cumulus or *stratus clouds that exhibit a ragged or shredded appearance (Plate 12). The term is not applied to any other *cloud genera.

**Framework Convention on Climate Change (FCCC)** The United Nations convention drawn up in 1992 that recognizes the significance of climatic change (*see feature*) and sets out a broad overall framework for research, discussion, agreement on measures to be taken, and implementation. Specific agreements on particular issues are introduced by amendments or protocols to the Convention, such as the *Kyoto Protocol.

**Franklin, Benjamin** (b. Boston, 17 January 1706; d. Philadelphia, 17 April 1790) American statesman and scientist, who, from a study of violent storms, discovered how such systems generally move towards the north-east, despite having strong winds in other directions. He came close to appreciating the pattern of circulation in *depressions. Through his hazardous kite experiment of 1752, he proved the electrical nature of *lightning and subsequently invented the lightning conductor. He made an extensive study of the *Gulf Stream, measuring its temperature and direction, estimating its velocity and effect upon the weather, and producing the first true chart of an ocean current.

**frazil** Tiny ice crystals and plates that form in a *supercooled layer of moderately fast-flowing or turbulent water. They are prevented from freezing into sheet ice by the motion of the water, and give rise to a layer known as *grease ice. When sufficient frazil has formed, the crystals may begin to aggregate into *pancake ice.

**free atmosphere** A general term for the upper levels of the atmosphere, away from the *planetary boundary layer, where the effects of surface heating and friction may be assumed to be absent. It thus approximately corresponds to the atmosphere above the level at which the wind becomes *geostrophic.

**free lift** The lifting force of an inflated hydrogen or helium balloon, i.e. the lift excluding the weight of the balloon and any equipment.

**freeze** **1.** To pass from the liquid to the solid state, known technically as undergoing a **phase transition**; the opposite of **melt** or **fuse**. Although it applies to any material, in meteorology it is nearly always restricted to the transformation of water into ice. **2.** A period when the air temperature remains below 0 °C for at least two days.

**freezing drizzle** *Drizzle that falls as liquid droplets, but freezes on impact with objects on the ground that have a temperature below 0 °C, thus producing a layer of *glaze.

**freezing fog** *Fog that freezes on contact with objects at sub-zero temperatures and produces a deposit of *rime or *glaze.

**freezing level** The level at which the air temperature is 0 °C. Note that this is frequently not the level at which water droplets turn to ice, because of the phenomenon of *supercooling, whereby droplets may remain liquid at temperatures as low as −40 °C. It may be more correctly termed the *melting level.

**freezing nucleus** A solid particle (*nucleus) that promotes the formation of an ice crystal from a water droplet.

**freezing point** The temperature at which a substance changes from the liquid to the solid phase. For pure water, this is 273.15 K (0 °C), which is slightly lower than the *triple point (273.16 K).

**freezing rain** *Rain that falls as liquid droplets, but which freezes on contact with the ground, trees, or other objects that are below 0 °C, creating a layer of *glaze.

**fret** *See* HAAR.

**friction** The mechanical resistance that occurs between two substances that are in motion relative to one another, between a fluid medium and any object in contact with it (including another fluid), and between adjacent layers in a single medium. In meteorology the effects of friction are particularly important in the flow of air at the surface, within the *planetary boundary layer, and at all scales, including the general circulation (*see feature*). Friction causes the *surface wind to deviate from the *geostrophic wind and the *gradient wind, spiralling out from the centres of *anticyclones and in towards the centres of *depressions. It is for this reason that *Buys Ballot's law is only an approximation. Friction is also an important factor in regions of *wind shear. *See also* EKMAN LAYER; EKMAN SPIRAL; SURFACE BOUNDARY LAYER.

**friction layer** *See* PLANETARY BOUNDARY LAYER.

**frigid zone** One of the two theoretical climatic zones, lying poleward of the Arctic and Antarctic Circles at 66° 33′ N and 66° 33′ S, respectively. *See also* TEMPERATE ZONE; TORRID ZONE.

**front** The boundary or zone of transition between two *air masses of different temperature or *humidity, which thus differ in density. Specific types of front are either features of the overall general circulation (*see feature*; *Antarctic Front, *Arctic Front, *Mediterranean Front, *Polar Front) or associated with *depressions (*anafront, *cold front, *katafront, *occluded front, *stationary front, *warm front) or other phenomena (*gust front). *See also* FRONTAL ZONE; FRONTOGENESIS; FRONTOLYSIS; SEA-BREEZE.

**frontal cyclone** *See* DEPRESSION.

**frontal fog** *Fog that commonly forms at a *front, when rain falls from the relatively warm air aloft into cooler, lower air, where it evaporates, raising the air to saturation and eventually forming fog droplets.

**frontal lifting** The forced ascent of the warmer air when two *air masses converge at a *front (or when one air mass is overtaking the other). Such lifting commonly occurs at a *warm front.

**frontal zone** The zone of transition between two *air masses, where mixing has occurred. In *depressions such zones may range in width from 10 km to as much as 200 km.

**frontogenesis** The formation or intensification of a *front or frontal zone, generally by an increase in the horizontal temperature gradient and vertical *wind shear within a restricted region.

**frontolysis** The weakening or dissipation of a *front or frontal zone. The converse of *frontogenesis.

**frost** **1.** A period when the air temperature is equal to or less than the *freezing point of water (i.e. 0 °C). *See* AIR FROST; GROUND FROST. **2.** A term for the icy deposit (*hoar-frost) that forms on the ground, vegetation, or other objects when the temperature falls below freezing and ice is deposited directly from *water vapour in the air. *See also* BLACK FROST; GLAZE; RIME; SILVER FROST.

**frost day** A period of 24 hours in which the minimum temperature in a *thermometer screen is equal to or below 0 °C.

**frost flowers** Individual ice crystals, sometimes over 200 mm in height and usually occurring in small clumps, that grow on fresh sea ice under calm conditions, when humidity above the ice is high. Frost flowers act as wicks, producing a higher concentration of dissolved ions than found in the subjacent sea-water. They appear to be a significant source of atmospheric bromine monoxide (BrO) in polar regions in spring.

**frost heave** The disturbance of the surface of the ground when water, freezing in the form of **ice lenses**, expands with consequent movement of the soil. The mechanism is involved in the formation of polygonal ground (regular patterns of stones) in Arctic and *permafrost regions, and (in an extreme form) in the creation of *pingoes.

**frost hollow (frost pocket)** Low-lying ground where *frost occurs more frequently than in the surrounding area. This is normally because cold air drains down neighbouring slopes into a localized pocket from which it is slow (or unable) to escape. In the case of the famous Rickmansworth frost hollow, a railway embankment prevents the natural drainage of cold air from the valley.

**frost point** The temperature to which a sample of air must be cooled, at constant pressure and humidity, to reach *saturation with respect to ice.

**frost-point hygrometer** An instrument for determining the *frost point of the air, usually achieved by cooling (or warming) a metal surface until a thin deposit of frost is observed to remain constant in thickness, at which the temperature is determined by a suitable *resistance thermometer.

**frost smoke** *See* STEAM FOG.

**Fujita, Tetsuya ('Ted')** (b. Kitakyushu City, Japan, 23 October 1920; d. Chicago, 19 November 1998) Japanese-American meteorologist, noted for his seminal work on *mesoscale phenomena such as *thunderstorms, *squall lines, *mesocyclones, and tornadoes (*see feature*). He developed the tornado damage scale (*see* FUJITA SCALE) and became the leading expert on tornadoes. His discovery of the existence of *microbursts was initially controversial, but is now accepted as being of great significance both meteorologically and for airline safety.

**Fujita-Pearson scale** *See* FUJITA SCALE.

**Fujita scale (Fujita-Pearson Scale, Enhanced Fujita Scale)** A scale (introduced in 1971) for describing the severity of a tornado (*see feature*) or other severe wind, based on the intensity of damage that is observed. The maximum wind speed is estimated from an analysis of the damage, and is thus not directly comparable with direct measurements by *Doppler radar. Class F5 tornadoes are rare.

| SCALE NUMBER | WIND SPEED | | DAMAGE |
|---|---|---|---|
| | mph | $kmh^{-1}$ | |
| F0 | ≤72 | ≤116 | light |
| F1 | 73–112 | 117–180 | moderate |
| F2 | 113–157 | 181–251 | considerable |
| F3 | 158–207 | 252–330 | severe |
| F4 | 208–260 | 331–417 | devastating |
| F5 | ≥261 | ≥418 | incredible |

Professor *Fujita proposed changes in 1992 to overcome shortcomings of the scale, including failure to consider different types of construction and the inability to classify a tornado if no damage was observed. Subsequent evaluation by meteorologists and engineers led to the introduction of the **Enhanced Fujita Scale**, which was formally implemented in February 2007. Note that the scale is still based on wind-speed estimates (not measurements), made using 28 damage criteria, and 3-second gust speeds.

| EF-SCALE NUMBER | WIND SPEED (3-sec. gust) | |
|---|---|---|
| | mph | $kmh^{-1}$ |
| EF0 | 65–85 | 105–137 |
| EF1 | 86–110 | 138–177 |
| EF2 | 111–135 | 178–217 |
| EF3 | 136–165 | 219–266 |
| EF4 | 166–200 | 267–322 |
| EF5 | ≥200 | ≥322 |

(The scales are defined in terms of miles per hour, so the conversion to kilometres per hour is approximate. For more accurate conversion, *see* WIND SPEED.) *See also* TORRO SCALE.

**fulgurite** [from Latin *fulgur*: 'lightning'] **1.** A tube of fused silica, with a glassy interior and a granular exterior, formed when lightning strikes relatively dry, sandy ground. Sand fulgurites often show a dendritic (branching) pattern, and may be between 1 and 5 cm in diameter and up to 20 m long. **2.** A thin sheet of fused material created by the action of lightning on rocky mountain peaks. Such rock fulgurites are less common than the type formed in sand. *See also* BALL LIGHTNING.

**fumigation** The condition in which pollutants trapped beneath a stable layer are transported down towards the ground. This often occurs with the onset of daytime heating and the resulting convective mixing beneath an *inversion.

**funnel cloud** A condensation funnel: a rotating, smooth column or cone of cloud that descends from the base of a *cumulus *congestus or *cumulonimbus cloud, but which does not reach the surface. *See* LANDSPOUT; TORNADO *feature*; TUBA; WATERSPOUT.

**funnelling** The process by which the local topography forces a wind to blow along a valley, with a corresponding increase in wind speed. *See also* RAVINE WIND.

**fusion** Another term for melting.

# G

**gale** A wind of force 8 on the *Beaufort scale, i.e. one with a mean speed of 34–40 *knots (17.2–20.7 m $s^{-1}$), or gusts reaching 43–51 knots, as measured at the standard *anemometer height of 10 m. In common usage, the term is applied to any high, damaging wind.

**galera** A cold north-westerly wind that affects the narrow coastal strip north of the Cantabrian Mountains, in Spain, and which is particularly frequent in winter.

**gale warning** An advisory notification issued by a forecasting office that wind speeds reaching force 8 on the *Beaufort scale, or gusts over 43 *knots, may be expected at sea or on exposed coasts.

**Galilei, Galileo** (b. Pisa, 15 February 1564; d. Arcetri, Italy, 8 January 1642) Italian astronomer and natural philosopher, who was one of the earliest true experimental scientists. He constructed, but did not necessarily invent, a *thermoscope and *thermometer. He had several gifted students, most notably Evangelista *Torricelli, who invented the *barometer.

**gamma radiation (gamma rays, γ-rays)** The portion of the electromagnetic spectrum beyond the X-ray region, where photons have the shortest wavelengths ($<$0.01 nm, approximately) and the highest energies (100 keV to $>$10 GeV). Gamma rays most commonly arise through radioactive decay of atomic nuclei (of *radon, for example), but may also be produced by other, extremely energetic acceleration mechanisms, primarily of an astrophysical nature.

**gamma-ray and X-ray events** Pulses of gamma rays and X-rays emitted from the region above *thunderstorms, and detectable only from space. The mechanism that produces the necessary exceptionally high energies is currently unknown, although it is suggested that *cosmic rays may be involved. Energetic cosmic rays that encounter a powerful electrical field, such as that generated by a positive cloud-to-ground discharge, could produce a shower of secondary electrons (and gamma rays) directed upwards. Such electrons may account for the blue tendrils seen in the lower part of *sprites. The gamma-ray events are also known as **terrestrial gamma flashes** (TGF).

**GARP** *See* GLOBAL ATMOSPHERIC RESEARCH PROGRAMME.

**GARP Atlantic Tropical Experiment (GATE)** A major international experiment, involving 70 countries, carried out between June and September 1974 under the *Global Atmospheric Research Programme. It studied physical processes in the tropics, particularly the role of cumulus convection and its relationship to the general circulation (*see feature*).

**garúa (camanchaca)** A dense *fog or low *stratus that affects the west coast of South America in winter. The *drizzle associated with it provides a significant amount of moisture in this arid region.

**GATE** *See* GARP ATLANTIC TROPICAL EXPERIMENT.

**GCM** *See* GENERAL CIRCULATION MODEL.

**GDPS** *See* GLOBAL DATA-PROCESSING SYSTEM.

**gegenschein** [German: 'counter glow'] A slightly brighter area of the *zodiacal band occasionally observed at night in the region immediately opposite the Sun. It arises from enhanced backscattering by interplanetary dust particles that lie close to the *antisolar point.

**general circulation model (GCM)** A complex computer program that simulates the behaviour of the general circulation (*see feature*) of the Earth's atmosphere or oceans (or both). Such models are used for the investigation of individual phenomena, and also for modelling climatic change (*see feature*). Using models of varying complexity, considerable success has been achieved in modelling phenomena such as the El Niño Southern Oscillation (*see feature*). Here, the most detailed programs are the **coupled general circulation models** (CGCM), which incorporate detailed circulation models for both the atmosphere and oceans but which require great computational resources. Similar models give a reasonable simulation of major features of present global climate and appear to give a satisfactory modelling of conditions during the last *ice age. This gives confidence that their predictions of global warming (*see feature*) in response to the increase in *greenhouse gases is generally correct. Current models do, however, show considerable sensitivity to small-scale errors, and there are uncertainties in the modelling of various fundamental factors, such as the exact influence of clouds upon the global heat balance.

**general inference** A term for the summary of weather conditions that is usually given in a forecast before the more detailed description of predictions for specific areas. It normally includes details of the current pressure distribution, and generalized information about the way in which the situation is likely to develop.

**generating head** A small clump or tuft of cloud in which *glaciation is occurring, creating *precipitation. Usually small *cirrocumulus or *altocumulus, particularly of the *floccus species, with trailing *virga. Present, but not conspicuous, in *cirrus *uncinus.

## General circulation

The overall circulation of the *atmosphere. Taken in its widest sense, this would incorporate small-scale detail, but the term is usually employed for some form of time-averaged representation of the circulation, often accompanied by an indication of the range of variations that are encountered.

The amount of *insolation is strongly dependent on latitude; on average, excess energy is received over a zone extending from the Equator to latitudes 40° N and S approximately, and a deficit from there to the poles. This energy imbalance is the basic cause of the overall circulation and leads to the formation of three major *meridional circulation cells in each hemisphere: the *Hadley cell, the *Ferrel cell, and the polar cell. Simultaneously, belts of high and low pressure are formed in each hemisphere: an *equatorial trough (low pressure) between the Equator and 5–10° N and S; *subtropical anticyclones centred on *c.*30° N and S; temperate lows at 40–70° N and S; and polar highs beyond *c.*70° N and S. The pattern tends to shift north and south with the changing seasons and there are considerable deviations from symmetry because of the distribution of continents and oceans around the globe. Certain pressure features are permanent or semi-permanent, however, and play an important part in the overall circulation. These *centres of action include the *Azores and *Pacific highs and the *Aleutian and *Icelandic lows. In the northern winter, the Canadian and *Siberian highs are important seasonal features.

Because of the *Coriolis effect, the expected meridional circulation becomes a predominantly *zonal flow, giving rise to specific wind systems at the surface: the *doldrums in the equatorial trough, with weak, intermittent winds; the *trade winds between the *subtropical highs and the Equator; the *westerlies between the subtropical highs and the temperate lows; and the polar easterlies flowing out of the polar highs. Again, there are considerable variations with the seasons, notably with the position of the *Intertropical Convergence Zone, and also with the *monsoons, particularly the Asian monsoon.

The *air masses produced at high latitudes encounter those spreading out from the subtropical anticyclones along the *Polar Front, producing a major temperature gradient and giving rise to the majority of *depressions. These are responsible for transferring large amounts of energy and angular momentum from low to high latitudes. Corresponding flows (the *jet streams) arise in the upper atmosphere where there are steep temperature gradients.

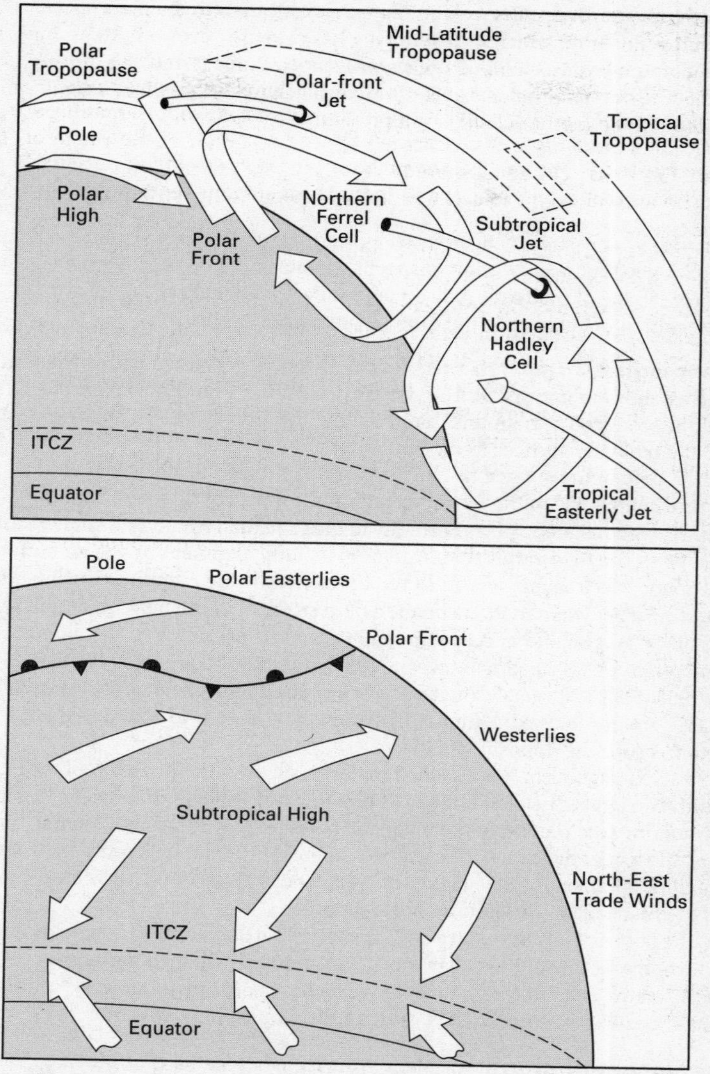


**General circulation** The overall global circulation is highly complex, but the basic *meridional circulation is shown (top), corresponding to the northern summer. A highly simplified representation of the surface-wind pattern is shown (bottom), which omits the wide variations in the position of the *Polar Front and the variable wind directions associated with the presence of *anticyclones and *depressions.

**genitus (-gen)** A suffix used in *cloud classification to denote a specific *cloud genus from which a current type has primarily evolved. An example is *altocumulus cumulogenitus (Ac cugen): altocumulus formed by the spreading out of cumulus. Unlike *mutatus, genitus implies that a significant amount of the mother cloud is still present.

**geocryology** The study of the physical processes governing glaciated regions, as well as *periglacial and *permafrost environments.

**geodesy** The study of the exact shape and size of the Earth, together with the precise geographical location of specific points on its surface. Nowadays, the precise shape of the Earth may be determined extremely accurately by observation of changes in the orbits of artificial satellites.

**geomagnetic field** The magnetic field that is generated within the Earth and extends out into space. The precise details of the mechanism by which the field is produced remain uncertain, but the dynamo action is believed to arise from circulating vortices in the liquid core.

The field at the surface resembles that of a dipole magnet, tilted at approximately 11° to the Earth's rotation axis and slightly offset from the centre, with magnetic poles currently in the Canadian Arctic and Antarctica. At any point the field varies from true north–south alignment by the amount of magnetic *declination, and from the horizontal by the amount of inclination (or **dip**). The declination, inclination, and strength of the field undergo slow secular changes, and also exhibit major reversals occurring at seemingly random intervals ranging between 40 000 and 35 000 000 years (average 300 000–1 000 000 years). The reversals are rapid on a geological time-scale, taking about 5 000 years, during which the field decays to zero and then reforms with the opposite polarity.

*Palaeomagnetism has revealed major changes in the direction of the Earth's magnetic field with time. At first this was thought to indicate *polar wandering (the motion of the magnetic poles relative to the continents), but it was subsequently shown to reveal *continental drift and thus form the basis of modern theories of *plate tectonics. Magnetic reversals were direct indication of the mechanism of **sea-floor spreading**.

The field extends into space and surrounds the Earth as the *magnetosphere, which interacts with the *solar wind. Radiation and charged particles released by *solar flares give rise to *aurorae, *geomagnetic storms, *sudden enhancement of atmospherics, and *sudden ionospheric disturbances.

**geomagnetic storm** A violent disturbance of the Earth's *geomagnetic field following a *solar flare or coronal mass ejection (*see* CORONA (3)), and generally ensuing about 24–36 hours after the event. Abrupt variations in the strength and direction of the interplanetary magnetic field, together with associated shockwaves, strongly perturb the *magnetosphere and may induce massive currents both in the ionosphere (where satellites may be damaged) and on the surface, leading to disruption or breakdown of power distribution grids. Major storms are usually accompanied by *aurorae, often visible at much lower geomagnetic latitudes than normal, with corresponding interference with short-wave radio communications. Conditions usually return to normal after a period of 2–3 days.

**geomagnetism** The study of the Earth's *geomagnetic field, including its origin, spatial extent, and variations with time. *See also* PALAEOMAGNETISM.

**geophysics** The study of all aspects of the physical properties of the Earth. The discipline involves such fields as geology, *geomagnetism, *meteorology, *oceanography, and seismology.

**geopotential height** A measure of the altitude of a point or layer in the atmosphere. Technically, it is expressed in terms of the *potential energy between the point and sea level, which depends upon the exact acceleration of gravity at that particular location. In practice, it does not differ from the actual altitude by more than a few per cent. It is, however, required for the definition of *isobaric surfaces, and thus *upper-level charts.

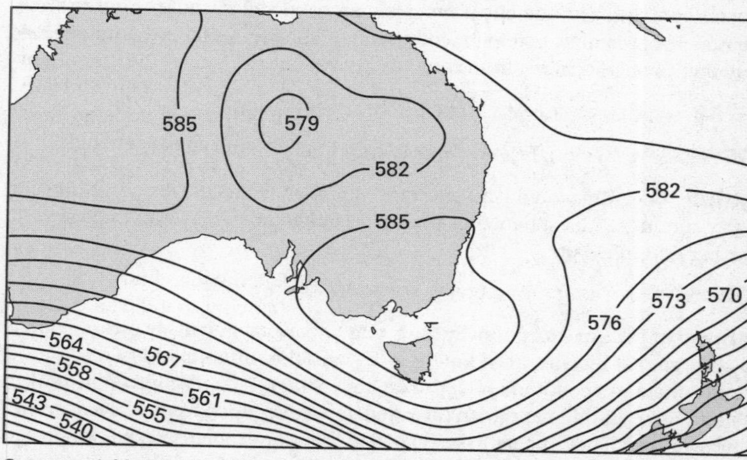


**Geopotential height** This chart of geopotential height may be compared with those for *pressure and *thickness for the same area and time.

**geosphere** The portion of the Earth that consists of liquid and solid material. It thus comprises the *hydrosphere (including the *cryosphere) and the *lithosphere, but excludes the *atmosphere.

**Geostationary Earth Radiation Budget (GERB)** An instrument carried by *Meteosat Second Generation satellites to monitor the Earth's Radiation Budget (ERB), in particular for an understanding of climatic change (*see feature*). It provides wide-band coverage of short-wave (reflected) solar *radiation and long-wave (emitted) thermal radiation. The short cycle time of 15 minutes means that it provides new insight into diurnal variations in the energy budget.

**geostationary satellite** An artificial satellite in an orbit 35 900 km above the Equator, at which altitude it completes one orbit in the same time as one rotation of the Earth (i.e. 24 hours). It thus remains (almost) perfectly stationary above a specific point and has a field of view covering approximately one-third of the Earth's surface (Plates 14 and 15). The *meteorological satellites in geostationary orbit include the *GMS, *GOES, and *Meteosat series. By international agreement, when a satellite reaches the end of its life it is moved to another (usually higher) orbit to allow the orbital site to be reused.

**geostrophic wind** A hypothetical wind that blows parallel to the *isobars or *contours, with low pressure to the left (in the northern hemisphere). The assumption is that there is exact balance between the *pressure-gradient force and the horizontal component of the Coriolis acceleration (*see* CORIOLIS EFFECT). Provided there is low isobaric *curvature, such a wind gives a good approximation to the motion of air in the free atmosphere above the *planetary boundary layer. The velocity is inversely proportional to the spacing of the isobars (or contours) and independent of air density, and may be determined from an *isobaric chart. *See also* GRADIENT WIND.

**GERB** *See* GEOSTATIONARY EARTH RADIATION BUDGET.

**GEWEX** *See* GLOBAL ENERGY AND WATER CYCLE EXPERIMENT.

**ghibli** A hot, dry, and dusty southerly wind that occurs in Libya in spring and early summer. A local name for the *sirocco, and known as the *khamsin farther east in Egypt.

**GHOST** *See* GLOBAL HORIZONTAL SOUNDING TECHNIQUE.

**gigantic jet** An exceptionally long, and apparently relatively rare, electrical discharge that reaches from the top of a *cumulonimbus cloud to the *ionosphere at an altitude of approximately 90 km. First discovered in 2001 in Puerto Rico, 30-odd have been recorded, most over open tropical oceans, with few over land. They appear to be related to the shorter *blue jets. It is suspected that the lower portions are blue, and the upper, red, like *sprites.

**glacial (glacial phase)** A period of time during an *ice age during which there is major accumulation of snow and ice on both *ice sheets and valley *glaciers, with consequent growth. The polar ice sheets, in particular, extend greatly towards the Equator. Such phases are normally accompanied by a fall in sea level, and alternate with *interglacial phases.

**glaciation** **1.** The process by which *supercooled water droplets change into *ice crystals in the top of a *cumulonimbus cloud as it reaches the *calvus stage. The cloud may subsequently develop *capillatus features or a *cirrus plume, or become an *anvil cloud (*incus). **2.** An alternative term for *glacial phase or *ice age. **3.** The state of being covered by a *glacier, *ice cap, or *ice sheet.

**glacier** A large body of ice that has resulted from the accumulation of snow over mountains or a central land mass, and which moves downhill under the influence of gravity. Glaciers transport large quantities of material derived from surrounding rocks, and carve out characteristic U-shaped valleys. At the edge of an *ice cap they may feed an *ice shelf or may give rise to *icebergs if they end at open water. Snow that falls in the *accumulation zone is transformed into *firn and eventually becomes compacted and consolidated into glacier ice. Below the *firn line, in the *ablation zone, mass loss through melting, *sublimation, or calving exceeds accumulation.

**glacier wind** A gravity wind that originates above a glacier. During the daytime, the ice is much colder than the surroundings and cools the layer of air immediately above it, which, being dense, then slides downhill. Such localized, *katabatic winds are often confined to a very shallow layer no more than 2–3 m deep.

**glaciology** The study of all aspects of ice and snow, especially as they occur in *glaciers, *ice caps, and *ice shelves, as well as past *ice sheets.

**Glaisher, James** (b. London, 6 April 1809; d. Croydon, England, 7 February 1903) British meteorologist and astronomer. After early meteorological observations for the Ordnance Survey, and astronomical work at Cambridge Observatory, he became Superintendent of the Magnetical and Meteorological Department at Greenwich Observatory in 1840. He was largely responsible for the introduction of scientific meteorology, partly through setting up a network of reliable observers with calibrated instruments. He introduced the first newspaper reports on the weather (*Daily News*, 31 August 1848), initially for two months, but subsequently resumed on 14 June 1849. Apart from numerous scientific meteorological investigations, Glaisher made extensive observations, primarily to measure *humidity, *lapse rates, and *temperature, during famous balloon ascents in the period 1862–6, reaching an estimated 37 000 feet on 5 September 1862. A founder, President, and long-serving Secretary of the British (now Royal) Meteorological Society.

**glaze (black ice, clear ice, glazed frost, verglas)** A transparent layer of ice that forms when *drizzle, *fog, or raindrops freeze on contact with a cold surface. If the droplets are *supercooled, rapid freezing may occur when the temperature is below or even slightly above 0 °C, but when there is no supercooling, glaze forms only when the object's temperature is well below 0 °C. *See also* BLACK ICE; RIME.

**glider** An experimental design of autonomous underwater probe, designed to obtain data about oceanic currents, which is propelled by changes in its buoyancy. The probe navigates using the *Global Positioning System and returns to the surface at specific intervals to transmit data over a satellite link. Unlike *profiling floats, it is able to move horizontally through the water column.

**Global Atmospheric Research Programme (GARP)** A major international programme (1967–82) organized by the *World Meteorological Organization and the International Council of Scientific Unions to study the processes and dynamics occurring in the atmosphere, with the specific aim of providing data to improve *general circulation models and thus weather forecasting.

**Global Data-Processing System (GDPS)** The global system for the processing, storage, and retrieval of meteorological data obtained by the *Global Observing System under the *World Weather Watch. The various meteorological centres are linked by the *Global Telecommunications System.

**global dimming** A decrease in the amount of solar *radiation reaching the Earth's surface, primarily caused by the haze produced by industrialization. This phenomenon, which occurred from the 1950s to the 1980s, and is now thought to have come to an end, partially offset global warming (*see feature*) during that period. Despite its name it was primarily a regional effect, with the greatest dimming between 10° N and 40° N, where industrial activity was concentrated, and also greatest near cities. About 60 per cent of the dimming arose from increased cloud cover. The decrease, even over sparsely populated areas, was about 4 $W/m^2$ in a quarter century. However, at some tropical sites *insolation rose during that period.

**Global Energy and Water Cycle Experiment (GEWEX)** A series of projects under the *World Climate Research Programme for the investigation of atmospheric radiation, *boundary-layer, cloud, *hydrological, and *precipitation processes. Specific individual projects include: the GEWEX Asian Monsoon Experiment (GAME), the Global Precipitation Climatology Project (GPCP), the Surface Radiation Budget (SRB) Project, the International Satellite Cloud Climatology Project (ISCCP), and the International Satellite Land Surface Climatology Project (ISLSCP).

**global horizontal sounding technique (GHOST)** The use of large numbers of *constant-level balloons, released simultaneously, to monitor large areas of the globe over a long period of time. The balloons are superpressure balloons, designed to float at a level of constant atmospheric density, i.e. at an *isopycnic surface.

**Global Observing System (GOS)** The overall system that obtains standardized observations from all parts of the world, coordinated by the *World Meteorological Organization on behalf of the 185 member countries. One of the three core components of the *World Weather Watch. Observations are received from a vast range of different types of facility, including manned sites and *automatic weather stations on land, ships and drift and anchored buoys at sea, aircraft and radiosondes in the atmosphere, and both *geostationary and *polar-orbiting satellites.

**Global Ozone Monitoring by Occultation of Stars (GOMOS)** An instrument carried by the *Envisat satellite to determine the distribution of *ozone in the *stratosphere. The instrument uses the absorption by ozone of *ultraviolet radiation from stars to obtain highly detailed three-dimensional maps of ozone distribution.

**Global Ozone Monitoring Experiment (GOME)** An instrument carried by the European satellite ERS-2 for mapping the global *ozone layer. Its two-dimensional mapping function was extended to three dimensions by the more sophisticated equipment carried by *Envisat (*see* GLOBAL OZONE MONITORING BY OCCULTATION OF STARS). GOME-2 is a greatly improved version that is being carried by *MetOp satellites and which monitors *ozone ($O_3$), *nitrogen dioxide ($NO_2$), *water vapour ($H_2O$), bromine oxide (BrO), and other atmospheric gases.

**Global Positioning System (GPS)** A worldwide navigation system consisting of 24 NavStar satellites, orbiting at an altitude of 19000 km. The orbits are so arranged that at least four satellites are in view from any point on the Earth's surface at all times. GPS receivers use information carried by the signals to compute their location, accurate (in commercial systems) to *c.*100 m. Such receivers are increasingly used in *constant-level balloons, *drift buoys, *radiosondes, and other autonomous sensor systems. One of a number of existing and proposed Global Navigation Satellite Systems (GNSS), which include the existing Russian GLONASS system and the forthcoming European Galileo system. *See also* LORAN.

**global radiation** The overall total of solar direct and indirect radiation as received by a horizontal surface.

**Global Sea Level Observing System (GLOSS)** An international programme aimed at obtaining accurate data on global and regional sea levels for various purposes, including long-term studies of climatic change (*see feature*). Satellite altimetry is employed to compensate for changes in the elevation of recording sites through geological processes, using signals from the *Global Positioning System satellites to define the position of sites in a global reference frame.

**Global Telecommunications System (GTS)** The worldwide system for the collection, transfer, and distribution of meteorological data as rapidly as possible to all potential users. Part of the overall *World Weather Watch system.

**global-warming potential (GWP)** A measure of the potential contribution of a gas to global warming (*see feature*). It is defined as the *radiative forcing produced by the release of 1kg of the substance over its lifetime, relative to the radiative forcing of 1kg of *carbon dioxide.

**glory** An optical phenomenon consisting of a series of coloured rings about the *antisolar point, visible against the background of a cloud, fog bank,

## Global warming

An increase in the average temperature of the Earth, primarily because of the rise in concentration of various *greenhouse gases in the *atmosphere. Although significant warming episodes have occurred in the past, particularly accompanying *interglacial phases, the term is generally used nowadays for the anthropogenic effect upon *climate. The principal source of concern is the great increase in *carbon dioxide released by the burning of fossil fuels. Various other gases also contribute, and all these substances are subject to the *Kyoto Protocol. Current results suggest that warming is occurring at *c.*1.5 deg C per century and an increase of 1.4–5.8 deg C by 2100.

Warming is primarily confined to the *troposphere and is balanced by a corresponding cooling of the *stratosphere and *mesosphere. Recent results indicate that this upper-atmospheric cooling is causing the altitude of the *ionosphere to decrease. Although the amount of warming at the surface is subject to some debate, its effects are apparently visible in the melting of *glaciers, and a general rise in sea level. Although modelling of global warming is incomplete, *general circulation models suggest that there will be significant latitudinal variations, particularly in increased cloudiness, frequency of severe storms, and changes in *precipitation.

or mist. The rings generally resemble those of a *corona, with similar coloration—violet on the inside, to red on the outside. Multiple rings may occur and the radii are inversely proportional to the droplet diameter. Glories are caused by *diffraction in which the droplets act as point-source reflectors. The intensities of the various rings are well explained by the theory of *Mie scattering, but a rigorous theoretical treatment of the colours and radii remains elusive.

**GLOSS** *See* Global Sea Level Observing System.

**GMS (Geostationary Meteorological Satellites)** A series of Japanese *geostationary satellites, stationed at 140° E.

**GNSS Receiver for Atmospheric Sounding (GRAS)** A *Global Positioning System (GPS) receiver, carried by *MetOp satellites, that functions as an atmospheric-sounding instrument. By receiving GPS signals, data from GRAS may be used to obtain temperature and humidity profiles at least 500 times a day. When combined with time and position data from the GPS signals, GRAS observations may be entered directly into *numerical weather prediction models.

**GOC** *See* Great Ocean Conveyor Belt.

**GOES (Geostationary Operational Environmental Satellites)** A series of *geostationary meteorological satellites launched by the United States, normally with two operational satellites at any time: GOES-E at longitude 75° W, and GOES-W at 135° W.

**GOME** *See* GLOBAL OZONE MONITORING EXPERIMENT.

**GOMOS** *See* GLOBAL OZONE MONITORING BY OCCULTATION OF STARS.

**GOS** *See* GLOBAL OBSERVING SYSTEM.

**Gothenburg Protocol** An international treaty, signed in 1999, governing the implementation of measures to reduce the long-range transport of atmospheric pollutants. *See also* CONVENTION ON LONG-RANGE TRANSBOUNDARY AIR POLLUTION.

**GPS** *See* GLOBAL POSITIONING SYSTEM.

**gradient** The variation in a specific parameter with respect to distance. In meteorology, generally taken to be a two-dimensional property, although three-dimensional gradients may become significant in the study of individual phenomena. Of particular importance are *humidity, *pressure, *temperature, and *thickness gradients. These are taken to be perpendicular to the corresponding *isopleths. *See also* PRESSURE GRADIENT.

**gradient wind** A hypothetical wind that follows a curved path parallel to curved *isobars or *contours. It is assumed to be a precise balance of the horizontal components of the *pressure-gradient force, the Coriolis force (*see* CORIOLIS EFFECT), and a centripetal force. It is more generally applicable than the *geostrophic wind at middle latitudes, but is still only an approximation to the true wind.

**GRAS** *See* GNSS RECEIVER FOR ATMOSPHERIC SOUNDING.

**grass minimum temperature** The temperature recorded by an accredited standard *minimum thermometer, exposed to the open air on short turf, with the bulb just in contact with the tips of the blades of grass. *See also* CONCRETE MINIMUM TEMPERATURE.

**graupel (soft hail)** Frozen *precipitation in the form of opaque *snow pellets produced when *supercooled water droplets collide and freeze. Typical diameters are 2–5 mm.

**gravity wave** A wave in which gravity—generally in the form of *buoyancy—plays the predominant role in restoring equilibrium in response to some external disturbance. Such waves are generated in the *troposphere above deep convective clouds, above mountains (where they create *lee waves), and in *jet streams. Gravity waves preferentially travel within stable layers. They appear to control circulation in the *mesosphere and may do so in the *stratosphere, but current knowledge of their exact role is limited. Note that this term must not be confused with **gravitational wave**, which is a disturbance of space-time, caused by the motion of gravitating bodies.

**grease ice** A stage in the formation of sea or lake ice, where a fairly dense concentration of separate *frazil ice crystals are prevented from freezing into a surface layer by the motion of the water. This gives rise to a characteristic greasy appearance.

**Great Ocean Conveyor Belt (GOC)** The principal subsurface oceanic circulation system. It is a *thermohaline circulation, primarily driven by the intense evaporation and cooling that occurs in the North Atlantic. This creates a *downwelling of dense, saline *bottom water, which flows south to Antarctica, where there is a very significant contribution of *Antarctic Bottom Water from the major source in the Weddell Sea. From here the flow continues through the Indian and Pacific Oceans, before eventually rising and flowing northwards in the Atlantic to complete the circulation.

**greco** [Latin: 'Greek'] A north-east wind over the Golfe du Lion in the western Mediterranean.

**green flash (sometimes erroneously known as the green segment)** A relatively infrequent, transient phenomenon when the last visible portion of the Sun's disk at sunset turns green. A similar effect may occur at the onset of sunrise, and an even rarer blue coloration has also been recorded. Reliable observers have also occasionally reported seeing a brilliant **green ray** that shoots up into the sky above the setting Sun. The green flash itself has been noted in light from the Moon and the planet Venus. It is caused by differential refraction of light of different wavelengths, combined with preferential weakening of the blue and violet colours by the normal *Rayleigh scattering in the atmosphere. Its occurrence appears to be enhanced by (or may even be contingent upon) *mirage conditions. Although it is often suggested that many visually observed green flashes are illusory, arising because sunlight has bleached the pigment in red-sensitive cones in the retina, this theory is disproved by the observations of green flashes as the Sun appears over the horizon at sunrise.

**greenhouse gas (GHG)** Any of the various gases that exert a greenhouse effect (*see feature*), particularly those present in the Earth's atmosphere, the most significant of which are *carbon dioxide, *methane, *nitrous oxide, *ozone, and *water vapour. The major GHGs that are of concern because they are causing climatic change (*see feature*) are carbon dioxide, methane, and nitrous oxide, together with *hydrofluorocarbons, *perfluorocarbons, and *sulphur hexafluoride. Emissions of these six gases or classes of gas are being restricted under the *Kyoto Protocol.

**Greenland Anticyclone** The semi-permanent area of high pressure that overlies the Greenland ice cap. The northern counterpart of the *Antarctic anticyclone.

**green thunderstorm** A term sometimes applied to systems where the area of *precipitation behind the *gust front has a bright green hue. Although

## Greenhouse effect

The increase in the temperature of a planet's atmosphere and surface that arises from the presence of certain gases in the atmosphere. Without an atmosphere, the Earth's surface would have an average temperature of *c.*256 K (–17 °C), at which the incoming radiation from the Sun would be balanced by the long-wave *infrared radiation emitted to space. Certain atmospheric gases, known as *greenhouse gases, particularly *water vapour, *carbon dioxide, and *methane, are almost transparent to incoming short-wave solar radiation, but strongly absorb long-wave radiation from the surface. Part of this absorbed energy is re-emitted back towards the surface, thus raising its temperature. The effect is named from its supposed similarity to that operating in a greenhouse. (In fact the increased temperature in the latter is largely caused by the fact that the glass prevents *convection from transporting heat to the outside.)

The presence of carbon dioxide (in particular) in an atmosphere has a major effect on planetary temperatures. It is believed that high concentrations of this gas caused high temperatures to prevail on the primordial Earth (*see* ATMOSPHERE (1)). In the case of Venus, the present-day abundance of carbon dioxide in the atmosphere (approximately 96.5 per cent) has raised the average surface temperature to about 460 °C.

Life on Earth would not exist without the natural greenhouse effect, but human activities are increasing its magnitude and thus contributing to global warming (*see feature*). The large quantities of carbon dioxide produced by the burning of fossil fuels, methane arising from increased livestock and rice production, and certain human-made gases (particularly the *chlorofluorocarbons) have a direct greenhouse effect.

often anecdotally associated with *hail, the coloration appears to occur with heavy *rain, as well as light and heavy hail. A full explanation is still wanting, but current suggestions appear to favour absorption of red light by the precipitation.

**gregale** A strong north-east wind over the western and central Mediterranean, most frequent in winter. It occurs when there is high pressure over Central Europe and low pressure over North Africa.

**grid** The network of points used to represent the atmosphere or oceans in *numerical weather prediction or *general circulation models, and generally taken at numerous atmospheric levels. For weather forecasting, typical grid resolutions are *c.*120 km and *c.*60 km for global models, and *c.*30 km for regional models. For certain studies, use may be made of a **nested-grid** model, in which an outer coarse grid surrounds a finer grid over the area of specific interest.

Similarly, use may be made of a **fine-mesh floating-grid** model, in which the grid moves with the system under investigation.

**Grosswetterlage** [German: 'overall weather pattern'] The relatively unchanging mean surface-pressure distribution over a specified interval of time. During this period the high- and low-pressure *centres of action remain essentially unchanging. Sometimes shortened to **Grosswetter**.

**ground, state of** A term used for observations made at *synoptic and certain *climatological stations on a patch of bare ground and a representative open, flat area typical of the site's location. *Frost and *dew are disregarded, and there are two scales, depending on whether the ground is with or without snow and ice cover. The scales provide for such observations as dry and dusty, moist, covered with *glaze, deep snow drifts, etc.

**ground discharge** A *lightning discharge from cloud to ground.

**ground frost** A temperature of 0 °C or less as recorded by a thermometer exposed to the open air on short turf, with the bulb just in contact with the tips of the blades of grass.

**growler** A small remnant of an *iceberg, almost completely submerged beneath the surface, and thus often a greater hazard than the larger, more visible *bergy bits.

**Guiana Current** A current that flows north-westwards along the northern coast of South America. It is an extension of the Atlantic Ocean's *South Equatorial Current, and subsequently merges with the *North Equatorial Current.

**Guinea Current** A warm current in the equatorial Atlantic which originates in the *Equatorial Counter Current and flows eastwards along the southern coast of north-west Africa and into the Gulf of Guinea.

**Gulf Stream** The portion of the North Atlantic *gyre that lies between the point where the *Antilles and *Florida Currents converge, and 40° N, 50° W, beyond which it is more properly known as the *North Atlantic Current. A relatively narrow, fast, warm current, east of the Grand Banks it is bounded on the north by the *Labrador Current. In popular usage, the term is often applied incorrectly to the *North Atlantic Current.

**Gulf Stream Extension** *See* NORTH ATLANTIC CURRENT.

**gust** A sudden, short-lived increase in the wind speed relative to the average speed at the time, generally regarded as lasting about 20 seconds, and thus much shorter than a *squall. The corresponding decrease in speed is a *lull. In the main, gusts are *turbulence created by obstructions on the ground and therefore the majority occur close to the surface. Vertical *wind shear and sudden changes in *lapse rate may, however, contribute under certain circumstances, such as those found in *clear-air turbulence.

**gust front** An outflow of cool air from the leading edge of a *cumulonimbus cloud. There may be strong gusts of wind, but no *precipitation is present.

**gustiness** A measure of the variation in wind strength. The ratio (expressed in per cent) of the difference between the maximum and minimum wind speeds to the mean wind speed over a specific interval. Typical values range between 25 and 100 per cent.

**gustnado** A North American term for a relatively weak, short-lived *whirl that is initiated by strong downdraughts on a *thunderstorm's *gust front. It does not produce a condensation funnel. Such tornadic vortices rarely exceed F2 on the *Fujita scale. *See also* LANDSPOUT; TORNADO *feature*; WATERSPOUT.

**guttation** The process by which plants exude excess water from the tips of leaves (as **guttation drops**) when transport within the plant exceeds the rate of *transpiration. The condition frequently occurs when the air cools at night while the soil remains relatively warm and moist. Guttation drops are often mistaken for *dew.

**gyre** The large-scale circulation formed by the wind-driven surface currents within the major ocean basins. The main gyres are clockwise in the northern hemisphere and anticlockwise in the southern. They are centred at approximately 30° N and 30° S and, because of the Earth's rotation, are stronger on the western sides of the basins. Smaller gyres with the opposite sense of rotation may occur within individual regions (such as that in the Greenland Sea). The various forces cause the centre of a gyre to be higher than the periphery: the centres of the North-Atlantic and North-Pacific gyres are *c.*1 m higher than the edges.

# H

**haar** In Scotland and north-eastern England, the local name for an *advection fog, formed over the North Sea, that has been carried inland by gentle onshore winds or a *sea-breeze. It is particularly frequent in spring and early summer, when water temperatures are low. Also known as **sea fret** in the north-east.

**haboob** [from Arabic *haab*: 'blow'] **1.** A severe *dust storm or *sandstorm in northern and central Sudan which transports and deposits large quantities of material. Haboobs are strongest in April and May, but occur in every month except November. The wind direction may be north (in winter) or east, south-east, or south (in summer). The storm front is extremely dense and turbulent and may be up to 1000 m high. **2.** A local Canadian term for a prairie dust storm.

**Hadley cell** A simple thermal cell originally proposed by George Hadley in 1735 to explain the circulation of the atmosphere in each hemisphere, and in particular, the origin of the *trade winds. In Hadley's model, air rises above the zone that is most strongly heated (i.e. the *tropics), flows at height towards the poles, where it sinks, and completes the circulation by an equatorward flow at low level. Hadley recognized that the rotation of the Earth was the reason for the fact that the trade winds did not blow directly from the north or south, but were instead northeasterlies and southeasterlies. The model was incorrect, however, in its assumption of a single circulation cell between the tropics and each pole. The concept of a Hadley cell does, nevertheless, correspond approximately to the circulation in the *troposphere between the Equator and latitudes 30° N and 30° S, so the term is now often applied to these cells. The atmospheric circulation on Venus has a slow meridional component between equator and poles which resembles that of a simple Hadley cell.

**Haidinger's brush** A faint polarization pattern visible to the human eye when viewing an evenly illuminated surface, such as a cloud or the sky. Not detectable by everyone, it normally consists of a yellowish hourglass-shaped region with two circular blue regions on each side of the waist. The long axis of the yellow area is at right angles to the light's plane of vibration. The figure arises because of dichroism (differences in the colour absorbed depending on the plane of polarization) in the region of the retina known as the macula lutea. *See also* POLARIZATION OF THE SKY.

**hail** Solid *precipitation in the form of balls or pellets of ice, generally taken to be 5 mm in diameter or larger. (Smaller particles are known as **small hail** or *ice pellets.) Hail originates as ice pellets or *graupel that are suspended in the *updraught within a *cumulonimbus cloud, and grow by the *accretion of *supercooled water droplets in alternating layers of *rime and *glaze, until the mass can no longer be supported. The largest sizes may exceed 50 mm across, and the heaviest individual hailstones of *c.*1 kg were recorded at Gopalganj, Bangladesh on 14 April 1986. Hailstone aggregates of 3.4 kg and 4 kg were reported from Hyderabad, India in 1939, and Yüwu, China in 1902, respectively.

**hail fog** Fog that is caused by the sudden, dramatic cooling of the surrounding air by *precipitation in the form of *hail.

**haize-beltza** [Basque: 'black wind'] A *katabatic wind descending the northern slopes of the western Pyrenees.

**half-life** The period of time taken for half the atomic nuclei in any given sample of a radioactive element to decay. Half-lives range from a fraction of a second to many thousands of years. *See* CARBON DATING.

**Halley, Edmond** (b. Shoreditch, near London, 8 November 1656; d. Greenwich, 14 January 1742) English astronomer and mathematician who made numerous contributions to various fields of the natural sciences. His meteorological and geophysical work includes the publication of the first chart showing the global wind circulation (1686), and a map of magnetic deviation (1701), which with its *isogonic lines was the first widely circulated chart to make use of *isolines.

**halo** One of the numerous different rings, arcs, and points of light that may be seen around the Sun or Moon, and which are caused by *refraction and *reflection by ice crystals either in high-altitude clouds (particularly *cirrostratus) or in smaller crystals suspended in the atmosphere, such as *diamond dust. The principal circular rings are haloes with radii of 22° and 46°, both of which show faint colours with red inside and violet outside. Other halo phenomena are: *anthelic arcs, *anthelion, *arcs of contact, *circumhorizontal arc, *circumzenithal arc, *parhelion, *paraselene, *parhelic circle, and *subsun.

**halocarbon** *See* CHLOROFLUOROCARBON; HALON; HYDROCHLOROFLUORO-CARBON.

**halocline** A layer in a body of water at which there is a marked change in *salinity, and hence in density.

**halon** A chemical compound, used in fire-fighting, derived from *methane or ethane by replacement of hydrogen atoms by fluorine, bromine, chlorine, or iodine. Halons are non-conductive and thus particularly valuable for use on fires of electrical or electronic equipment. They contribute about 40 per cent of atmospheric bromine, however, thus depleting the *ozone layer, and many were banned under the *Montreal Protocol.

Two halons are causing particular concern. Halon-1202, previously ignored, has recently been found to pose a considerable threat because its concentration is increasing rapidly, although its source remains unknown. Halon-1211 ($CBrClF_2$) is mainly manufactured in China (90 per cent of current declared world production), and its atmospheric abundance is 50 per cent more than expected from manufacturing figures officially reported. If uncorrected, it alone will cause many years' delay in restoring the ozone layer.

**hammada** [from Arabic *hamid*: 'dead, lifeless'] One of the three principal desert landforms, with extensive bare rock surfaces and little gravel or sand. *See also* ERG; REG.

**harmattan** A dry, dust-laden, but relatively cool north-easterly or easterly wind in north-west Africa, part of the *trade-wind system. In the wet (summer) season it tends to give way to *monsoon winds from the Gulf of Guinea and the Atlantic Ocean.

**Hawaiian High** *See* NORTH PACIFIC HIGH.

**haze** **1.** Atmospheric obscuration by dry particles that are sufficiently small to remain suspended in the air for long periods and which give a pearly quality to the daylight. The term does not imply any specific upper or lower limits to visibility.

**2.** Less commonly, obscuration by water droplets or by moist *hygroscopic particles, smaller than cloud particles. Such conditions are more frequently classed as *mist or *fog.

**hazemeter** *See* TRANSMISSOMETER.

**HCFC** *See* HYDROCHLOROFLUOROCARBON.

**head wind** In aviation, any wind that reduces the distance an aircraft covers over the ground, i.e. one that causes the ground speed to be less than the air speed. *See also* EQUIVALENT HEAD WIND; TAIL WIND.

**health resort station** A *climatological station that carries out observations at 09:00 and 18:00 GMT, the latter report covering temperature, rainfall, sunshine, and past weather during the day, specifically for release to the press and other media.

**heap clouds** A common name for any *cumuliform clouds, but most often applied to *cumulus and *cumulonimbus.

**heat burst (heatburst)** A rare, sudden, and short-lived temperature rise that occurs behind a decaying severe *thunderstorm when dry warm air is forced to descend from the middle *troposphere, temporarily displacing the pool of rain-cooled air at the surface. Heat bursts may be accompanied by extremely strong gusts. During an event that occurred near Oklahoma City on 22–23 May 1997, the temperature rose from 87.6 °F to 101.9 °F (30.9 °C to 38.8 °C) between 21:00 and 21:25 at Chickasha, and a gust of about 170 km $h^{-1}$ was recorded at Tipton. On 14 June 1960, a heat burst at Kopperl, Texas raised the temperature to nearly 140 °F (60 °C), roasting corn on the plant, and turning grass into hay.

**heat capacity** *See* SPECIFIC HEAT.

**heat island** A localized pool of warm and often polluted air that occurs over an urban area, with a temperature that may be several degrees higher than in the surrounding region.

**heat lightning** A popular term for *lightning that is too distant for the associated thunder to be heard, and which may appear red because of *scattering and absorption of other wavelengths by the intervening atmosphere. Often incorrectly believed to be lightning caused simply by the existence of high atmospheric temperatures.

**heat low** *See* THERMAL LOW.

**Heaviside layer** *See* IONOSPHERE.

**hectopascal (hPa)** A multiple (100) of the *pascal, and a useful unit for the expression of atmospheric *pressure. In all scientific and technical work it has now replaced the earlier derived unit, the *millibar, being more rigorously defined. It is equally convenient, because 1 hPa ≡ 1 mb.

**heiligenschein** [German: 'holy light'] A bright, colourless fringe of light surrounding the shadow of an observer's head, i.e. around the *antisolar point. There are two separate mechanisms, the strongest being when sunlight falls on dew-covered grass. The dewdrops and blades of grass act together as retro-reflectors, returning the light preferentially towards the source. The second, weaker effect (known as a **hot spot** in aerial photography) occurs with a rough surface, such as dry grass or leaves. An observer looking at the antisolar point does not see the shadows of the leaves, which are masked by the objects themselves. Every visible surface is fully illuminated, so the immediate vicinity appears brighter than the surrounding area, where the shadows come into view.

**Heinrich event** A period of increased iceberg calving from the Laurentide (North American) *ice sheet into the North Atlantic during the last *ice age. At least six of these events have been recognized from the increased frequency of large-sized, ice-rafted debris deposited in marine sediments. The greatly increased input of light fresh water to the surface layer probably caused major disruptions to oceanic circulation, specifically to the *Great Ocean Conveyor Belt, with consequent effects upon the climate in the northern hemisphere, and particularly in areas around the North Atlantic. The correlation with the more frequent *Dansgaard–Oeschger events is unclear.

**helikite** A newly developed combination of kite and helium-filled balloon for carrying meteorological instrumentation. It should remain aloft whatever the prevailing conditions, unlike the *kite balloon, which may lose its shape and descend when subjected to gusts of cold air.

**heliostat** An instrument mounting that is driven to keep any devices that it carries (for example, a *pyrheliometer) oriented towards the Sun, or a mirror that is mounted and driven in such a way that sunlight falls on a specific device throughout the day.

**helium (He)** A colourless *inert or noble gas, atomic number 2, that is a minor atmospheric component (*c*.0.000524 per cent by volume), but may be present in significant amounts (up to 7 per cent) in natural gas. Because helium is non-flammable and has a lifting power 92 per cent of that of hydrogen, it is generally preferred for the inflation of airships and research balloons. Helium-3 ($^{3}He$) is used in fusion experiments and magnetic resonance imaging.

**helm wind** A strong, blustery wind that descends the western side of the Cross Fell range in Cumbria. It is a *lee wave produced by a prevailing north-easterly flow and exhibits a *föhn effect. A bank of cloud, the **helm**, lies along the mountain-tops or just above them. At times, a further narrow, stationary, but rotating roll cloud forms roughly parallel to the first and a few kilometres downwind. This **helm bar** is a *rotor cloud and the surface wind dies away beneath it. Farther away from the fells there may be a gentle westerly flow towards the mountains.

**heterosphere** The region of the atmosphere, above about 85 km, in which the types and relative proportions of the various atmospheric gases (*see* AIR) diverge from the constant ratios that exist in the underlying *homosphere. These changes mainly result from the *dissociation of oxygen molecules into atomic oxygen by solar *ultraviolet radiation and also because any mixing occurs through *diffusion, which is less efficient than the *turbulence that takes place at lower levels.

**HFC** *See* HYDROFLUOROCARBON.

**high** *See* ANTICYCLONE.

**high-precipitation supercell** A *supercell thunderstorm with intense *precipitation that creates an opaque curtain to the rear of the *wall cloud and *flanking line in an area that would usually be precipitation-free in 'normal' and *low-precipitation supercells.

**High Resolution Infrared Sounder (HIRS)** A 20-channel infrared radiometric sounder carried by *NOAA *polar-orbiting meteorological satellites. Data, when combined with observations from the *Advanced Microwave Sounding Units, are used to determine atmospheric temperature and pressure profiles to altitudes of *c*.40km. HIRS data also measure *sea-surface temperatures, *ozone levels, precipitable water, and cloud properties (height, coverage, and *albedo). Legacy units are being fitted to *Eumetsat *MetOp satellites (*see* INITIAL JOINT POLAR SYSTEM).

**high-resolution picture transmission (HRPT)** The continuous broadcasting of imagery from *polar-orbiting *meteorological satellites, in digital format and with a higher resolution than *automatic picture transmission.

**hill fog** *Fog that arises when low cloud covers high ground, but which has not been formed by the uplift and *condensation of moist air. (The latter form being classed as *upslope fog.)

**hindcasting** The retrospective *analysis of the *synoptic situation to determine the causes of a specific weather event. Although such analysis is usually undertaken for unforeseen extreme events, it may also be carried out to discover why less dramatic circumstances were poorly forecast. The term is similarly applied to modelling past climates.

**HIRS** *See* HIGH RESOLUTION INFRARED SOUNDER.

**hoar-frost** Soft ice crystals that form on vegetation or other objects that cool below 0 °C. Such frost often consists partly of *dew, originally deposited as liquid droplets that have subsequently frozen with a drop in temperature (to form *silver frost), and partly of ice that has formed directly (i.e. by *sublimation) on the surfaces concerned. *See also* GLAZE; RIME.

**hodograph (more correctly known as a hodogram)** A representation, on a polar diagram, of the distribution of horizontal wind vectors (and thus *wind shear) with respect to height.

**homosphere** The region of the atmosphere, below about 85km, in which (neglecting certain minor components such as *carbon dioxide, *ozone, and *water vapour) the types of atoms and molecules and their relative proportions remain constant. *Compare* HETEROSPHERE.

**Hooke, Robert** (b. Freshwater, England, 18 July 1635; d. London, 3 March 1703) British physicist, who made many contributions to a wide range of sciences, including meteorology. His practical skills enabled him to devise improved versions of *barometer, *rain gauge, and *thermometer, together with the first practical *hygrometer and a **weather clock** for recording several measurements every 15 minutes. He is sometimes described as the 'father of meteorology' because he was the first to advocate a standardized format for recording observations, and promoted national and international networks of observers.

**hook echo** A characteristic radar echo associated with some (but not all) tornadoes (*see feature*). It shows *cyclonic curvature (and thus resembles the numeral '6'), and appears as an appendage to the larger echo returned by the parent *multicell or *supercell thunderstorm. It arises from backscattering from *precipitation in a flow of air being drawn into a strong circulation, alongside a similar stream of precipitation-free air. Hook echoes may last just a few minutes and, although not infallible indicators of tornado development, do reveal extreme *turbulence.

**horizon** The apparent boundary between the Earth's surface and the sky. The actual horizon depends on the presence of hills or mountains, but the theoretical (or sea) horizon is a small circle centred on the observer. Its radius (i.e. the distance of the horizon) is mainly determined by the altitude of the observer, but is also affected by *refraction and is thus dependent on the temperature profile of the atmosphere. For an observer 1.5 m above sea level, the apparent horizon is *c.*4.5 km distant. At an altitude of 30 m, the distance increases to *c.*24 km.

**horizontal visibility** The maximum distance at which an observer can see an object situated in essentially the same horizontal plane. It is, for example, the visibility given in shipping forecasts and reports from coastal stations. *See also* OBLIQUE VISIBILITY; VERTICAL VISIBILITY.

**horse latitudes** Regions of light, variable winds and generally fine weather, associated with the *subtropical anticyclones at latitudes 30–35° N and S.

**hot-film/hot-wire anemometer** *See* THERMAL ANEMOMETER.

**Howard, Luke** (b. London, 28 November 1772; d. London, 21 March 1864) A British pharmacist and later manufacturing chemist, with a lifelong interest in meteorology. He was the first to develop a classification of clouds, similar to the scheme used by Linnaeus for plants and animals. This was presented to the Askesian Society in 1802, under the title 'On the Modifications of Clouds', and published in his later work *The Climate of London* (1818–20). His principal classes *cirrus, *cumulus, and *stratus are still used today, together with the term 'nimbus', which he also introduced.

**HRPT** *See* HIGH RESOLUTION PICTURE TRANSMISSION.

**Humboldt, Alexander von** (b. Berlin, 14 September 1769; d. Berlin, 6 May 1859) German naturalist and explorer who made major contributions to various sciences, including geology, geomagnetism, and meteorology. Together with Aimé Bonpland he undertook an expedition to South and Central America in 1799–1804, during which he studied the *Peru Current, and correctly ascribed altitude sickness to lack of *oxygen. He is the first person known to have prepared weather maps showing *isobars and *isotherms. He later advocated a worldwide network of magnetic observatories to study *magnetic storms, data from which eventually proved the link with *solar activity.

**Humboldt Current** *See* PERU CURRENT.

**humidity** A measure of the *water vapour content of the air, and generally taken to be the *relative humidity. *See also* DEW POINT; SPECIFIC HUMIDITY.

**humidity mixing ratio** The ratio of *water vapour to the mass of dry air with which it occurs, normally expressed as grams per kilogram. *See also* SPECIFIC HUMIDITY.

**humilis** [Latin: 'humble'] A *cumulus *cloud species that appears flattened and which, when viewed from the ground, has a greater horizontal than vertical extent.

**hurricane** [from Spanish *huracán*, from Taino *hurakán, hura*: 'wind'] The term used for a tropical cyclone (*see feature*) that occurs in the Atlantic, Caribbean, or eastern Pacific.

**hurricane force** A surface wind whose average speed exceeds 64 *knots (force 12 on the *Beaufort scale).

**hurricane wave** The increased height of sea level (as much as 6–7 m) that may be caused by the extremely low pressure in the centre of a tropical cyclone (*see feature*). *See* STORM SURGE.

**hydrochlorofluorocarbon (HCFC)** A chemical compound of *hydrogen, chlorine, fluorine, and carbon. In general, HCFCs are less destructive of atmospheric *ozone than *chlorofluorocarbons (CFCs), have shorter atmospheric lifetimes, and are partly degraded in the *troposphere. They are therefore regarded as acceptable short-term alternatives to CFCs, but because they do still have destructive potential are subject to the *Montreal Protocol, with production ceilings. At a meeting in Montreal in 2007, agreement was reached for the dates for cessation of production and use to be brought forward by ten years. HCFCs will be phased out in industrialized countries by 2020 and in developing countries by 2030.

**hydrofluorocarbon (HFC)** A chemical compound of *hydrogen, fluorine, and carbon (also known as a fluorinated hydrocarbon). Unlike *chlorofluorocarbons, HFCs contain no chlorine (or bromine), and thus do not destroy atmospheric *ozone. They are, however, *greenhouse gases and are specifically covered by the *Kyoto Protocol. As with *HCFCs, under the new Montreal agreement of 2007 they will be discontinued by 2030.

**hydrogen (H)** A flammable gas, the lightest element, consisting in its simplest form of one electron and one proton. There are two further *isotopes, *deuterium and *tritium. Although extremely abundant throughout the universe, it exists only in minute quantities ($5.0 \times 10^{-5}$) in the atmosphere, because any free hydrogen diffuses upwards and is lost to space.

## Hurricanes Katrina, Rita, and Wilma

The North-Atlantic hurricane season of 2005 was the most active ever recorded, with 28 named storms and 15 hurricanes, seven of which were Category 3 (*see* SAFFIR-SIMPSON SCALE) or higher. Of these, three, Katrina, Rita, and Wilma, were of particular interest.

Katrina (23–30 August) was the storm that created such destruction along the Gulf Coast of Mississippi and Louisiana, and inundated New Orleans, with some 1 500 deaths in four states. Regrettably, the various authorities had not taken heed of the evacuation problems with Hurricane Floyd in 1999, nor the extreme vulnerability of New Orleans to a major *storm surge that was revealed by a simulation with a theoretical model ('Hurricane Pam'). Katrina initially made landfall in Florida as a Category 1 storm, but decreased to *tropical-storm strength overland. It intensified over the warm waters of the Gulf of Mexico, crossing western Cuba as Category 3 and becoming Category 5. It also became very large, with storm-force winds out to *c.*370 km, and hurricane-force winds to *c.*170 km from the centre. Its central pressure of 902 hPa on September 28 was then the fourth lowest on record. It made landfall in Louisiana as a Category 3 storm with estimated wind speeds of *c.*110 knots. Storm-surge measurements were complicated by the destruction of many instruments, but heights were 7.3–8.5 metres over a stretch of coastline at least 32 km long.

Hurricane Rita (18–26 September) became hurricane-force southeast of Key West, Florida on 20 September and rapidly intensified to Category 3 and, gaining energy from the *Loop Current, went from tropical storm to Category 5 in less than 36 hours. The central pressure fell by the extreme amount of 70 hPa in 24 hours up to 00:00 UTC on 22 September, when the pressure was estimated at 897 hPa or slightly less, displacing Katrina to become the fourth-lowest pressure recorded. Rita weakened, however, becoming Category 3 by the time it made landfall on extreme southwestern Louisiana, near the Texas border. The maximum storm surge amounted to *c.*4.7 metres and further major destruction occurred in certain areas previously affected by Katrina.

Hurricane Wilma (15–25 October) was first classed as a *tropical depression in the Caribbean Sea southwest of Jamaica on 15 October, slowly becoming a tropical storm two days later, and a hurricane on 18 October. It then underwent explosive intensification. It went from a 60-knot tropical storm to a 150-knot Category 5 hurricane in 24 hours, an unprecedented strengthening. Peak wind speed reached 160 knots on 19 October, when the *eye contracted to *c.*3.7 km, the smallest diameter ever recorded, and the estimated central pressure was 882 hPa, the lowest known for an Atlantic hurricane. (The 6-, 12- and 24-hour pressure drops are all the greatest ever recorded.) At Category 4 on 21 October, Wilma made landfall on the Yucatan peninsula, where the storm surge is thought to have been

3.7–4.6 metres. Although it created extensive damage, no figures are available for possible casualties. Reaching the Gulf of Mexico at Category 2, Wilma was accelerated by a powerful *steering flow northeastwards. It intensified to Category 3 and made landfall in southwestern Florida on 24 October, crossing to the Atlantic coast in about 4.5 hours, decreasing slightly in intensity. It intensified for the final time shortly after leaving Florida, raced northeast at 40–50 knots (75–93 $km^{9-1}$), and eventually combined with another extratropical depression off Nova Scotia on 27 October.

**hydrogen sulphide ($H_2S$)** A colourless, poisonous gas, produced by decaying vegetable and animal matter, and by anaerobic bacteria. In deep seas (such as the Black Sea) and oceans with anoxic zones, the hydrogen sulphide diffuses upwards as far as the *chemocline, the boundary between oxygenated surface waters and the anoxic zone. It has been suggested that some of the past mass extinctions were the direct result of the release of large quantities of hydrogen sulphide into the atmosphere. Large-scale volcanic activity, such as the eruption of the Siberian Traps in the late Permian, would have greatly increased the atmospheric *carbon dioxide content, with an accompanying reduction in free oxygen and enhanced global warming (*see feature*). Warm oceanic waters would have absorbed less oxygen, leading to an expansion of anoxic conditions, destabilizing the chemocline, and resulting in an eruption of $H_2S$ into the atmosphere, directly killing bacteria that metabolize oxygen, animals and plants, and simultaneously destroying the *ozone layer. The concept is supported by *carbon-isotope determinations from the various geological periods.

**hydrography** The study of large bodies of water, such as the oceans, lakes, and rivers, with regard both to their physical properties and also—as the term is most commonly understood—to navigation.

**hydrological cycle** The transport of water in all its forms within and between the atmosphere, *cryosphere, *hydrosphere, and the Earth's surface. It is a closed cycle, although very small quantities of *juvenile water may be added by volcanic action. There may also be some slight losses when water is *dissociated in the upper atmosphere to hydrogen and oxygen, with some hydrogen being lost to space.

**hydrology** The study of water in that portion of the *hydrological cycle that excludes specifically atmospheric processes. It therefore covers the effects and properties of water as *precipitation in all its forms, *evaporation, run-off in streams and rivers, groundwater, soil moisture, and ocean waters.

**hydrometeor** Atmospheric water in any form, whether as solid, liquid, or vapour. Apart from the obvious types of *precipitation, it also includes forms such as clouds themselves, cloud and fog particles, blowing spray, and *dew. The term is more frequently used than the companion terms *electrometeor and *lithometeor, but is probably obsolescent.

**hydrometer** An instrument for determining the density of liquids, such as that of sea-water.

**hydrosphere** The portion of the Earth's surface that consists of water, as distinct from the solid, rocky *lithosphere and the gaseous *atmosphere. It is also generally taken to include the *cryosphere.

**hydroxyl** A molecule consisting of one hydrogen and one oxygen atom, created in the upper atmosphere by the *dissociation of water vapour into the hydroxyl radical and hydrogen.

**hyetograph** **1.** A map or diagram showing the distribution of rainfall, either by area or by time. **2.** A proprietary form of recording *rain gauge.

**hygrograph** A recording form of *hygrometer, the chart from which is known as a **hygrogram**. The active sensitive element is most frequently human hair, which changes in length depending on the humidity.

**hygrometer** An instrument for determining atmospheric *humidity. There are various forms, including those that employ dry- and wet-bulb thermometers (*psychrometers), electrical sensors that use changes in capacitance or resistance, dew-point hygrometers that determine the temperature at which dew forms on a mirror, and sensors that measure the absorption of radiation by the water vapour in the air.

**hygroscope** A device for giving an indication of whether the air is dry or damp, which forms the basis for simple weather predictors. The indicator may be a change in colour of hygroscopic material or the variations in torsion of a substance such as catgut, the latter being used in weather houses to cause the appropriate figure to appear.

**hygroscopic** Having a tendency to capture water molecules from the air, normally through *absorption (2). Hygroscopic particles play an important part in many atmospheric processes.

**hygrothermograph** An instrument that combines the functions of the *hygrograph and the *thermograph, i.e. by recording both *humidity and *temperature on a single chart.

**hypercane** A poorly defined term for a hypothetical, intense tropical cyclone (*see feature*) that it has been suggested might result from extreme heating produced by the impact of a kilometre-sized *meteorite in the ocean. The term is also occasionally used in the media for an exceptionally violent tropical cyclone.

**hypsometer** [from Greek *hypsos*: 'height' + meter] Originally, a device for ascertaining height, but, in meteorology, an instrument for determining atmospheric *pressure by measuring the temperature of the *boiling point of water. This may then be converted to pressure from the known boiling-point/pressure relationship.

# I

**IASI** *See* INFRARED ATMOSPHERIC SCANNING INTERFEROMETER.

**ICAO standard atmosphere** A standard model of the atmosphere adopted by the International Civil Aviation Organization (ICAO). The principal features are the assumption of a standard mean sea-level pressure of 1013.25 hPa and an overall *lapse rate of 6.5 deg C $km^{-1}$ from the surface to the *tropopause, which is taken to lie at a height of 11 km.

**ice** The solid phase of *water ($H_2O$), which occurs in many forms throughout the atmosphere and on the Earth's surface. Its density is generally *c.*917 kg $m^{-3}$, less than that of water, which is why it floats. The processes in which it is involved—*deposition, *freezing, melting, and *sublimation—require the transfer of large amounts of *latent heat, so these processes are of great significance in the Earth's overall energy budget.

Specific forms of ice include: *diamond dust, *droxtal, *glaze, *graupel, *hail, *hoar-frost, *ice crystals, *ice pellets, *ice prisms, *rime, and *snow. It occurs in various clouds, particularly *cirrus, *cirrocumulus, and *cirrostratus, and plays a part in many phenomena, among them *haloes (including many associated optical phenomena) and *ice fog. It is involved in the creation of *rain through the *Bergeron process.

When ice forms on the surface of water, various types occur, including *frazil, *grease, and *pancake ice. In greater masses, ice occurs as *glaciers, *ice caps, *ice sheets, and *ice shelves, as well as *bergy bits, *growlers, *icebergs, and *pack ice. *See also* PERMAFROST.

**ice age** A period when large areas of the Earth's surface were covered with *ice caps, also known as a glacial period or *glaciation (2). Several such episodes are known to have occurred in the Precambrian (*see* SNOWBALL EARTH), with others in the Cambrian, Ordovician, Carboniferous–Permian, and, most recently, the Pleistocene. Within ice ages, which typically last 100 000 years, there are stages of maximum ice advance (*glacials) and ice retreat (*interglacials). During the Pleistocene, the edge of the northern polar cap advanced down to approximately 50° N. It is not known whether the last retreat, approximately 10 000 BP, represents the end of the Pleistocene Ice Age, or an interglacial.

It is now accepted that there are several significant mechanisms that contribute to causing ice ages (and climatic change in general—*see feature*), although their relative importance remains uncertain. Some of these are: *continental drift and mountain building (driven by *plate tectonics), which alter the distribution of land masses and their topography; the various astronomical effects suggested by the Milankovich theory (*see feature*); changes in the concentration of the *greenhouse gases *carbon dioxide and *methane; and changes in the value of the *solar constant. *See also* CLIMATIC OPTIMUM; LITTLE ICE AGE.

**ice apron** A mass of ice adhering to a steep rock face high on a mountain, and which is commonly the source of *ice avalanches.

**ice avalanche** A large mass of ice that separates from an *ice apron or the snout of a hanging glacier and crashes down a mountainside. Ice avalanches are often extremely destructive and may reach sufficient velocity to climb part-way up the opposite side of a valley.

**iceberg** A large mass of ice that has calved from a *glacier, or parted from an *ice shelf, and which may be either floating or grounded in shallow water. Icebergs derived from an ice shelf frequently have large, relatively flat upper surfaces and are consequently known as **tabular bergs**. Some may be exceptionally large: one that was sighted in November 1956 near Scott Island (67° S, 179° E), and which had broken away from the Ross Ice Shelf, measured approximately $100 \times 330$ km, with a height above water of *c.*40 m.

**ice blink** A yellowish-white illumination of the underside of low clouds, caused by light reflected from an ice-covered surface. It is not as bright as *snow blink, but is easily distinguished from the darker appearance of clouds over water (*water sky) or above ice- and snow-free land.

**ice cap** A permanent deposit of ice and snow, such as those found in mountainous and glaciated regions. Technically, ice caps have areas of less than 50 000 $km^2$ and are thus smaller than *ice sheets, such as those covering Greenland and Antarctica, although the terms are often used interchangeably, with 'ice cap' commonly used for the polar ice on both Earth and Mars. During *ice ages, these (or similar) semi-permanent caps formed the nuclei of far greater ice sheets that extended over a significant fraction of the Earth's surface.

**ice crystal** One of the various crystalline forms that water adopts in its solid phase, and which grows on an *ice nucleus at temperatures below freezing. Depending upon the exact temperature or range of temperatures present, crystals may occur as hollow columns, flat plates, sector plates, prisms, needles, or dendritic crystals of varying complexity. Most display hexagonal symmetry. Because of the fixed angular relationship between faces, ice crystals refract light at specific angles, and are responsible for a wide range of *halo effects.

Ice crystals are present in various clouds, particularly *cirrus, *cirrostratus, *cirrocumulus, and *cumulonimbus. They are important in the formation of 'cold' rain (*see* BERGERON PROCESS) and in *atmospheric electricity. *See also* DIAMOND DUST; DROXTAL; SNOW.

**ice-crystal cloud** Any of the *cloud genera *cirrus, *cirrostratus, or *cirrocumulus, which consist of ice crystals rather than water droplets. *See also* MIXED CLOUD.

**ice-crystal theory** *See* BERGERON PROCESS.

**ice day** A period of 24 hours, beginning at the standard *synoptic time of 09:00, throughout which the maximum air temperature remains below 0 °C.

**ice fog** A form of *fog that consists of suspended ice crystals, 20–100 $\mu$m in diameter, or, when dense, of even smaller crystals (*droxtals). It occurs at high latitudes in calm, clear weather at very low temperatures (below −30 °C), and close to a source of water vapour. It may give rise to *halo phenomena.

**Icelandic low** A semi-permanent low-pressure area, predominantly located over the Denmark Strait between Iceland and Greenland, and particularly intense in winter, when the average pressure is around 994 hPa. It actually represents the mean position of many individual deep *depressions. It is a major *centre of action in the northern hemisphere, and its fluctuations in strength are associated with the *North Atlantic Oscillation.

**ice nucleus** A generic term embracing *freezing nucleus and *sublimation nucleus. *See* NUCLEUS.

**ice pellets** A form of *precipitation produced by *stratiform clouds and consisting of translucent pellets of ice (frozen raindrops) that do not exceed 5 mm in diameter. They may be spherical, irregular, or (rarely) conical in shape.

**ice prisms** Small, unbranched ice crystals in the form of needles, columns, or plates that form at low temperatures (below −30 °C) from cloud or clear sky. *See* DIAMOND DUST.

**Ice Saints** St Mamertus, St Pancras, and St Gervais, whose feast-days fall on 11, 12, and 13 May, respectively. In continental Europe these days are widely regarded as being subject to frosts or forming part of a cold spell.

**ice sheet** A massive, permanent ice dome, exceeding 50 000 km$^2$, and in particular the enormous areas glaciated during *ice ages. The only surviving ice sheets are those covering Antarctica and Greenland, which have significant effects on the Earth's weather and climate. They are on a larger scale than the smaller *ice caps. During the Quaternary Period the great North American Laurentide ice sheet had a maximum extent of *c.*13.4 million km$^2$. *See also* POLAR ICE.

**ice shelf** An extensive area of thick ice, fed by glaciers from a neighbouring *ice cap or *ice sheet, that extends outwards from the coast, and the outer portion of which is floating rather than being grounded on the sea-floor. At the seaward edge, calving produces large tabular *icebergs. At present, the main ice shelves are those in Antarctica, in particular the Ross, Ronne/Filchner, Larsen, and Amery Ice Shelves.

**ice sizzle** An informal North American term for the noise produced when melting glacier ice releases trapped, pressurized air bubbles.

**ice storm** A popular term for a sudden, widespread, and highly destructive incidence of *glaze, such as the event that occurred over a large area of Canada and New England on 5–9 January 1998.

**icing** The deposition of ice on aircraft, in the form of *frost, *glaze, *rime, or a mixture of the two last. Icing most commonly occurs at temperatures just below freezing, and its rate increases with the *supercooled water content of clouds and with droplet size. It is more significant in *cumuliform than *stratiform clouds.

**IGC** *See* INTERNATIONAL GEOPHYSICAL COOPERATION.

**ignis fatuus (jack-o'-lantern, will-o'-the-wisp)** [Medieval Latin: 'foolish fire'] A pale light or phosphorescence occasionally seen over bogs and marshy ground at night. Although no rigorous scientific explanation has been advanced, it is widely believed to arise from burning *methane, ignited by spontaneous oxidation of diphosphane, $P_2H_4$. Both methane and phosphane ($PH_2$) are produced by decaying vegetation.

**IGY** *See* INTERNATIONAL GEOPHYSICAL YEAR.

**IJPS** *See* INITIAL JOINT POLAR SYSTEM.

**illuminometer** *See* DAYLIGHT RECORDER.

**incus (inc)** [Latin: 'anvil'] An *accessory cloud specifically associated with cumulonimbus *capillatus, where the upper portion of a *cumulonimbus has spread out on reaching an *inversion to give a large, overhanging layer of cloud (Plate 9). This may be smooth, striated, or fibrous in appearance.

**index** **1.** A part of a measuring instrument that indicates the reading by its position relative to a graduated scale. **2.** A term sometimes used for the zero point of such a scale. **3.** A numerical value, often mathematically derived, that indicates the intensity of a particular phenomenon (such as the *Cold Tongue Index or the *North Atlantic Oscillation index).

**index correction** The quantity that must be added to or subtracted from an instrumental reading to compensate for the *index error. It is equal to, but opposite in sign to the latter.

**index error** The error in an instrument reading that results from incorrect positioning of the zero point during manufacture.

**Indian Ocean Dipole** A tentatively identified and named *oscillation of *sea-surface temperatures and atmospheric *pressure over the Indian Ocean which resembles a lesser version of the El Niño Southern Oscillation (*see feature*).

**Indian summer** A popular term for a period in mid to late autumn of clear skies, warm sunny days, and cool nights. Normally taken to have been preceded by moderately severe *frost. Such intervals do not occur every year, although in some years there may be more than one.

**indirect circulation** A closed, vertical circulation in which warm air sinks and cold air rises. In such a circulation, *kinetic energy is converted into *potential energy. In the general circulation (*see feature*), the cells between the *subtropical highs and the *Polar Front exhibit this form of indirect circulation.

**inert gas** Any of the unreactive gaseous elements (*helium, *neon, *argon, *krypton, xenon, and *radon) which form Group 0 of the periodic table. Also known as the **rare** or **noble gases**, they are not, in general, chemically active in nature, although some compounds (notably of krypton, xenon, and radon) may be created in the laboratory. Xenon, in particular, will react with biological molecules and is used in magnetic resonance imaging.

**inferior mirage** *See* MIRAGE.

**inflection point** In satellite images of the cloud system of a developing *depression, the point on the trailing edge at which the curvature changes from *cyclonic to *anticyclonic. *See* COMMA CLOUD; LEAF CLOUD.

**Infrared Atmospheric Scanning Interferometer (IASI)** A sophisticated instrument carried by the *MetOp polar-orbiting satellites, able to provide vertical profiles of atmospheric temperature and humidity of unprecedented accuracy. Temperature is measured to an accuracy of 1 K and humidity to within 10 per cent, both with a vertical resolution of 1 km. It also monitors various atmospheric components, such as *carbon monoxide (CO), *ozone ($O_3$), *methane ($CH_4$), and bromine oxide.

**infrared radiation** Electromagnetic radiation beyond the red end of the visible spectrum, with a nominal wavelength range of 700 nm (0.7 μm) to 1 mm, perceived by humans as heat. For convenience, the range is often divided into the **near-infrared** (0.7–35 μm)—although the long-wave boundary is ill-defined—and the **far-infrared** (approximately 35–300 μm). The latter range is blocked by *atmospheric absorption. The range between 300 μm and 1 mm is frequently known as the sub-millimetre band.

**Initial Joint Polar System (IJPS)** A cooperative agreement between *Eumetsat and *NOAA for the replacement of the current series of morning NOAA satellites by an improved European system, the *Eumetsat Polar System. The United States is supplying legacy equipment (*Advanced Microwave Sounding Units, *Advanced Very High Resolution Radiometers, and *High Resolution Infrared Sounders) for incorporation into most of the *MetOp satellites to provide continuity of observations to ensure compatibility of the observational datasets.

**Insat** A series of satellites launched by the Indian Space Research Organisation, including satellites that incorporated both communications and scientific (meteorological) instrumentation. The first dedicated, geostationary meteorological satellite was MetSat-1 (*see* KALPANA).

**insolation** The solar radiation that is received at the Earth's surface per unit area. It varies according to latitude and, over the course of a year, with the season and also slightly because of the eccentricity of the Earth's orbit. There are also long-term changes because of suspected variations in the *solar constant and through the fluctuations proposed by the Milankovich theory (*see feature*). Apart from this general use, the term is also employed for the specific amount of radiation measured by appropriate instruments such as a *pyranometer.

**instability** The property of a system such that any perturbation to the steady state will increase in amplitude (*compare* STABILITY). In meteorology, it is the property of an air parcel that continues its motion if it is forced to ascend or descend, rather than returning to its original level. A parcel will be unstable if, on ascending, the *environmental lapse rate is greater than the *dry adiabatic lapse rate or the *saturated adiabatic lapse rate (as appropriate), so the parcel remains warmer than its surroundings. *See also* ABSOLUTE INSTABILITY; CONDITIONAL INSTABILITY.

**instability index** One of various numerical indices of the probability of *thunderstorm activity, generally based upon determination of the degree of instability in the atmosphere.

**instability line** In *synoptic meteorology, a line or band of convective activity, often several hundred kilometres long, that is not associated with any form of surface *front. When the activity is intense and results in a line of *thunderstorms, the term *squall line is more appropriate.

**intensification** The increase, with time, of the overall *pressure gradient around the centre of a system, such as the increase in pressure in an *anticyclone or the deepening of a *depression. The opposite process is *weakening.

**interdiurnal variation** The change in any given atmospheric property, such as *humidity or *temperature, from one day to the next.

**interglacial** A period of increased temperatures that occurs between major *glacial phases, and which lasts between 10 000 and 20 000 years. During interglacials, major *ice sheets are confined to high latitudes. If it is assumed that the Pleistocene glaciation has not yet ended, the last eleven millennia constitute the Holocene interglacial. *See also* GLACIAL; STADIAL; INTERSTADIAL.

**Intergovernmental Panel on Climate Change (IPCC)** A panel of experts established in 1988 by the United Nations Environment Programme and the *World Meteorological Organization with a mandate to assess knowledge of the climate system and climatic change (*see feature*); the environmental, economic, and social impacts; and possible response strategies. Its activities are covered by three principal Working Groups, which may be broadly summarized as: I—scientific assessment of processes operating to cause climate change and modelling of these processes; II—assessment of climate-change impacts; III—determination of possible response strategies. In addition to holding major scientific conferences, the IPCC issues a number of publications (Special Reports and Technical Papers) dealing with specific aspects of its work, but the most significant are its Assessment Reports. The First was issued in 1990 and provided the basis for negotiations on the *Framework Convention on Climate Change. The Second Assessment Report was issued in 1995; the Third in 2001; and the Fourth in 2007. The level of scientific expertise and credibility exhibited by the IPCC in its various activities is, in itself, a major achievement.

**International Geophysical Cooperation (IGC)** The period from 1 January to 31 December 1959, during which much of the scientific work started during the *International Geophysical Year was continued or extended.

**International Geophysical Year (IGY)** The period from 1 July 1957 to 31 December 1958, during which an extensive, international programme of geophysical observations was carried out over all parts of the globe. The period was chosen to coincide with a *sunspot maximum. Apart from the wide degree of international cooperation, it was notable for the establishment of various scientific stations in Antarctica, and the launch of the first artificial satellites.

**International Meteorological Organization (IMO)** A non-governmental organization, consisting of the directors of various national meteorological services, founded in 1873. A conference in 1947 adopted the World Meteorological Convention, which led to the formation of the *World Meteorological Organization.

**International Polar Year** Periods devoted to the scientific study of the polar regions. During the first two periods, 1 August 1882 to 1 September 1883 (the First Polar Year, FPY), and 1 August 1932 to 31 August 1933 (the Second Polar Year, SPY), extensive meteorological, magnetic, and auroral observations were carried out. The dates were chosen to coincide with times of *sunspot maximum and minimum, respectively. These were the forerunners of the *International Geophysical Year, which is regarded as the Third Polar Year. The Fourth Polar Year, organized by the International Council for Science (ICSU) and the *World Meteorological Organization, runs from March 2007 to March 2009, the extended period providing two full annual cycles and equal coverage of both the Arctic and Antarctic. More than 200 projects involve thousands of scientists from more than 60 nations.

**International Years of the Quiet Sun (IQSY)** The period 1 January 1964 to 31 December 1965 (chosen to coincide with a *sunspot minimum), during which an observational programme was carried out similar to that made during the earlier *International Geophysical Year.

**interpluvial** *See* PLUVIAL.

**interstadial** A relatively short period of warming, lasting around 2 000 years, that occurs within a major *glacial phase. *See also* STADIAL; INTERGLACIAL.

**Intertropical Convergence Zone (ITCZ)** The region over the tropical oceans where the *trade winds from the northern and southern hemispheres *converge, generally marked by a more or less continuous band of *cumuliform clouds. The ITCZ tends to migrate north and south with the seasonal shift in the greatest *insolation. The extent of this migration and the general position of the ITCZ vary around the world. Between Asia and Australia it ranges between 30° N and 20° S; farther east, over the central and eastern Pacific, it remains closer to the Equator. In the Atlantic, it does not move south of the Equator. Over Africa, during the wet season, it marks the convergence of the *monsoon winds and the north-easterly *harmattan. *See also* SOUTH PACIFIC CONVERGENCE ZONE.

**Intertropical Front** A term sometimes applied to the region over the continents where the north-east and south-east trades converge. Although significant temperature and humidity contrasts sometimes occur, especially over Asia and West Africa, these do not often produce significant weather patterns, unlike *fronts at higher latitudes.

**intortus (in)** A *cloud variety: *cirrus in which the individual filaments appear irregularly curved and entangled.

**Intraseasonal Oscillation** *See* MADDEN-JULIAN OSCILLATION.

**inversion** A point or layer at which the *lapse rate becomes zero, or changes sign. Note that technically, contrary to common usage, an inversion does not solely indicate an increase of temperature with height. There is an inversion at the *tropopause, for example, where the positive lapse rate (decline in temperature with height) in the *troposphere at first becomes essentially zero in the lowest layer of the *stratosphere and then begins to increase with height (negative lapse rate). A further inversion occurs at the *stratopause, where the increase in temperature with height becomes a decrease (reverting to a positive lapse rate). Inversions may occur at any height, but the major ones at the tropopause, stratopause, and *mesopause are permanent features of the atmosphere.

**inversion layer** A layer in which there is a temperature *inversion. Inversions frequently occur in *anticyclones, but are also common in *depressions when air in the middle troposphere subsides. They cause an increase in *stability and tend to limit the upward growth of cloud. When particularly strong, with high *potential temperatures that suppress small-scale convection in the layers beneath them, they are often termed **capping inversions**. The lowermost layer of air frequently becomes an inversion layer (a **surface inversion**) when there is strong *nocturnal radiation from the ground on a *radiation night.

**ion** An electrically charged atom or molecule, created through the loss or gain of one or more electrons. *See* IONIZATION.

**ionization** The process by which an atom or molecule gains or loses one or more electrons and becomes negatively or positively charged. In the atmosphere, ionization occurs predominantly above 50 km, where it is caused by *cosmic rays and solar X-ray and *ultraviolet radiation. At lower levels, the main contributions are from *radioactivity and *cosmic rays, together with more localized sources such as the break-up of raindrops, fragmentation of *ice crystals, and the friction produced by wind-driven sand particles.

**ionosphere** The region of the atmosphere, consisting of the upper *mesosphere and the *thermosphere (i.e. from *c.*60–70 km to 1000 km or more), where *ultraviolet and X-ray radiation from the Sun ionizes atoms, producing free electrons and high electrical conductivity. The ionosphere reflects many radio waves back towards the Earth's surface, but also blocks some incoming radiation from space.

For historical reasons, the ionosphere is often regarded as consisting of various 'layers' with specific radio-propagation characteristics: the D-layer, 50–90 km; E-layer (Heaviside or Heaviside–Kennelly layer), 90–150 km; and F-layer (or Appleton layer), above 150 km. The term 'regions' is more appropriate, because there is an overall increase in electron density with height, rather than sharply distinct bands. Great variations occur with solar activity, and also diurnally. The D- and E-regions are prominent during daytime, when the F-region splits into the $F_1$ and $F_2$ regions at *c.*150 km and *c.*350 km. At night *recombination occurs in the D- and E-regions, which weaken or disappear. The F-region remains strong, but the two components merge at an altitude of *c.*350 km.

Recent observations suggest that the ionosphere is slowly decreasing in altitude by *c.*0.2 km per year, almost certainly as a result of the cooling in the upper atmosphere that accompanies global warming (*see feature*).

**IPCC** *See* INTERGOVERNMENTAL PANEL ON CLIMATE CHANGE.

**IQSY** *See* INTERNATIONAL QUIET SUN YEAR.

**iridescence (also known as irisation)** [from Latin *iris*: 'rainbow'] Brilliant colours that occur in certain clouds, most markedly as red and green tints, with occasional yellow and blue. Although arising from the same mechanism (*diffraction by cloud particles) as a *corona (1), iridescence generally appears as bands that run parallel to the edge of the clouds, which are normally within 30–35° of the Sun. The colours are purest and strongest when the particles cover a restricted range of sizes. Most commonly seen in *cirrocumulus, *cirrostratus, and *altocumulus, iridescence also occurs strongly in the rarer *nacreous clouds.

**Irminger Current** A warm branch of the *North Atlantic Current that flows north through the Denmark Strait, the channel between Greenland and Iceland.

**isallo-** A prefix used to indicate a change with time (a tendency) in a particular meteorological element as plotted on a *synoptic chart. Two commonly used forms are *isallobar, indicating *pressure tendency, and isallotherm, indicating temperature change with time.

**isallobar** A line joining points of equal change in atmospheric *pressure over a specified interval, usually three hours (i.e. points of equal *pressure tendency).

**isallobaric low** The region on a *synoptic chart which shows the greatest decrease in pressure over a specific time interval (usually three hours), as shown by *isallobars. It is the centre into which a *depression will tend to move.

**isentropic analysis** The analysis of various processes in the atmosphere (in the context of both forecasting and research), mapped on to *isentropic surfaces. Any *adiabatic ascent or descent takes place along such surfaces, in contrast to *isobaric analysis, where ascending or descending air parcels pass through the *isobaric surfaces.

**isentropic surface (constant-entropy surface)** A surface at which all points are at a constant *potential temperature (or *entropy) at a given time. *See also* ISOBARIC SURFACE.

**iso-** [from Greek *isos*: 'equal'] A prefix added to indicate that the value of the specific parameter under consideration does not vary, e.g. *isobar. *See also* ISOGRAM.

**isobar** A line joining points of equal atmospheric *pressure. On surface *synoptic charts, the isobars represent pressure readings reduced to sea level to eliminate the effects of the different heights of the reporting stations. For this purpose, it is assumed that the mean temperature of the (generally hypothetical) air column between the station's site and sea level is the same as the temperature reading at the station.

**isobaric analysis** An *analysis of conditions carried out for a specific *isobaric surface, i.e. one at a constant pressure. Such an analysis has the advantage that the *geostrophic wind is related to *contour spacing irrespective of the actual air density.

**isobaric chart (sometimes known as a constant-pressure chart)** A *synoptic chart drawn for a specific *isobaric surface, typically produced for

1000, 850, 700, 500 hPa, etc. Such charts normally show observational data and analysis of specific parameters such as the *geopotential height of the surface, *humidity, *temperature, and wind (*see feature*).

**isobaric surface (sometimes known as a constant-pressure surface)** A surface of equal atmospheric *pressure at a given time.

**isoceraunic line** A line joining points that have the same frequency of audible thunder, taken as a percentage, averaged over a year.

**isochasm** A line joining points at which the frequency of visible *aurora is constant.

**isochrone** A line joining sites at which a given phenomenon was observed to occur at the same time.

**isodynamic line** A line joining points at which the intensity of the Earth's magnetic field is constant.

**isogon** A line joining points at which the wind direction is constant.

**isogonic line** A line joining points with a constant *magnetic declination.

**isogram** *See* ISOPLETH.

**isohaline** A line joining points of constant *salinity.

**isohel** [iso + Greek *helios*: 'sun'] A line joining points that have the same duration of sunshine.

**isohyet** A line joining points of equal *rainfall for a particular period or for a particular storm.

**isohypse** A line connecting points on an *isobaric surface that have a constant altitude. *See* CONTOUR.

**isokinetic sampling** Atmospheric sampling for particulates using equipment that ensures that the velocity of the airflow from which the sample is collected is the same as that in the outside environment. This ensures that the distribution of particle sizes is truly representative.

**isoline** A line joining points on a chart at which a physical property has the same numerical value, examples being *isobar, *isotherm, etc.

**isomer (isomeric line)** A line that joins points that have recorded the same specific percentage of their annual *rainfall in any one particular month.

**isoneph** A line joining sites with equal cloud amounts.

**isopleth (sometimes known as an isogram)** A generic term for a line joining points at which the value of some parameter is constant.

**isopycnic surface** A surface of constant air (or water) density.

**isotach (also known as an isovel)** A line joining points of equal wind speed.

**isotherm** A line joining points of equal or constant temperature.

**isothermal** Having a constant temperature. The term is applied to any atmospheric layer where the *lapse rate is zero.

**isothermal chart** A *synoptic chart that shows the distribution of temperature at a particular level, normally by means of *isotherms.

**isotope** A chemical element that exists in two or more forms with different atomic masses, i.e. as atoms that have identical numbers of protons within their nuclei, but differing numbers of neutrons. The majority of elements consist of a mixture of isotopes, e.g. carbon, which has six protons, but six, seven, or eight neutrons, thus occurring as carbon-12, carbon-13, and carbon-14 (more correctly written as $^{12}C$, $^{13}C$, and $^{14}C$). The last of these is radioactive and forms the basis of *carbon dating. Although isotopes have identical chemical properties, some processes preferentially involve specific isotopes and may be used to determine climatic conditions during past epochs. *See* CARBON-ISOTOPE DETERMINATION; OXYGEN-ISOTOPE METHOD.

**isovel** *See* ISOTACH.

**ITCZ** *See* INTERTROPICAL CONVERGENCE ZONE.

J

**Jacob's ladder** *See* CREPUSCULAR RAYS (1).

**Jason-1** A *polar-orbiting satellite, launched in December 2001, designed specifically to monitor the Earth's oceans. Primarily a successor, with similar but improved instrumentation, to the very successful *TOPEX-Poseidon satellite.

**jet** A narrow, fast current of air at low levels, which is distinguished from the higher *jet streams. *See* BARRIER JET; COLD CONVEYOR BELT; CONVEYOR BELT; LOW-LEVEL JET; NOCTURNAL JET; WARM CONVEYOR BELT.

**jet streak** A localized region of exceptionally high wind speeds, not exceeding a few hundred kilometres in length, and generally located within a *jet stream. The upper-level *divergence associated with jet streaks is a major factor in *explosive cyclogenesis, and also in the creation of *split fronts.

**jet stream** A narrow current of high-speed winds, typically thousands of kilometres long, hundreds of kilometres in width, and a few kilometres in

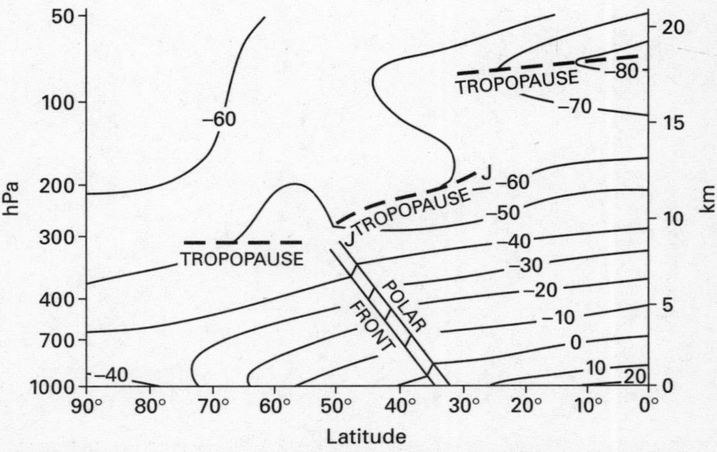


**Jet stream** The westerly jet streams (J) in the northern hemisphere lie where significant breaks occur in the *tropopause. Typical temperatures are also shown. Note how the temperature increases sharply across the *Polar Front.

depth, that occurs in the upper *troposphere and lower *stratosphere. Depending on context, an arbitrary lower limit of 25–30 m $s^{-1}$ is generally used to determine whether a particular flow should be described as a jet stream. The maximum speed recorded was 656 km $h^{-1}$ above South Uist in the Outer Hebrides, Scotland, on 13 December 1967. Jet streams are an essential element in the overall general circulation (*see feature*; including that of the upper atmosphere), and often occur where there are breaks in the level of the *tropopause. (*See* EQUATORIAL JET STREAM; POLAR FRONT JET STREAM; POLAR-NIGHT JET STREAM; SUBTROPICAL JET STREAM.) They are closely associated with *fronts and *depressions, and significantly affect the growth and decay of such systems (*see* ENTRANCE REGION; EXIT REGION). A jet stream may sometimes break or branch (a condition known as **split flow**), creating a region of *diffluence, below which convection is enhanced, often leading to the development of intense *thunderstorm systems. Jet streams are also regions of considerable *wind shear and *clear-air turbulence.

**juvenile water** Water that is reaching the Earth's surface for the first time from the deep interior through volcanic activity. Most of the water expelled by volcanoes is part of the overall *hydrological cycle, but small quantities may arise from deep in the mantle.

PL1 Typical altocumulus clouds, consisting of flattened 'pancakes' of cloud with clear sky visible between them.

PL2 Smooth lens-shaped altocumulus lenticularis clouds produced by wave motion over hills some distance away, together with lower *cumulus clouds.

PL3 A sheet of altostratus cloud, showing extensive billows, producing a *vertebratus-like appearance.

PL4 Cirrus clouds in which the ice crystals initially fell almost vertically, before encountering a layer of *wind shear, causing an abrupt change in direction.

PL5 Cirrocumulus is often so high that the individual cloud elements appear very small to the naked eye, and the lack of contrast makes them difficult to distinguish.

PL6 An extensive sheet of cirrostratus, with a well-defined 22° *halo and, on the left, a *parhelion just outside the halo.

PL7 Cumulus clouds, here arranged in *cloud streets, with higher-level cirrus indicating an advancing *warm front.

PL8 A typical *shower, or *cumulonimbus cloud, showing several active cells, with the most distant one (left centre) having spread out into an anvil.

PL9 A pair of cumulonimbus incus clouds. These cells had reached the low-altitude winter *tropopause and then spread out into the characteristic anvil shape.

PL10 Nimbostratus with a typical ragged base, associated with a *depression and bringing persistent rain to the hills of the Peak District in Derbyshire.

PL11 An extensive sheet of stratocumulus, with narrow gaps between the cloud elements, through which sunlight is penetrating to give rise to *crepuscular rays.

PL12 A sheet of stratus, ragged (*fractus) in places, shrouding the tops of the Alps. Such cloud does not normally give rise to any precipitation, except occasionally very slight *drizzle.

PL13 A *high-resolution picture transmission (HRPT) image from the NOAA 17 *polar-orbiting satellite on 27 May 2007. The deep depression produced extensive flooding in central England and damaging winds.

PL14 A composite image from three channels (visible and near-infrared), from Meteosat 8 (a *Meteosat Second Generation satellite) on 9 May 2003. The higher-resolution images greatly improve coverage at high latitudes.

PL15 A Channel-4 (infrared) image from Meteosat 8 on 13 December 2004. The darkest areas are hottest. Note the ribbon of *jet-stream cirrus across the Red Sea, and that the *Intertropical Convergence Zone is well north of the Equator.

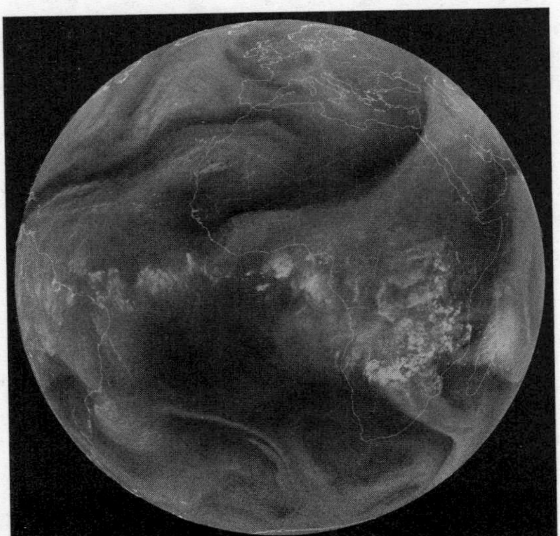

PL16 A Channel-5 (water vapour) image from Meteosat 8 on 6 March 2004. Dark areas represent low humidity. The band of dry air stretching across North Africa is particularly marked, as is the tropical cyclone Gafilo off Madagascar.

# K

**Kalpana** An Indian, geostationary meteorological satellite, stationed at 74° E. Originally named MetSat-1, and launched on 12 September 2002, it was renamed Kalpana on 6 February 2003 in honour of Indian-born Kalpana Chawla who died on 1 February 2003 in the Space Shuttle Columbia disaster.

**Karman vortex street (sometimes known as a von Karman vortex street)** A characteristic train of alternating clockwise and anticlockwise eddies that are shed by a bluff (i.e. not streamlined) obstacle in a flow. Such streets of vortices are often visible in satellite images of the cloud layer immediately behind an island, for example.

**katabatic wind** A wind consisting of dense air that has cooled by radiative cooling over upland areas or contact with snow and ice fields and which then drains down into the valleys. Also known as a **drainage wind** and a **mountain breeze**. Examples of strong katabatic winds are the *bora, *mistral, and the extreme winds that drain from the Antarctic ice cap.

**katafront** A warm or cold *front above which the warm air is descending (generally by subsidence at nearly all levels) relative to the cold air. Such a front is subdued and less active than a corresponding *anafront.

**kelvin (K)** The unit of heat on the *Kelvin temperature scale, defined such that the *triple point of pure water is at 273.16 K. The magnitude of one kelvin is precisely equal to one degree *Celsius ($1\,K \equiv 1\,°C$). *See also* ABSOLUTE ZERO.

**Kelvin effect** The effect whereby the *vapour pressure over a curved surface (e.g. that of a water droplet) is greater than that over a flat plane. The smaller the radius, the greater the vapour pressure. The effect is particularly important for the process of *condensation and for small droplets (such as those in clouds).

**Kelvin–Helmholtz instability** An instability caused by high values of vertical *wind shear that occur within a shallow layer, such as an *inversion in which there is a strong temperature or density gradient. In the atmosphere, the resulting waves (**Kelvin–Helmholz waves**) are often revealed by the presence of *billow clouds, including forms that resemble breaking water waves.

**Kelvin temperature scale** The international standard scale for scientific temperature measurement, defined in terms of the *triple point of water and a unit of heat (the *kelvin). Its zero point is at *absolute zero. Many physical laws and equations are more easily expressed by using such an absolute scale, rather than a relative scale such as the *Celsius temperature scale.

**Kelvin wave** An atmospheric or oceanic *gravity wave normally generated by an imbalance between the Coriolis force (*see* CORIOLIS EFFECT) and *pressure gradients, and which travels eastwards (at approximately 2–3 $m\ s^{-1}$ in the ocean). Kelvin waves propagating along the *thermocline are particularly important in El Niño Southern Oscillation events (*see feature*).

**Kelvin wedge** The angle included between the two arms of a vessel's wake. Because of the properties of wave propagation in water, this angle is constant for all velocities and is almost exactly 39°.

**Kern's arc** An extremely rare *halo phenomenon, appearing as a faint white circular arc, centred on the zenith, with the same radius as the *circumzenithal arc, but strongest on the opposite side of the sky.

**khamsin** A hot, dry, dusty wind occurring in late spring and summer around the eastern Mediterranean. A counterpart of the *sirocco, it is a southerly wind over Egypt, and an easterly over the Negev Desert and parts of Saudi Arabia. Like the sirocco, it is drawn northwards or westwards ahead of an approaching *depression. Known as the *sharav in Israel.

**kinetic energy** The energy that any mass has as the result of its motion. *See* DIRECT CIRCULATION; INDIRECT CIRCULATION.

**kite balloon** A tethered balloon shaped to have significant aerodynamic lift, and thus to provide a stable platform for specific instrumentation and observations. *See also* HELIKITE.

**knot (kn)** A speed of one international *nautical mile per hour, precisely equivalent to 1.852 $km\,h^{-1}$ or approximately 0.514 $m\ s^{-1}$. A unit used for reporting and forecasting wind speed.

**kona storm** [from Polynesian *kona*: 'leeward'] A severe south-westerly wind in the Hawaiian Islands which is associated with the passage of a secondary *depression to the north. It brings heavy rain to the dry, south-western side of the islands normally in the lee of the prevailing north-easterly *trade winds.

**konimeter** *See* DUST COUNTER.

**Köppen, Wladimir** (b. St Petersburg, 25 September 1846; d. Graz, Austria, 22 June 1940) German meteorologist and climatologist who produced a map of climatic zones, differentiated by temperature regime, in 1884. His pioneering scheme for *climate classification (*see* KÖPPEN CLASSIFICATION) was introduced in 1900.

**Köppen classification** A *climate classification system that is relatively simple, yet allows expansion to describe small climatic units. It is based primarily upon annual and monthly means for temperature, rainfall (particularly the seasonal distribution of rain), and wind, which together determine the types of flora and fauna within particular regions. It is similar to the classification of *air masses in using a capital letter to denote the main climatic type, followed by a lower-case letter to indicate subtype. The principal classes and subclasses are:

| | |
|---|---|
| A | tropical rain climates |
| B | *arid climates |
| C | warm *temperate rain climates |
| D | *boreal forest and snow climates |
| E | treeless cold snow climates |
| EF | perpetual frost (*ice cap) |
| ET | *tundra |
| BS | *steppe grassland climates |
| BW | warm *desert climates |
| s | summer dry season |
| w | winter dry season |
| x | wet early summer, dry late summer season |
| f | generally wet, with no specific dry season |
| m | *monsoon climate, with a short dry winter season |

**kosava** A *ravine wind in the valley of the Danube, where it cuts through the Carpathian Mountains to the east of Belgrade.

**krypton (Kr)** An *inert gas, atomic number 36, that occurs at low concentrations (1 part in 670 000) in the atmosphere.

**Kuro Shio** [Japanese: 'black salt-water current'] Part of the northern Pacific *gyre, a rapid, warm-water current with a characteristic dark blue colour that flows north-eastwards roughly parallel to the coast of Japan. Beyond Japan at latitude 35° N, it splits into two branches, both known as the **Kuro Shio Extension**, the major, southern flow becoming the *North Pacific Current, and the minor branch the *Aleutian Current. Also known as the **Kuroshio** and the **Japan Current**.

**Kyoto Protocol** The treaty drawn up at the Kyoto Conference in December 1997, as part of the *Framework Convention on Climate Change, for the limitation of emissions of six *greenhouse gases or classes of gases (*carbon dioxide, *methane, *nitrogen dioxide, *hydrofluorocarbons, *perfluorocarbons, and *sulphur hexafluoride). A significant omission from the treaty was the question of aircraft emissions, which are a major source of *ozone in the upper *troposphere, where ozone acts as a greenhouse gas. The concepts of *emissions credits and emissions trading were introduced at this conference.

# L

**laawan** [Arabic: 'the helper'] A west wind in the Middle East, so named because it helped farmers to winnow grain at harvest-time.

**labile** [from Latin *labilis*: 'liable to slide'] The condition in which a parcel of air has a *lapse rate equal to or greater than the *adiabatic lapse rate. The parcel therefore either has neutral equilibrium or is unstable.

**Labrador Current** A cold current flowing southwards from Baffin Bay and along the coast of Labrador and Newfoundland. After encountering the *Gulf Stream east of the Grand Banks, the two currents flow eastwards, separated by an extremely steep temperature gradient (sometimes known as the *North Wall or **cold wall**). *See also* Arctic Polar Front.

**lacunosus (la)** [Latin: 'having holes'] A *cloud variety in which *cumuliform clouds are arranged in a sheet or layer with moderately regular breaks or holes. The cloud elements are often fairly thin and their edges tend to be ragged. Normally applies to *altocumulus and *cirrocumulus and, occasionally, to *stratocumulus.

**lahar** [Javanese: 'lava'] A mudslide consisting of water-borne volcanic debris, often highly destructive and the cause of much loss of life. Lahars may be triggered by intense rainfall, the melting of snow or glacier ice through volcanic activity, or the collapse of the wall surrounding a crater lake. They solidify when they stop moving, and may remain a threat for many years after an eruption.

**lake-effect snow** Heavy snowfall downwind of large bodies of open water, such as the Great Lakes in North America. When a cold airstream of *continental polar air (cP) crosses relatively warm water (as in the autumn and early winter) it becomes unstable, leading to *mesoscale bands of cloud over the water and shoreline. Although the clouds are relatively shallow (extending to heights of perhaps 4 km), they produce significant precipitation on the farther shore. This effect is largely responsible for the extremely heavy snowstorms experienced by Buffalo, New York, for example.

**Lamb catalogue** A classification of daily weather types, introduced in 1950 by Professor Hubert Lamb. A classification is available for every day from 1 January 1861 to the present day, and the system has been extended to certain earlier periods. There are eight directional types, i.e. the principal compass directions (N, NE, E, etc.), and three non-directional: *cyclonic, *anticyclonic, and unclassifiable. The information provided by this listing is of great value in climatological studies, particularly with regard to climatic change (*see feature*).

**laminar flow** Non-turbulent flow, in which a fluid moves steadily and smoothly across a surface along *streamlines that everywhere lie in parallel sheets.

**land-breeze** A localized, night-time wind that blows off the land, and is created by the temperature (and hence pressure) differential between land and sea. It is the counterpart to the *sea-breeze and normally begins around midnight or the early hours of the morning, reaching a maximum strength around dawn. It often gives rise to a land-breeze front that may extend well out to sea, and which is frequently visible on satellite images as a line of *cumuliform clouds.

**land-lash** An obsolescent, British provincial term for heavy rain accompanied by high winds.

**LANDSAT** A series of satellites launched by the United States beginning in 1972, known as Earth Resources Technology Satellites (ERTS), but renamed when in orbit. Although primarily designed for studies of land use and resources, the satellites provided useful data for meteorological and oceanographic studies.

**landspout** A form of *vortex resembling a tornado (*see feature*), but weaker and shorter-lived, which extends down from a vigorously growing *cumulus, cumulus *congestus, or *cumulonimbus, and makes contact with the ground. It often appears similar to a *dust devil. Landspouts are not associated with a *gust front (*compare* GUSTNADO) nor with the organized rotation found in a *mesocyclone (i.e. in a *supercell thunderstorm). The method of formation (through strong *downdraughts and compensating *updraughts) is similar to that of a *waterspout and differs from the mechanism that creates true tornadoes.

**Langmuir, Irving** (b. Brooklyn, New York, 31 January 1881; d. Woods Hole, Massachusetts, 16 August 1957) American physicist and chemist, awarded the Nobel Prize for Chemistry in 1932 for his work on surface chemistry. A major contributor to several scientific fields, in meteorology Langmuir proposed that *dry ice and *silver iodide could be used for *cloud seeding to induce *precipitation. He developed what is now known as the Langmuir chain-reaction theory (*see* COALESCENCE) to account for the growth of large raindrops in warm clouds.

**Langmuir chain-reaction theory** *See* COALESCENCE.

**La Niña** A relatively recent term for the 'cold' phase of the El Niño Southern Oscillation (*see feature*), and the counterpart to the *El Niño phase. There is a steeply sloping *thermocline, giving deep warm water in the western Pacific and accentuated *upwelling of cold water off the coast of South America.

**lapse rate** The rate of change in a property with an increase in height. In meteorology, the term normally applies to the change in temperature. In the *troposphere this is an overall decrease from the ground to the *tropopause. Note that this is defined as a positive lapse rate. A negative value indicates that temperature increases with height. The most important theoretical lapse rates are the *dry adiabatic lapse rate and the *saturated adiabatic lapse rate (which apply to dry and saturated air parcels, respectively). The actual conditions prevailing within the atmosphere are described by the *environmental lapse rate. *See also* ADIABATIC; PSEUDO-ADIABATIC LAPSE RATE.

**latent heat** The heat absorbed or released by any substance that undergoes a change of phase (i.e. with no change in temperature); *compare* SENSIBLE HEAT. In meteorology, the term is exclusively understood to apply to the three phases of water (*ice, liquid *water, and *water vapour) and to the relevant amounts of energy involved in the transitions. These may be quite substantial, which is why the various phases of water are of primary significance in meteorological processes. The latent heat of fusion (ice to water) is $334 \times 10^3$ J kg$^{-1}$; and of *sublimation (ice to water vapour) is $2834 \times 10^3$ J kg$^{-1}$. The latent heat of vaporization (water to water vapour) depends upon temperature and is $2501 \times 10^3$ J kg$^{-1}$ at 0 °C, reducing to $2406 \times 10^3$ J kg$^{-1}$ at 40 °C.

**law of storms** A formulation of the general characteristics of the wind strengths and directions found in a *cyclone, and of the way in which the system propagates as a whole. First described by *Brandes in 1826, H. Dove in 1826, and W. Redfield in 1831, it was of extreme practical importance for mariners, being the first scientifically based rule on which they could base their decisions.

**layer clouds** A common name for any *stratiform clouds, more specifically for *stratus, *nimbostratus, *altostratus, and *cirrostratus.

**lead** A long, navigable stretch of open water in *pack ice, which may be several kilometres wide, as distinct from a narrow **lane**, or the extensive area known as a *polynia.

**leader** The initial streamer in each phase of a *lightning discharge that establishes an ionized path along which the main charge will subsequently pass. The very first channel is established by the *stepped leader, and subsequent strokes in the discharge are preceded by a *dart leader.

**leaf cloud (baroclinic leaf, cloud leaf)** An elongated cloud system, roughly resembling a leaf, visible in satellite images in the early stages of the formation of a *depression. It arises within the air of the *warm conveyor belt. The poleward edge of the leaf cloud has the shape of a flattened letter 'S', and the western edge often shows a notch, which will later develop into the *dry slot. With further *cyclogenesis, the pattern becomes a *comma cloud.

**lebeche** *See* LEVECHE.

**lee depression** *See* OROGRAPHIC DEPRESSION.

**lee eddy** *See* PILLOW EDDY.

**leeward** Towards or on the downwind side. *Compare* WINDWARD.

**lee wave (also known as a mountain wave)** A wave or series of waves that forms above and downwind of an obstacle, such as a range of mountains. A stationary ('standing') *gravity wave that develops in a stable layer between less stable higher and lower layers. The amplitude is strongly dependent on both the height of the barrier and the wind speed, and is normally greatest at some intermediate level, decreasing in both higher and lower layers. In certain cases the wave train may extend hundreds of kilometres downwind, and be extremely persistent while conditions remain constant. The downward motion of the initial wave behind the obstacle may occasionally be very large and bring air from high levels down close to the surface, creating *föhn conditions. The crests of the waves often exhibit *wave clouds, such as *altocumulus *lenticularis. *See also* ROTOR.

**lenticular clouds** Any of the *cloud types that occur as the *lenticularis species. *See also* WAVE CLOUDS.

**lenticularis (len)** A *cloud species in which the elements are approximately lens or almond shaped (and commonly known as **lenticular clouds**). The *cloud types most often encountered are *altocumulus (Plate 2) and *cirrocumulus, although *stratocumulus occurs occasionally. In most cases, the clouds are of *orographic origin and lie at the crests of *lee waves, remaining essentially stationary in the sky while the conditions (including wind speed and direction) remain constant. Depending on circumstances, however, they may sometimes lie far downwind of any mountain range or obvious source of vertical motion.

**leste** A hot, dry, dusty easterly or south-easterly wind from Morocco that reaches the Canary Islands and Madeira. It is closely related to the *sirocco.

**levanter** A humid easterly wind that affects the Straits of Gibraltar and the eastern coasts of Spain, most frequent in summer and early autumn. It generally occurs in *stable air under an *inversion and may produce a *banner cloud behind the Rock of Gibraltar. It is particularly strong over the Straits of Gibraltar, where there is a funnelling effect between the Atlas Mountains and the Spanish highlands. *See also* VENDAVAL.

**leveche (occasionally known as the lebeche)** A hot, dry wind on the narrow coastal strip of south-eastern Spain. It brings dusty *continental tropical air from the Sahara, and is particularly frequent in summer. *See* GHIBLI; KHAMSIN; SIROCCO.

**Le Verrier, Urbain Jean Joseph** (b. St Lô, 11 March 1811; d. Paris, 23 September 1877) French astronomer, who specialized in celestial mechanics, and whose work led to the discovery of Neptune in 1846. Appointed Director of the Paris Observatory in 1854 and, prompted by the storm of 14 November 1854 which caused a large loss of vessels engaged in the Crimean War, he was instrumental in setting up the first specialized meteorological network. Stations throughout France and in other countries reported regular observations to Paris using the telegraph. Publication of standardized data from the reporting stations first occurred in 1857.

**libeccio** The local name for a blustery south-westerly or westerly wind that occurs over Corsica and the central Mediterranean. It is particularly strong in winter, and may bring heavy rain or thunderstorms in summer.

**lidar (also incorrectly known as laser radar and optical radar)** [light detection and ranging—by analogy with 'radar'] A pulsed laser system that, in meteorological applications, may be used to detect the presence and motions of *aerosols, *cloud droplets, and other particles. Pulses of light are emitted by the laser, and the backscattered return signal from suspended particles is detected using a telescope. Specific applications include *cloud-base recorders, and the detection of *clear-air turbulence.

**life cycle** The various stages in the development and decay of a *depression, involving formation, *deepening, *occlusion, and *filling.

**lifting condensation level** The level to which a parcel of air must be raised before expansion and cooling (at the *dry adiabatic lapse rate) cause it to reach saturation.

**lightning** The natural electrical discharge that has its origin in the separation of charge that occurs in *cumulonimbus clouds. This discharge may take various forms: cloud-to-ground, cloud-to-cloud (inter-cloud), within a cloud (intra-cloud), or cloud-to-air. The process of a cloud-to-ground stroke (for example) occurs in several stages: an initial *stepped leader, a *return stroke, followed by one or more *dart leaders and their respective return strokes. In the majority of cases, the stepped and dart leaders bring negative charge from the cloud to the ground. The small percentage (*c*.5 per cent) of positive flashes are associated with much higher return currents.

Lightning is associated with several other atmospheric phenomena, notably *blue jets, elves (*see* ELF), *sprites, and *trans-ionospheric pulse pairs. *See also* BALL LIGHTNING; BEAD LIGHTNING; FORK LIGHTNING; HEAT LIGHTNING; ROCKET LIGHTNING; SHEET LIGHTNING; SFERICS; STREAK LIGHTNING.

**limb** The edge of the visible disk of any astronomical body, such as the Sun, Moon, or a planet. In meteorology, the term is sometimes applied to the outer edge of the atmosphere.

**limb-scanning radiometer** A satellite-borne *radiometer that is designed to scan the Earth's *limb rather than sounding the atmosphere vertically beneath the satellite.

**line convection** A narrow belt of *convection, often less than 5 km in width, with a corresponding restricted *rainband, that occurs on some *cold fronts where the warm air is rising (i.e. at an *anafront).

**line squall** A line or belt of instability at which a set of characteristic changes are encountered. The phenomenon may be frontal (as in an active *cold front) or non-frontal (in a *squall line). Typical changes are: a rapid rise in pressure and drop in temperature; a sudden rise in wind speed, accompanied by a *veer (in the northern hemisphere); a line of low, dark cloud; and strong convective activity in the form of *thunderstorms (frequently accompanied by *hail).

**lithometeor** An obsolescent term for any solid particles, such as sand, dust, smoke, or *haze particles, that are suspended in the atmosphere.

**lithosphere** The outer layer of the solid Earth, consisting of the oceanic and continental crust, and also the uppermost, brittle layer of the mantle. Its thickness varies greatly from just 1–2 km over the hot magma plumes at oceanic ridges to *c.*60–120 km for oceanic crust, and perhaps as much as 300 km below the oldest parts of the continents.

**Little Ice Age** A period from about 1500 to 1850 with low winter temperatures in north-western Europe and a general advance of *glaciers. The term should be used with caution, because, contrary to an earlier view, it was not consistently cold throughout the period and, although certain short episodes (particularly of cooling) show both hemispheric and global correlations, it was primarily a regional phenomenon. There is evidence that the *zonal westerly flow in winter frequently lay hundreds of kilometres further south than normal.

After generally deteriorating conditions, extremely cold winters and cold wet summers occurred in the 1590s. Although summer temperatures then recovered, winters remained *c.*0.5 deg C below the average for later years. Extreme conditions returned in the 1690s, again with exceptionally cold winters. There was then a sudden warming, with temperatures by 1730 similar to those of the warmest years of the 20th century. A swing back to colder conditions, particularly during the winter, occurred after an exceptionally cold winter in 1740. Winter temperatures began to rise from 1850, but it was not until the 20th century that average annual figures did the same. By contrast, spring and summer temperatures have remained remarkably steady since 1850.

**local and named winds** *See* ABROHOLOS; ANDHI; BAD-I-SAD-O-BIST-ROZ; BAGUIO; BARAT; BISE; BORA; BRICKFIELDER; BURAN; BURGA; CHILI; CHINOOK; CIERZO; CORDONAZO; DESCUERNACABRAS; DIABLO; DOCTOR; ETESIAN WINDS; GALERA; GHIBLI; GRECO; GREGALE; HABOOB; HAIZE-BELTZA; HARMATTAN; HELM WIND; KHAMSIN; KOSAVA; LAAWAN; LESTE; LEVANTER; LEVECHE; LIBECCIO; MAESTRO; MATACABRAS; MELTEMI; MISTRAL; NORTE; NORTHEASTER; NORTHER; PAMPERO; PAPAGAYO; PONENTE; PONIENTE; PUELCHE; PURGA; SANTA ANA; SEISTAN; SHAMAL; SIMOOM; SIROCCO; SOLANO; SQUAMISH; SUKHOVEI; TEHUANTEPECER; TERRAL; TORNADO *feature* (1); TRAMONTANA; VENDAVALES; VERANILLO; VERANO; VIRAZON; WILLIWAW; ZONDA.

**long-range forecast** A weather *forecast for a period more than ten days after the day on which it is issued, sometimes termed an **outlook**. *See also* MEDIUM-RANGE FORECAST.

**long wave** *See* ROSSBY WAVE.

**long-wave radiation** Electromagnetic radiation emitted by the surface and atmosphere at wavelengths between 3 and 100 μm, i.e. within the range of *infrared radiation. This corresponds to radiation from a body at a temperature of 200–300 K. *Compare* SHORT-WAVE RADIATION.

**looming** A *refraction effect in which objects that are usually below the horizon become visible, caused when the air density decreases more rapidly with height than normal. Rays of light from the object that would usually pass over the observer's head are refracted downwards more strongly, and thus become visible. The term is sometimes erroneously applied to the effect known as *towering. The opposite effect is *sinking. *See also* MIRAGE.

**Loop Current** A subsidiary current that circulates within the Gulf of Mexico, with a period of about one year.

**LORAN (LOng-range Aid to Navigation)** A radio-navigation system using synchronized signals emitted by 'master' and 'slave' stations. Extensively employed for meteorological purposes by drift buoys, *radiosondes, *GHOST balloons, etc. The earliest version (LORAN-A) was introduced in World War II, and the later version, LORAN-C, using a much lower frequency, became widely used. Despite the challenge from satellite-based Global Navigation Satellite Systems (GNSS), such as the *Global Positioning System, it has recently been appreciated that a ground-based system has certain specific advantages. An advanced form, known as 'Enhanced LORAN' (eLORAN, E-LORAN or LORAN-E), is therefore being introduced.

**low** *See* DEPRESSION.

**lowering** *See* WALL CLOUD.

**Lowitz, arcs of** A rare *halo phenomenon, normally appearing as parabolic arcs convex towards the Sun and tangent to the 22° halo at points below the position of any *parhelion. The theoretical explanation is uncertain, but the arcs may arise by refraction in crystals that are rotating as they fall. If so, two other components, concave towards the Sun, and above and below it, appear to be unrecorded.

**low-level jet** A narrow band of high winds that occurs well below the level of the *tropopause, where the major *jet streams occur. Often associated with severe weather. Both *barrier jets and *nocturnal jets fall within this category, as do some *warm conveyor belts.

**low-precipitation supercell** A tornadic *supercell thunderstorm (i.e. one containing a rotating *mesocyclone) that produces a limited amount of *hail, but effectively no rain at the ground. The *precipitation core is thus translucent (*compare* HIGH-PRECIPITATION SUPERCELL). In addition, low-precipitation supercells produce lightning within the anvil and in the clear air beneath it rather than beneath the cloud base and in association with the precipitation column. Such a situation may give rise to little or no audible thunder at the ground. A cloud-to-ground stroke beneath the anvil may be totally unexpected and thus regarded as a *bolt from the blue.

**LRTAP** *See* CONVENTION ON LONG-RANGE TRANSBOUNDARY AIR POLLUTION.

**lull** A brief reduction in wind speed below the average for the time in question. The counterpart of a *gust.

**lustrum** [Latin: a purification ceremony] A period of five years, often used in climatological studies.

**lysimeter** An instrument for measuring the moisture content of the soil and other parameters such as the rates of *percolation and *evapotranspiration. It generally consists of a section of soil and typical vegetation (e.g. grass) isolated from the surroundings and mounted so that changes in weight may be monitored, and from which the required information may be derived.

# M

**Mach number** The ratio between the speed of a body immersed in a fluid to that of the speed of sound in the surrounding fluid. When the Mach number exceeds 1, the speed is supersonic, and when above 5, it is hypersonic.

**mackerel sky** A popular term for a sky with extensive *cirrocumulus or high *altocumulus that resembles the pattern on mackerel, i.e. the *cloud varieties *radiatus and *undulatus. In common usage the term is frequently extended to any form of regularly patterned cloud layer, even *stratocumulus, regardless of the size of the cloud elements.

**macroburst** *See* DOWNBURST.

**macroclimate** The climate of a large region of the Earth, either part of or the whole of a country, as distinguished from *mesoclimate and *microclimate.

**macroclimatology** The study of *macroclimates, both as individual systems and in relationship to one another.

**macrometeorology** The study of planetary-scale processes, such as the general circulation (*see feature*), that occur in the atmosphere.

**Madden-Julian Oscillation (MJO; also known as 30–60 day oscillation, 30–60 day wave, 40–60 day oscillation, and intraseasonal oscillation)** An equatorial pattern of enhanced and reduced *convection and *precipitation that travels eastwards across the tropical oceans. Particularly marked over the Indian and western Pacific Oceans, it becomes less distinct over the colder eastern Pacific, but strengthens again over the Atlantic. It moves at *c.*5 m/s over the eastern hemisphere and *c.*10 m/s over the western. Taken around the globe, there are typically one or two areas of enhanced convection and precipitation, and one or two regions where the effects are reduced. MJO activity is linked to the El Niño Southern Oscillation (*see feature*) events, being typically strong about 6–12 months before the onset of an El Niño event, absent during an event, but becoming strongest during any *La Niña phase.

**maestro** [Italian: 'master'] A north-westerly wind that occurs in summer over the western part of the Adriatic. The name is also given to a similar wind on the coast of Corsica and Sardinia.

**magnetic storm** *See* GEOMAGNETIC STORM.

**magnetopause** The boundary between a planet's *magnetosphere and the magnetic field carried by the *plasma of the *solar wind and which prevails in interplanetary space.

**magnetosphere** The region surrounding a planet within which its magnetic field predominates over external fields. Because of the effect of the *solar wind, the magnetosphere is compressed on the sunward side, and drawn out into a long *magnetotail downwind, giving an overall shape resembling an elongated teardrop, with cavities (**magnetic cusps**) occupying the regions above the magnetic poles.

**magnetotail** The portion of a planet's *magnetosphere, drawn out by the *solar wind, that may extend many tens or hundreds of planetary radii away from the Sun. In the centre of the magnetotail is a **neutral sheet**, separating the regions of opposite magnetic polarity. When major disturbances occur in the solar wind, reconnection of the opposing fields may take place, accelerating charged particles towards the magnetic poles and giving rise to an *aurora or a *geomagnetic storm.

**mamma (mam)** [Latin: 'udder'] A *supplementary cloud feature consisting of bulges or pouches beneath the lower surface of a cloud. They are the visible signs of *downdraughts bringing colder air into the lower layers. Beneath *cirrus, *cirrocumulus, *altocumulus, *altostratus, and *stratocumulus they are generally irregular in shape, but beneath *cumulonimbus anvils they are often bulbous, and may appear particularly striking when illuminated by low sunlight.

**mammatus clouds** An obsolescent term for clouds (particularly *cumulonimbus) with *mamma.

**mares' tails** A popular name for moderately dense wisps of *cirrus, particularly those clumps that are thicker at one end than at the other.

**marine climate** *See* MARITIME CLIMATE.

**maritime air** Any *air mass that has a *source region over the oceans. Such air always has high *humidity in its lower layers. *See* MARITIME ARCTIC AIR; MARITIME POLAR AIR; MARITIME TROPICAL AIR; EQUATORIAL AIR.

**maritime Arctic air (mA)** An *air mass that has its *source region over the Arctic or Antarctic regions. In winter, when there is largely unbroken ice cover across the Arctic Ocean, it initially has *continental characteristics (cA), but is rapidly modified on crossing open water, being warmed from beneath, and thus becoming unstable. It is generally similar to *maritime polar air, but colder throughout its depth. In summer, it originates directly over the open *pack ice in both the Arctic and Antarctic, and there are additional sources at high latitudes over Canada and Eurasia (particularly Siberia), where the extensive marshy ground means that the regions more closely resemble *maritime than continental areas, and affect the air mass accordingly. *See* ARCTIC AIR.

**maritime climate (occasionally known as a marine climate)** A form of *climate that is governed by the area's proximity to the sea, and which prevails over islands and the windward side of continents. Characteristic features are restricted diurnal and annual ranges of temperature and, at least on windward coasts, moderate to heavy precipitation.

**maritime polar air (mP)** A cool, moist *air mass, which in the northern hemisphere often originates as *continental polar air over the North American or Asian land masses and is modified as it moves out over the Atlantic and Pacific Oceans. In the southern hemisphere, there is a continuous zone of mP air between approximately 40° and 60° S that tends to expand towards Antarctica in summer. Heating by the relatively warm oceans leads to instability, with convective clouds and *showers over the sea and on windward coasts. Convection is greatest in the rear of *depressions and tends to be suppressed towards lower latitudes, when subsidence may limit *cumulus growth and promote extensive *altocumulus or *stratocumulus.

**maritime tropical air (mT)** A warm, moist, and generally stable *air mass that originates in the *subtropical anticyclones over the oceans. The principal *source regions are the North Atlantic and North Pacific in the northern hemisphere, and the South Atlantic, South Pacific, and south Indian Ocean in the southern. The last three together cover practically half of the southern hemisphere in a broad belt down to about latitude 40° S. (In the southern summer, mT air covers most of southern Africa and South America, with only very small source regions of *continental tropical air in both continents.) Because of its warmth and humidity, *stratiform cloud frequently develops as mT air moves towards the poles. When the sea-surface temperature is low, it may produce extensive *advection fog in the South-Western Approaches, the South China Sea (*see* CRACHIN), off Newfoundland, and off northern California.

**Mason's hygrometer** *See* PSYCHROMETER.

**matacabras** [Spanish: 'kills goats'] An uncomfortable, cold wind, not necessarily northerly, in the mountainous regions of Spain. The term is also associated with *sleet (i.e. melting snow) and *small hail, sometimes accompanied by wind. *See also* DESCUERNACABRAS.

**Maunder Minimum** A period from about 1645 to 1715 when very few *sunspots or *aurorae were recorded from Western Europe. Recent research on Chinese and Korean records suggests that sunspot numbers were not as low as previously thought. The Maunder Minimum did, however, coincide with an increase in the $^{14}$C level in trees (*see* CARBON DATING), caused by an increased flux of *cosmic rays, and came towards the end of the period known as the *Little Ice Age. *See also* SPÖRER MINIMUM.

**Maury, Matthew Fontaine** (b. Spotsylvania, Virginia, 14 January 1806; d. Lexington, Virginia, 1 February 1873) American meteorologist, hydrographer, and early oceanographer. Gathered information from seafarers on currents and winds, published the first modern chart of global wind fields in 1848, and later studied the Atlantic, Indian, and Pacific Oceans in detail. Published the first bathymetric chart of the Atlantic in 1854, and the first true oceanographic work, *The Physical Geography of the Sea*, in 1855.

**maximum thermometer** A thermometer that is designed to register the maximum temperature obtained during the interval between readings. Generally a mercury-in-glass thermometer with a constriction immediately above the bulb, which allows the mercury column to expand but which breaks the column on cooling. The maximum temperature is indicated by the far end of the column. Resetting is effected by shaking the mercury past the constriction. *See also* SIX'S THERMOMETER.

**MCC** *See* MESOSCALE CONVECTIVE COMPLEX.

**MCS** *See* MESOSCALE CONVECTIVE SYSTEM.

**mean free path** The average distance travelled by an atom or molecule between collisions with another similar atmospheric component. The mean free path increases with decreasing pressure, and in the *exosphere becomes hundreds of kilometres, allowing particles with *escape velocity to be lost to space.

**measurement** *See* OBSERVATION.

**mediocris (med)** [Latin: 'average'] A *cumulus *cloud species in which the individual clouds have similar (and moderate) horizontal and vertical extents. The top of the cloud normally shows signs of upward growth.

**Mediterranean air** A subsidiary form of *air mass, characterized by significant warming of *maritime polar, mP (or even *maritime Arctic, mA) air that penetrates to the Mediterranean region. The resulting instability gives rise to *depressions with very heavy *precipitation on the *warm front, and intense *showers and *thunderstorms behind the *cold front.

**Mediterranean climate** A climatic type characterized by hot, dry summers and mild, moist winters (*Köppen classification Cs). The influence of the *subtropical anticyclones is dominant in summer, whereas in winter there is a westerly airflow. The type region is the largest, extending from Portugal and Morocco to Turkmenistan, but smaller regions of similar type are found in California, Chile, South Africa, and southern Australia.

**Mediterranean Front** The boundary between *Mediterranean air to the north and *continental tropical air from the Saharan region to the south. The temperature difference across it may amount to as much as 15 deg C in winter.

**medium-range forecast** A weather *forecast for a period of 3–10 days after the date on which it is issued. *See also* LONG-RANGE FORECAST.

**Megacity Initiative: Local and Global Research Observations (MILAGRO)** A research project to investigate air pollution in and around certain megacities (any urban area with more than ten million inhabitants). The initial study of Mexico City not only revealed an extremely high concentration of *secondary organic aerosols but that these compounds are a major component of air pollution, whose significance was previously unsuspected, and which will require substantial adjustment of pollution models.

**meltemi** [Turkish: 'bad-tempered'] The Turkish name for the *etesian winds.

**melting band** *See* BRIGHT BAND.

**melting level** The altitude at which the air temperature is 0 °C, at which ice crystals and *snowflakes melt as they descend in the atmosphere. *See also* FREEZING LEVEL.

**melting point** The temperature at which a substance changes from the solid to the liquid phase. For pure water, this occurs at 273.15 K (0 °C), slightly below the *triple point (273.16 K) by which the modern temperature scale is defined.

**meniscus** The (curved) upper surface of a column of liquid. In reading barometers or thermometers that contain a mercury column, for example, allowance must be made for the curvature of the meniscus.

**mercury (Hg)** A metallic element used extensively in *thermometers and previously in mercury-column *barometers. It has a melting point of −38.87 °C, and cannot therefore be employed in thermometers required for very low temperatures. It is then replaced by alcohol, although the latter provides less accuracy.

**meridian** **1.** The great circle on the *celestial sphere that passes through both celestial poles and the observer's *zenith and *nadir, intersecting the horizon at the north and south points. **2.** Any of the great circles on the surface of the Earth that pass through both geographical poles and are used to denote longitude.

**meridional cell** Large-scale circulation in either the atmosphere or ocean that occurs predominantly in a north–south direction, such as the atmospheric *Hadley cells. Most such cells form part of a *direct circulation, driven by *convection.

**meridional circulation** Atmospheric circulation that occurs essentially in a north–south plane. Although the general circulation (*see feature*) was originally believed to be of this basic form, it is now accepted that it involves large-scale *zonal flow, and that *jet-stream motions are essential components.

**meridional extension** A very pronounced extension of *ridges or *troughs in a north–south direction.

**meridional flow** A situation in which the winds exhibit a large north–south component, unlike the normal *zonal flow. Such conditions generally precede *blocking.

**meso-** [from Greek *mesos*: 'middle'] A prefix commonly used in meteorology to indicate that a phenomenon is intermediate in size (*mesoscale); position (*mesosphere); or applies at intermediate scales (*mesoclimatology).

**mesoclimate** The *climate of a geographically restricted area, such as that of a forest, upland area, or valley, which is not necessarily representative of the overall climate of the region as a whole.

**mesoclimatology** The study of *mesoclimates.

**mesocyclone** A deep, vertical column of air, with *cyclonic rotation, that develops within a severe *thunderstorm. Mesocyclones generally have a diameter of 3–10 km, but are known to exceed 16 km on occasions. Present in all *supercell thunderstorms, and associated with the formation of all strong and violent tornadoes (*see feature*).

**mesometeorology** The study of weather on a smaller scale than that covered by *synoptic meteorology, and thus of a restricted geographical area, such as a valley, watershed, etc. Although in certain cases the area may be studied by installing a number of individual weather stations, it is generally carried out with satellite instrumentation, with perhaps some ground sites to provide 'ground truth' and instrumental calibration.

**mesopause** The boundary between the *mesosphere and *thermosphere. Modern observations suggest that it may lie at two distinct levels, $86 \pm 3$ km, and $100 \pm 3$ km, the higher value being found towards the poles in summer, when there is upwelling at high latitudes. It is an *inversion, at which the temperature of the atmosphere reaches its minimum value. This temperature is now known to be more variable than previously thought, but lies in the range 110–173 K (–163 to –100 °C).

**mesoscale** A term applied to weather systems that are intermediate in size between *synoptic-scale and *storm-scale systems. They are generally too small to be adequately studied by the normal *synoptic network and to be shown clearly on a standard synoptic chart, and thus lie between about 80 km and 250 km across. Some typical mesoscale systems are the cloud bands that produce *lake-effect snow, *mesoscale convective complexes, and *squall lines. Some authors use the term for rather smaller features that have a spatial scale of between 1 and 50 km.

**mesoscale convective complex (MCC)** A particularly large and vigorous *mesoscale convective system, defined from the properties of the cloud shield as revealed by satellite infrared imagery. The area with a temperature below –52 °C must exceed 50 000 $km^2$, and there must be a continuous area, exceeding 100 000 $km^2$, below –32 °C. This larger area must be elongated, with an eccentricity (ratio of minor axis to major axis) of 0.7 or more. These criteria must be satisfied for a period of more than 6 hours for the classification to apply.

**mesoscale convective system (MCS)** A large-scale system characterized by deep convective clouds (*cumulus *congestus and *cumulonimbus), various associated *stratiform clouds, heavy precipitation, and a vigorous circulation. The last becomes progressively more organized as the system develops. In particular, the individual downdraughts tend to combine to produce a vigorous outflow, and a deep layer of cold air in the lower troposphere. At upper levels, the individual cirrus plumes combine to produce a large *anvil or cirrus shield. The overall lifetime of such a system exceeds four hours, much longer than that of a normal, individual cumulonimbus cell.

**mesosphere** The atmospheric layer between the *stratosphere and the overlying *thermosphere (and thus lying between *c.*50 and 83–103 km), in which temperature decreases with height, reaching the atmospheric minimum of approximately 150 K (–123 °C) at the *mesopause. Circulation within the mesosphere is poorly understood, but (together with stratospheric circulation) appears to be a single *Hadley cell, with rising air in the summer hemisphere and descending air in the winter hemisphere. There is a pair of jet streams located approximately above the *subtropical anticyclones at altitudes of *c.*60–65 km, westerly in the winter hemisphere, and easterly in the summer hemisphere. These somewhat resemble the stratospheric *Polar-Night Jet Stream and *Equatorial Jet Stream.

**meteor** **1.** An old-fashioned term (although one that is still used in *World Meteorological Organization definitions) for any liquid or solid particles, apart from a cloud, that may be suspended in the atmosphere, precipitated from it, or deposited on the surface. It also applies to any electrical or optical phenomena. *See* ELECTROMETEOR; HYDROMETEOR; LITHOMETEOR. **2.** A small fragment of solid material of cometary or minor-planetary origin that enters the Earth's atmosphere and is vaporized at altitudes of *c.*110–70 km. (Before encountering the atmosphere, such particles are termed *meteoroids.) The term is also applied to the luminous trail that may be visible at night. Larger bodies may be classed as *fireballs, *bolides, or, if fragments reach the ground, *meteorites. Meteors and *micrometeorites may contribute some of the *freezing nuclei found in *noctilucent clouds.

**METEOR** A series of *polar-orbiting meteorological satellites, launched by the USSR and then Russia. They are placed in an eccentric orbit (rather than one close to circular) so that they spend a greater proportion of their time in a suitable position to obtain images of northern latitudes.

**meteorite** A solid, stony, or metallic body, in most cases of minor-planetary origin, that has survived passage through the Earth's atmosphere and reached the surface. From chemical analysis and spectroscopic studies, certain meteorites may be matched to specific solar-system bodies, including the Moon, Mars, and various minor planets, such as Vesta. Others (those known as carbonaceous chondrites) consist of extremely primitive material dating back to the formation of the solar system ($4.6 \times 10^9$ BP) and are found to contain inclusions (such as microscopic diamonds) and the decay products of radioactive elements formed at a still earlier epoch.

**meteorograph (formerly called an aerograph)** An instrument for obtaining a chart recording of various meteorological elements; particularly the now obsolete form that was carried aloft by early kites and balloons.

**meteoroid** The general term for any small body moving on an independent orbit around the Sun in interplanetary space, before it strikes an atmosphere or a solid body. Meteoroids have two main sources: they are fragments of minor planets, or dust particles shed by comets. The majority are very small, giving rise to *micrometeorites and *meteors, rather than resulting in *meteorites.

**meteorological optical range (MOR)** A formally defined distance, essentially equivalent to *visibility: the length of path in the atmosphere over which the light from a known source is reduced to 0.05 of its original intensity.

**meteorological reconnaissance flight** A flight made by a suitably equipped aircraft, either regularly to obtain data from an area inadequately covered for *synoptic purposes, or for a study of a specific phenomenon, such as a tropical cyclone (*see feature*). In addition, research flights may be carried out (often by several aircraft simultaneously) to obtain data on particular atmospheric processes, such as the three-dimensional structure and evolution of *cumulonimbus clouds.

**meteorological rocket** An unguided solid-fuel rocket, launched to altitudes of 60–100 km. In addition to standard instrumentation, such as temperature, pressure, and *ozone sensors, some versions may be equipped to obtain air samples, or to trap particles found in *polar stratospheric clouds or *noctilucent clouds.

**meteorological satellite** An artificial satellite specifically designed and operated to obtain meteorological observations in real time for forecasting purposes. It is thus distinguished from earth-resource and other satellites that may carry instrumentation to monitor specific phenomena for climatological or other purposes.

Two types are in use: *geostationary satellites such as *GOES, *GMS, and *Meteosat; and *polar-orbiting satellites, such as the *NOAA and *METEOR series.

**meteorology** [from Greek *meteoros*: 'high' + *logos*: 'discourse'] The study and science of all aspects of the atmosphere, and thus including both *weather and *climate. It aims to understand the physical and chemical nature of the atmosphere, its dynamical behaviour, and its complex interactions with the surface. It also includes both short- and long-term weather forecasting, and the determination of past and future climatic change (*see feature*). In modern times the discipline is also extended to include the study of the atmospheres of other planets. The term was first employed by *Aristotle.

**Meteosat** A series of spin-stabilized *geostationary meteorological satellites, launched by the European Space Agency (ESA), and operated by *Eumetsat. There have been two generations of satellites, and a third is under development. Of the first generation, Meteosat-1 was launched on 23 November 1977, and coverage has been essentially continuous ever since the last in the series, Meteosat-7, in 1997. The satellites are normally stationed at longitude 0°, but for a period, when a *GOES-E satellite failed, Meteosat-3 (a redundant unit) was moved west to provide the missing coverage. The primary (service and backup) satellites have now been replaced by *Meteosat Second Generation units. Two first-generation Meteosats are currently stationed at 63° E over the Indian Ocean to act as primary and backup units for the Indian Ocean Data Coverage service.

Meteosat's scanning *radiometer returns images of the whole of the Earth's visible disk in three different spectral channels (visible: 0.5–0.9 μm; infrared water-vapour absorption: 5.7–7.1 μm; thermal infrared: 10.5–12.5 μm). The repeat rate is 30 minutes. After processing by the ground station, images are transmitted to the satellite, and then rebroadcast to users within the satellite's communications footprint (normally within a great-circle radius of 55° from the subsatellite point). Additional data channels are available to relay environmental information from surface stations to the system's data centres.

**Meteosat Second Generation (MSG)** An advanced series of geostationary satellites, the first of which (Meteosat-8) became operational in January 2004. The series carries advanced instrumentation, primarily the *Spinning Enhanced Visible and Infrared Imager *radiometer, with twelve spectral channels, an improved ground resolution of *c.*1 km, an image repeat rate of 15 minutes, and a delay time of just 5 minutes. The satellites also carry the experimental *Geostationary Earth Radiation Budget instrument. The instrumentation, and the data returned, have been optimized to be used with the latest, most advanced, analysis and *numerical weather prediction methods (Plates 14–16). Meteosat Third Generation satellites are under development. *See also* Satellite Application Facility.

**methane ($CH_4$)** A gas that is a trace component of the atmosphere (*c.*$2.0 \times 10^{-4}$ per cent by volume), but which makes a significant contribution to the greenhouse effect (*see feature*), as well as tending to deplete atmospheric *ozone, by which it is destroyed. Extensively produced by the decay of organic matter and from rice paddies, ruminant animals, and termites, it is also associated with all forms of fossil fuel as natural-gas deposits. From its occurrence in swampy areas it obtains its common name of **marsh gas**.

**methane hydrate** A *clathrate, consisting of *methane trapped within a water-ice matrix, found mainly on and below the sea-floor at the lower edge of continental slopes, where low temperatures and high pressures permit it to exist. Some deposits also occur in *permafrost regions. Enrichment in the $^{12}C$ *isotope shows the methane to be of organic origin. Estimates suggest that methane hydrate deposits are a potential reserve twice the size of the combined total for coal, oil, and conventional natural gas. Commercial exploitation is likely to be difficult, but preliminary tests have already begun.

Methane evolved through melting of methane hydrate has been proposed as one factor in sudden climatic change (*see feature*), including a surge in global temperatures at $55 \times 10^6$ BP, and changes at the end of the last *ice age around 15000 BP. It is also suspected to have caused major underwater slumping and consequent *tsunami, particularly the Storrega submarine landslide off Norway, in *c.*8000 BP, when around 5 600 $km^3$ of sediments slid downslope for about 800 km.

**methyl bromide ($CH_3Br$; bromomethane)** A pesticide used extensively as a soil fumigant, which is an important source (*c.*25 per cent) of atmospheric bromine. It contributes to the destruction of the *ozone layer, and, after protracted negotiation, was included in the *Montreal Protocol and phased out of use in developed countries by 2001. It is scheduled to be discontinued in developing countries by 2015. However, in 2004, the United States demanded that it be given critical-use exemption, and the chemical is used extensively in that country, particularly on Californian fields. Natural sources also exist, it being produced, most notably, by algae and kelp in the oceans, and by certain plants, including brassicas.

**methyl chloroform ($CH_3CCl_3$)** A halogenated hydrocarbon with various industrial uses, but which contributed to destruction of the *ozone layer. Under the *Montreal Protocol, its use was abolished by January 1996.

**MetOp (Meteorological Operational)** A series of three *polar-orbiting meteorological satellites developed by the European Space Agency (ESA), *Eumetsat, *NOAA, and the French space agency (CNES), and forming the space segment of the *Eumetsat Polar System (EPS). MetOp-1 was launched on 19 October 2006 and became fully operational by 15 May 2007. The satellites will eventually become the morning series of a NOAA-Eumetsat system that will replace the ageing NOAA (2) series. MetOp satellites (orbiting at an altitude of *c.*800 km) will continue some of the monitoring carried out by the *European Remote Sensing Satellites and will also complement the observations made by *Envisat. Apart from a newly developed suite of sophisticated instruments (*see* ADVANCED SCATTEROMETER; GLOBAL OZONE MONITORING EXPERIMENT; GNSS RECEIVER FOR ATMOSPHERIC SOUNDING; INFRARED ATMOSPHERIC SCANNING INTERFEROMETER; and MICROWAVE HUMIDITY SCANNER) they carry legacy instruments identical to those used on NOAA polar-orbiting satellites (*see* ADVANCED MICROWAVE SOUNDING UNIT; ADVANCED VERY HIGH RESOLUTION RADIOMETER; and HIGH RESOLUTION INFRARED SOUNDER) to ensure compatibility and continuity between datasets obtained with different instruments.

**Mexican plume** A plume of potentially warm air (*see* POTENTIAL TEMPERATURE) that originates over Mexico and flows towards the United States at an intermediate level, where it forms a *capping inversion. Extremely high humidity builds up in the lower layers, partly through *advection of moist air from the Gulf of Mexico. When an upper-air *trough approaches from the west, the capping inversion may break down completely, giving rise to exceptionally powerful *multicell and *supercell thunderstorms over Texas, Arkansas, and Oklahoma. *See also* SPANISH PLUME.

**MHS** *See* MICROWAVE HUMIDITY SENSOR.

**microbarograph** A *barograph specifically designed to record small and rapid fluctuations in atmospheric *pressure.

**microburst** An intense, sharply defined *downburst that affects a relatively small area (less than 4 km across) and produces abrupt *wind shear. Microbursts generally last for about 5 minutes, but some are more persistent, with durations of up to 30 minutes. Wind speeds may be as high as 150 knots (280 $km h^{-1}$). The sudden wind reversals and violent downdraught constitute a major hazard for aircraft when landing or taking off, and are known to have been responsible for around thirty accidents in the United States. They also pose considerable danger to sailing vessels. Dry microbursts occur in dry environments when *precipitation falls from clouds with high bases, producing *virga, but not reaching the ground. The extreme evaporative cooling intensifies the downdraught. Wet microbursts occur in extremely moist environments when the downdraught draws in air from outside the storm.

**microclimate** The conditions found in the atmosphere at very small scales, particularly those where small features of the surface significantly affect temperature, humidity, and other factors. Often associated with the study of specific animals, insects, plants, bacteria, etc., including extremophiles, bacteria that exist in extreme environments, such as hot springs and submarine oceanic vents (**black smokers**), and algae that are found beneath the surface of rocks.

**microclimatology** The study of *microclimates.

**micrometeorite** A *meteoroid that is so small (<0.1mm) and has such a low mass ($<10^{-6}$g) that it is not vaporized as a *meteor but slowly sinks to the Earth's surface. Micrometeorites may be collected in the stratosphere or recovered from ice cores. They are interplanetary dust particles that originally formed part of comets or minor planets.

**micrometeorology** The study of small-scale phenomena that take place in the atmosphere, particularly within the *planetary boundary layer. The study also extends to *turbulence and certain other processes that may also take place at higher levels.

**micrometre (μm)** An SI submultiple of the metre ($10^{-6}$m), frequently used for infrared and microwave wavelengths and the diameter of *aerosols, cloud droplets, etc. Previously known as the **micron**, a term that remains in common use.

**micron** *See* MICROMETRE.

**micropulsations** Rapid fluctuations in the strength, direction, and polarity of the local *geomagnetic field, often showing an approximate periodicity. The latter may range from *c.*0.1s to 300s or more. Linked to short-term variations in the *ionosphere and *magnetosphere.

**microseisms** Low-amplitude, quasi-periodic motions of the ground that do not originate in earthquakes but have various other artificial and natural causes. These include storm conditions at sea, and microseisms may be used for the early detection of tropical cyclones (*see feature*) and other disturbances.

**microthermal climate** A low-temperature climate with very short summers (average temperature around 10 °C) and long, cold winters (average temperature −3 °C or below). Generally *Köppen classification D, and found between latitudes 40–65° N and S.

**Microwave Humidity Sensor (MHS)** A five-channel, microwave scanning *radiometer carried by *MetOp satellites. It is particularly suitable for the detection of *humidity at various altitudes, variations in cloud cover, the presence of atmospheric ice, and all forms of *precipitation. It also returns data on surface temperatures.

**microwave sounding unit** *See* ADVANCED MICROWAVE SOUNDING UNIT.

**middle atmosphere** A broad term for the region of the atmosphere that consists of the *stratosphere and the *mesosphere.

**Mie scattering** A generalized theory of the *scattering of radiation by particles, which may be employed irrespective of the relative sizes of the particles and the wavelength of the radiation. It is thus more widely applicable than *Rayleigh scattering. In general, *aerosol and *fog particles (for instance)

## Milankovich theory

The hypothesis that the basic cause of long-term climatic change (*see feature*), and particularly of the *ice ages, lies in quasi-periodic variations in the Earth's orbit around the Sun. The idea was originally proposed by Sir John Herschel in 1830, extended by J. A. Adhémar and James Croll, and revived by *Milankovich in 1920. Current analysis suggests that astronomical factors alone are insufficient to account for the major ice ages, but that they probably have a *forcing effect, being reinforced by other mechanisms, notably *carbon-dioxide concentration.

There are three principal effects: (a) Variation in the **eccentricity** of the Earth's orbit (between *c.*0.005 and 0.05), with a primary period of *c.*100 000 years and secondary periods of *c.*400 000 and *c.*125 000 years. This causes a significant variation in the amount of radiation incident on the Earth. (b) Alteration of the inclination of the Earth's axis (known as the **obliquity**) between 21.6° and 24.4°. This varies with a period of *c.*41 000 years, and periods of minimum inclination cause a reduction in the average intensity of radiation incident at high latitudes. (c) Variation in the orientation of the Earth's axis in space that causes the axis to sweep out a cone (**precession of the equinoxes**). This causes a change in the timing of the seasons relative to *aphelion and *perihelion. When the (northern) winter solstice and aphelion occur together, for example, *insolation is then at an absolute minimum in the northern hemisphere. Precession has two periodicities of *c.*19 000 and *c.*23 000 years, which combine to give an overall period of *c.*22 000 years.

whose diameter is comparable to that of the incident light will give rise to white, rather than coloured, scattered light.

**MILAGRO** *See* MEGACITY INITIATIVE: LOCAL AND GLOBAL RESEARCH OBSERVATIONS.

**Milankovich (Milanković, Milankovitch), Milutin** (b. Dalj, Austria-Hungary, 28 May 1879; d. Belgrade, 12 December 1958) Yugoslav scientist who carried out extensive studies of the effects of astronomical factors on the amount of solar radiation incident on the Earth, and hence on climate. He determined the length of the various cycles (*see feature* MILANKOVICH THEORY) and vigorously championed the idea that the variations were the basic cause of *ice ages.

**mile, nautical** *See* NAUTICAL MILE.

**millibar** A commonly used unit of atmospheric *pressure, a submultiple ($10^{-3}$) of the *bar, a c.g.s. unit of pressure, and introduced by William Napier *Shaw in 1909. In most meteorological works it has now been superseded by the hectopascal (hPa): 1 mb = 1 hPa. The latter is itself a multiple of the SI unit the *pascal (Pa).

**minimum deviation** The smallest angle through which light is deviated when undergoing *refraction in particles such as water droplets or *ice crystals. This angle depends upon the refractive index of the material: for a simple pathway through a raindrop it is *c.*42°, and for a hexagonal ice crystal *c.*22°. These correspond to the radii of the primary rainbow (*see feature*) and 22° halo, respectively. Rays emerging at the minimum deviation make the greatest contribution to the brightness of the final image.

**minimum temperature** The lowest temperature reached during any given period, normally taken to be that of the air, as recorded by a *minimum thermometer in a *thermometer screen. *See also* CONCRETE MINIMUM TEMPERATURE; GRASS MINIMUM TEMPERATURE.

**minimum thermometer** A thermometer designed to register the minimum temperature attained between readings. Generally an alcohol-in-glass thermometer, with an *index that is contained within the alcohol column. Contraction of the alcohol drags the index back towards the bulb, but expansion does not alter its position, thus allowing the minimum temperature to be read from the index. Resetting is normally by tilting the thermometer so that the index slides until it meets the *meniscus at the end of the alcohol column. *See also* SIX'S THERMOMETER.

**mirage** An optical phenomenon caused by *refraction in the lowest layers of the Earth's atmosphere that affects the image of a distant object. The term is usually applied to a whole range of effects caused by differential refraction. Depending on the exact density profile (which in turn depends upon *temperature and *humidity), the image may be compressed or elongated vertically (*see* STOOPING; TOWERING), appear elevated above or depressed below the horizon (*see* LOOMING; SINKING), be inverted, or split into multiple images.

Two specific types of mirage are the **inferior mirage** and the **superior mirage**, named according to whether the virtual image appears below or above its 'true' position, i.e. its apparent position if it were seen through a homogeneous layer of air. An inferior mirage occurs when the lowest layer of air is heated, as over a hot road or desert sands, or when cold air overlies a warm sea. Rays of light from the sky are strongly refracted upwards and reach the eye on paths that make them appear to come from closer to the surface, leading to the familiar sight of 'pools of water' on a hot road. Commonly, a direct image appears above a **vanishing line**, below which there is a vertically compressed inverted image of the distant objects above a band of apparently clear sky (or 'water') along the horizon. A superior mirage occurs when the lowest layer is much colder than an overlying layer, such as when warm air crosses an ice- or snow-covered surface. Rays of light from distant objects are strongly refracted downwards and reach the eye along paths suggesting that they come from higher in the sky. Under certain circumstances, an inverted image of distant objects may appear hanging above the objects themselves. *See also* FATA BROMOSA; FATA MORGANA; MOCK MIRAGE.

**MISR** The multi-angle imaging spectroradiometer carried on the *Terra satellite for obtaining data on the three-dimensional distribution of *aerosols and particulates from sources such as volcanoes, industrial emissions, fires, etc. It also provides information on their contribution to the overall radiation budget.

**mist** A suspension of small water droplets or moist *hygroscopic particles that slightly obscures visibility. Mist is reported when visibility exceeds 1 km; below that limit the obscuration would be classed as *fog. Obscuration by dry particles is defined as *haze.

**mistral** [Provençal: 'masterly'] A strong, cold, dry, and squally northerly wind that blows down the Rhône valley in south-eastern France and over the Golfe du Lion into the Mediterranean. In extreme cases it may reach as far as Sicily, where it has been recorded at force 10 on the *Beaufort scale. It may be a purely *katabatic wind, but also arises when a low-pressure centre lies over the Ligurian or Tyrrhenian Seas. Its strength is increased by the funnelling effect of the Rhône valley. It is most violent in winter and spring, and has been known to reach 130 $km h^{-1}$ over the Rhône delta.

**mixed cloud (mixed-phase cloud)** Any cloud that contains both water droplets and ice crystals, particularly when both are present at the same level. *Altostratus, *nimbostratus, and *cumulonimbus usually fall in this category.

**mixing ratio** The ratio of the mass of some specific gaseous component to the mass of the air with which it is combined. The term is most frequently used with respect to *water vapour (*see* HUMIDITY MIXING RATIO), and is usually expressed in grams per kilogram.

**mizzle** A term used in Devon and Cornwall for *Scotch mist.

**MJO** *See* MADDEN-JULIAN OSCILLATION.

**moat** The clear annulus surrounding a tropical cyclone (*see feature*) that shows plainly on satellite images. Air lifted by the storm descends around the periphery (as well as in the *eye), giving rise to cloud-free skies and fine weather.

**MOC** *See* MERIDIONAL OVERTURNING CIRCULATION.

**mock mirage** A term proposed for a recently recognized optical effect in which an observer, sited above an *inversion layer that gives rise to *anomalous propagation (ducting), sees a section of the image of a distant object as inverted. *See also* MIRAGE.

**mock moon** *See* PARASELENE.

**mock sun** *See* PARHELION.

**mock sun ring** *See* PARHELIC CIRCLE.

**model** Any theoretical representation of a process or system, including conceptual, as well as numerical, models. In general, for practical reasons, a number of simplifying assumptions must be made, which limit the extent to which a model truly represents actual conditions, the time period over which any conclusions are valid, and the application for which that model may be employed. In meteorology and climatology, numerical models of varying complexity are used, ranging from **intermediate coupled models**, in which a simple model for the atmosphere is linked to a simple model of the oceans, to the most complex **coupled general circulation models** that incorporate highly sophisticated modelling of both atmosphere and oceans. *See also* GENERAL CIRCULATION MODEL; NUMERICAL WEATHER PREDICTION.

**model atmosphere** A theoretical or mathematical representation of conditions within the *atmosphere, particularly with regard to the vertical *temperature and *pressure distribution. Examples are the various *standard atmospheres, and a model in which the simplifying assumption is made that the atmosphere is fully *barotropic. More complex *baroclinic models require conditions at various *isobaric surfaces to be specified.

**MODIS** The Moderate-Resolution Imaging Spectroradiometer carried on the *Terra satellite. Designed to be sensitive to a broad range of frequencies from visible to long-wave infrared radiation. Giving coverage of the whole globe every one or two days, this sensor provides information about a large number of factors, including cloud and surface temperature, cloud amounts (particularly *cirrus clouds, which have been poorly monitored previously), snow and ice cover, smoke, and *aerosols. It also indirectly monitors *carbon dioxide by providing data on photosynthetic activity in plants and photoplankton in the oceans.

**moist adiabatic lapse rate** *See* SATURATED ADIABATIC LAPSE RATE.

**moisture content** *See* SPECIFIC HUMIDITY.

**Mono** A *katabatic wind that affects parts of California. It originates in the Great Basin, streams down the western slopes of the Sierra Nevada, and may reach 160 $km^9h^{-1}$ in the Yosemite Valley.

**monsoon** [from Arabic *mausim*: 'season'] **1.** A wind that is persistent in direction throughout a season and also displays a pronounced change in direction from one season to another. The basic cause is strong differential heating of land and sea areas, most marked in the case of Asia, where air flowing out from the *Siberian anticyclone in winter (the north-easterly monsoon) is replaced by the south-west monsoon in summer. Similar seasonal reversals occur over West Africa, Australia, and, although less pronounced, over the south-western United States. **2.** By association, the rainy season that occurs with the onset of the south-west monsoon.

**monsoon air** A term sometimes used for the *air mass involved in the *monsoons, although it is essentially the same as *maritime tropical air.

**monsoon climate** The climate found in regions affected by *monsoon winds, particularly those in the tropics (Köppen classification Am). Such regions typically have a dry season, which includes the winter period of greatest cold followed by a transition to extreme heat, which occurs just before the arrival of the monsoon rains. A second warm spell usually follows the end of the rainy season.

**Montreal Protocol** The treaty on 'Substances that Deplete the Ozone Layer', drawn up in 1987 by the United Nations Environment Programme (UNEP) to phase out the production and use of various substances, notably the *chlorofluorocarbons (CFCs) and *hydrochlorofluorocarbons (HCFCs), which have a deleterious affect on atmospheric *ozone and thus create the *ozone holes. The initial list of substances has subsequently been increased through various amendments. Revisions agreed in 2007 bring the dates for the elimination of HCFCs forward by ten years: to 2020 for industrialized countries and 2030 for developing countries.

**Moon** The Earth's only natural satellite, which is astronomically significant because of its large size relative to its parent body. The Moon is responsible for the major component of oceanic tides (*see feature*), but although it does contribute to *atmospheric tides, the effect of solar heating is far greater.

**MOPITT (Measurements of Pollution in the Troposphere)** An instrument borne on the *Terra satellite, which monitors the distribution of two significant *pollutants, *carbon monoxide and *methane (which is also an important *greenhouse gas).

**morning glory** A characteristic sudden *squall often experienced in the southern Gulf of Carpentaria in Australia, but suspected to exist in less extreme form elsewhere in the world. Often accompanied by one or more long *roll clouds that typically travel at 40–50 $km\,h^{-1}$, each with associated subsidiary gusts. The strongest morning glories arise through the interaction of *sea-breezes that initially propagate towards one another from both sides of the Cape York Peninsula. When they interact, they create a hydraulic jump, similar to a river *bore, with an abrupt uplift at the leading edge, and a corresponding pressure jump of about 1 hPa. The morning glory itself propagates in the same direction as the stronger sea-breeze (usually that from the Pacific Ocean). Wave motion behind the front gives rise to the bands of cloud, but after the glory has passed there is a persistent increase in the depth of the low-level cold air and in atmospheric pressure.

Similar interaction may occur between a sea-breeze front and a *cold front, and such **southerly morning glories** have been observed in association with cold fronts moving north or north-east over northern Australia. *See also* UNDULAR BORE.

**mother cloud** The *cloud genus from which another cloud has evolved, i.e. the type to which the suffixes *genitus or *mutatus are applied.

**mother-of-pearl cloud** *See* NACREOUS CLOUD.

**mountainado** A vertical *vortex, usually associated with intense *katabatic winds. It is suggested that irregularities in the topography may cause the horizontal *vorticity associated with the normal, vertical *wind shear to turn vertically, in a manner similar to the way in which the intense *updraught in a *supercell thunderstorm causes a horizontal vortex to become a pair of counter-rotating vertical vortices, leading to the formation of a tornado (*see feature*).

**mountain breeze** *See* KATABATIC WIND.

**mountain-gap wind** A localized wind that blows through a mountain pass, its strength being accentuated by the funnelling effect. *See also* RAVINE WIND.

**mountain shadow** The shadow of a hill or mountain, cast by the rising or setting Sun on to the atmosphere or a layer of cloud. Regardless of the shape of the mountain, the shadow always appears as a dark cone. In reality the edges of the shadow are parallel, but appear to converge because of perspective. *See also* SHADOW OF THE EARTH.

**mountain wave** *See* LEE WAVE.

**mountain wind** A *katabatic wind that descends into a valley at night from the surrounding mountains, and which is caused by the greater cooling of the air above the higher slopes (*compare* VALLEY WIND), resulting in a gravity flow of cold air down the valley.

**MSG** Abbreviation for *Meteosat Second Generation.

**MSU** *See* ADVANCED MICROWAVE SOUNDING UNIT.

**multicell storm** A *thunderstorm in which two or more cells are active simultaneously. When an initial cell is sufficiently strong, the inflow may give rise to a daughter cell that becomes active and produces *lightning discharges before the parent cell has died away. Such a system is less organized than a *supercell thunderstorm, but, through the repeated generation of new cells, may persist for a long time. More frequently encountered in tropical (rather than temperate) regions, and where such storms often occur as *squall lines. Multicell storms may sometimes give rise to weak tornadoes (*see feature*).

**mutatus (-mut)** A suffix used in *cloud classification to denote the *cloud genus from which a currently observed type has evolved. An example is stratus stratocumulomutatus (St scmut): *stratus that originated as *stratocumulus. Unlike *genitus, mutatus implies that substantially all of the *mother cloud has been altered.

# N

**nacreous cloud** A rare *polar stratospheric cloud, occurring at altitudes of 15–30 km, that displays extremely strong *iridescence, hence the alternative name of **mother-of-pearl cloud**. Rapid uplift in a *lee wave causes water vapour to deposit on *nitric acid trihydrate particles, with even growth to a diameter of about 2 μm. Because of the extreme dryness at such altitudes, temperatures below 190 K (−83 °C) are required for the cloud particles to form. These conditions occur more frequently in winter over the Antarctic than over the Arctic. The regular distribution of particle sizes gives rise to the pure colours through *diffraction effects.

**nadir** The point on the *celestial sphere immediately below the observer's feet. Diametrically opposite to the *zenith.

**Nakaya diagram** A diagram, developed by Ukichiro Nakaya, that relates the specific form of ice crystals produced to the prevailing environmental conditions (humidity, temperature, and wind).

**NAT** *See* NITRIC ACID TRIHYDRATE.

**National Oceanic and Atmospheric Administration (NOAA)** The national meteorological and oceanographical organization of the United States. NOAA has overall responsibility for both forecasting and research on all aspects of weather and climate, including the provision of *meteorological satellites, such as the *GEOS and *NOAA series.

**nautical mile** A unit of distance, now defined internationally as precisely equal to 1 852 metres (originally one minute of arc as measured along a great circle of the Earth). A speed of one nautical mile per hour is defined as one *knot, a unit often used to express wind velocities.

**nebulosus (neb)** A *cloud species consisting of a uniform, featureless veil or layer of cloud. Applies primarily to *stratus and *cirrostratus.

**neon (Ne)** A colourless *inert gas, atomic number 10, that occurs at low concentrations ($1.8 \times 10^{-3}$ per cent by volume) in the atmosphere.

**nephanalysis** The analysis of a satellite image or a *synoptic chart to determine the type and extent of specific forms of *cloud and the associated *precipitation.

**nephoscope** An instrument for determining the direction in which a cloud is moving and its angular displacement relative to a point vertically beneath it. From the angular displacement and the cloud height, its velocity may be determined.

**NESIS** *See* NORTHEAST SNOWFALL IMPACT SCALE.

**neutral point** *See* POLARIZATION OF THE SKY.

**névé** **1.** The original term for granular snow, now more generally known as *firn. **2.** A term increasingly used for an area of firn, i.e. for the *accumulation zone of a glacier.

**NEXRAD (Next Generation Weather Radar)** A network of pulsed *Doppler radar units installed by the United States' National Weather Service to provide quantitative data on precipitation, wind speeds and direction, and other factors. The system is specifically designed to monitor severe weather and dangerous situations, to track storms, and to give early warning of potentially damaging hail, flooding, and extreme wind speeds (including tornadoes—*see feature*). Velocities may be determined to distances of about 200 km, and precipitation to about 450 km.

**nightglow** The faint *airglow emission that is detectable at night. Under moonless conditions it amounts to about half of the light from the sky (the remainder being starlight). The overall spectrum is highly complex, but the dominant components are the **auroral green** light from oxygen at 557.7 nm, the *sodium D lines at 589.3 nm, and various *hydroxyl bands. The principal emission occurs in specific layers of the *ionosphere: most at 80–100 km, but some at 200–300 km. *See also* DAYGLOW; TWILIGHT GLOW.

**nimbostratus (Ns)** [Latin: 'rain' + 'layer'] One of the main *cloud genera: a thick grey or dark grey, middle-level (*see* ÉTAGE) layer cloud that produces more or less continuous *precipitation in the form of *rain or *snow (Plate 10). It is always thick enough to hide the Sun, and the bottom often appears diffuse because of the precipitation. Ahead of approaching *warm fronts, *altostratus normally lowers and thickens to become nimbostratus, which in turn may extend down almost to ground level.

**Nimbus** A series of *polar-orbiting meteorological satellites, launched by the United States beginning in 1969, and following on from the earlier *Tiros series, of which they were a development. Later, more advanced satellites that continue the same basic observing programme are generally known as *NOAA satellites.

**nitric acid trihydrate (NAT; $HNO_3 \cdot 3H_2O$)** A compound consisting of one nitric acid molecule ($HNO_3$) and three water molecules. It is the principal component of one type of *polar stratospheric cloud, where it forms solid crystals at temperatures below 195 K. These also act as *freezing nuclei in two other types of such clouds.

**nitric oxide (NO; nitrogen monoxide)** A colourless gas that occurs in extremely low quantities in the atmosphere (approximately $10^{-9}$ per cent by volume), but which is nevertheless regarded as a significant pollutant and contributor to *acid rain. It is produced by the burning of fossil fuels and also by *photochemical processes in the upper atmosphere.

**nitrogen (N)** A colourless and odourless gas that is the most abundant atmospheric gas, amounting to 78.08 per cent by volume in the *homosphere (*see* AIR). Although relatively inactive as a molecular species ($N_2$) in the atmosphere, it participates in numerous important reactions in the *nitrogen cycle. *Nitrogen oxides are significant pollutants.

**nitrogen cycle** An important chemical cycle, consisting of a complex series of processes that circulate *nitrogen between the atmosphere and animals, plants, and soil. Nitrogen in the atmosphere may be converted into *nitrogen oxides by *lightning discharges; these are taken up by the soil and form nitrates that are then absorbed by plants and eventually by animals. In addition, nitrogen-fixing bacteria associated with the roots of certain plants convert the element into a form that may be absorbed. Other bacteria break down nitrogen compounds in dead animals and plants, or their wastes, and return it to the soil or, in some cases, to the atmosphere. *See also* CARBON CYCLE.

**nitrogen dioxide ($NO_2$)** A gas (a *nitrogen oxide) that is present in small quantities both near the surface and in the *stratosphere. In the lower atmosphere it arises from burning fossil fuels and industrial processes; and in the upper atmosphere through the *dissociation of *nitrous oxide.

**nitrogen oxides ($NO_x$)** Chemical species containing *nitrogen and oxygen, particularly the reactive species *nitric oxide (NO) and *nitrogen dioxide ($NO_2$), which are regarded as pollutants. Their average lifetime is relatively short at about two days before they react with other substances. The burning of fossil fuels is the primary source of $NO_x$ (with approximately one-third deriving from cars and other vehicles), but *lightning and the burning of biomass are also significant sources. Approximately 25 per cent of the overall total is broken down by *photochemical dissociation in the stratosphere and some is removed as nitric acid ($HNO_3$) by snowfall. $NO_x$ species are also converted into nitrate *aerosol particles. In a motor vehicle's catalytic converter, rhodium particles catalyse the breakdown of nitrogen oxides into nitrogen and oxygen.

**nitrous oxide** A colourless gas ($N_2O$) used as an anaesthetic and aerosol propellant, which primarily resides in the *stratosphere. It is a *greenhouse gas, 200 times as effective as *carbon dioxide at absorbing energy from the Sun, and is subject to restrictions imposed by the *Kyoto Protocol. The principal natural source is from nitrate ions produced by soil bacteria.

**NOAA** **1.** *See* NATIONAL OCEANIC AND ATMOSPHERIC ADMINISTRATION. **2.** A series of *polar-orbiting meteorological satellites provided by that organization, the latest being NOAA 18, launched on 20 May 2005.

**noble gas** *See* INERT GAS.

**noctilucent clouds (NLC)** The highest clouds in the atmosphere, occurring just below the *mesopause at altitudes of 80–85 km. Visible occasionally from the ground at high latitudes (above *c.*45°), around the summer *solstice, when the observer is in darkness but the clouds are illuminated by sunlight. Observations from orbit suggest that a separate layer occurs above the equatorial region and this may extend north and south. Here the clouds are visible from the ground only fleetingly around sunrise and sunset. Noctilucent clouds have a distinctive bluish-white or yellowish coloration and occur as a tenuous layer, in which high-altitude waves create various undulations. The differing lengths of the line of sight through this thin layer give rise to apparent structures, somewhat similar to those seen in the much lower *cirrus and *cirrostratus. Various specific structural forms (called **Types**) are recognized (*see* BAND (2), BILLOW (2), VEIL (2), WHIRL (2)). Noctilucent clouds appear to be essentially identical to *polar mesospheric clouds, but do not occur closer to the poles than latitudes 60° N and S.

The clouds are carried towards the west-south-west (in the northern hemisphere) by upper-atmosphere winds, but the characteristic structures (thought to arise from high-altitude waves) often move in the opposite direction. Noctilucent clouds consist of particles of ice *c.*0.1 μm in diameter, but there is currently no agreement on the nature of the *freezing nuclei, which may be particles of dust from *meteors, or ion clusters created by *cosmic rays. (The earlier suggestion that the nuclei might be volcanic dust is now generally discounted, for lack of any adequate mechanism that could transport dust to such altitudes.) *Nucleation occurs at the *mesopause itself, a few kilometres above the observed cloud levels.

**nocturnal jet** A localized layer of strong winds, a few hundred metres above the surface, which sometimes occurs at night. Strong radiational cooling leads to the formation of an *inversion and this effectively decouples the higher flow from the friction associated with the *boundary layer. The effect may be so marked that the nocturnal jet exceeds *geostrophic wind speeds.

**nocturnal radiation** The radiation emitted by the Earth's surface at night. This normally greatly exceeds the amount of radiation received from the *atmosphere during the hours of darkness, particularly when the sky is free from cloud and relatively dry. It leads to cooling of the surface, and often to a *radiation inversion. *See also* RADIATION NIGHT.

**normal** In meteorology, the average value of a specific parameter, such as temperature or rainfall, taken over a number of years. For practical purposes, a period of 30 years is usually taken as providing a representative sample.

**norte** *See* NORTHER.

**North Atlantic Current (Gulf Stream Extension, North Atlantic Drift)** Often popularly referred to as the *Gulf Stream. The portion of the North-Atlantic *gyre that lies east of 45° W, flowing towards Europe. A branch passes between Iceland and the British Isles northeastward into the Arctic Ocean, where it becomes the *Norwegian Current. Its temperature is some 5 deg C above average for its latitude and, by warming the prevailing southwesterly winds, it is thus a major factor in the mild climate of northwestern Europe. It has been suggested that it weakened in the late 17th and early 18th centuries. This, together with an extension of Arctic surface water farther south—Iceland was surrounded by *pack ice—may have contributed to the *Little Ice Age.

**North Atlantic Drift** *See* NORTH ATLANTIC CURRENT.

**North Atlantic Oscillation (NAO)** A fluctuation of the pressure pattern prevailing over the North Atlantic that has a marked influence on the winter weather of bordering land masses, particularly of Europe. It is similar to the better-known, better-understood (and thus more predictable) El Niño Southern Oscillation (*see feature*). Its two states are defined by comparison of the pressures over the Azores and Iceland. When the *Azores high is strong and extends towards Portugal (high NAO index), there is a deep low-pressure region centred near Greenland (the *Icelandic low). As a result there are strong westerly winds over the Atlantic, causing mild winters in northern Europe, but dry conditions further south. There is intense cooling and deep convection of the water in the Labrador Sea, but less cooling in the Greenland Sea, and weakened production of warm Sargasso water.

In the opposite state (low NAO index), the Azores high collapses and a high-pressure region over Greenland replaces the Icelandic low. Circulation around this high introduces a northerly and easterly flow of Arctic air over north-western Europe, leading to a cold winter. Depressions, which might normally cross the British Isles, take a southerly route across the Iberian Peninsula, giving increased rainfall to the Mediterranean region and as far as the Middle East. The Greenland Sea becomes the dominant sink of cold water in the North Atlantic, while heating and convection in the Sargasso Sea become strong.

The system tends to remember the prevailing state from one winter to the next, but there is an approximately decade-long oscillation between high-index and low-index states. This may be linked to the period of circulation of regions of anomalous *sea-surface temperature and to the strengthening and weakening of the vertical oceanic circulation. *See also* ARCTIC OSCILLATION; ATLANTIC MULTIDECADAL OSCILLATION.

**northeaster (nor'easter)** A strong or gale-force north-easterly wind, particularly the cyclonic storms that affect the northern portion of the East Coast of the United States during the winter and which bring the greatest snowfalls.

**Northeast Snowfall Impact Scale (NESIS)** A scale that ranks snowfall events in the northeastern United States in terms of their impact on society as a whole. The formula used is a function of the area covered by the snowstorm, the amount of snow, and the total population living within the path of the storm. The ranking is: Extreme, Crippling, Major, Significant, and Notable. Events such as those of 12–14 March 1993 and 6–8 January 1996 rate as 'Extreme'.

**North Equatorial Current** The warm current that flows westwards across the southern North Atlantic Ocean. It is the southern portion of the North Atlantic *gyre, completing the flow between the cold *Canary Current and the warm *Gulf Stream.

**norther** A cold, northerly wind that often brings a drastic drop in temperature, especially the wind experienced over the southern region of the Great Plains in North America. Such a sudden outbreak of polar air often crosses the Gulf of Mexico and reaches Central America, where it is known as a norte.

**North Pacific Current (North Pacific Drift)** A warm current (part of the North Pacific *gyre) that flows eastwards across the northern Pacific Ocean from about 25° N to 45° N. It is a continuation of the *Kuro Shio Current, but does not extend right across the Pacific, turning south at about the longitude of Hawaii. *See also* ALEUTIAN CURRENT; ALASKA CURRENT; CALIFORNIA CURRENT.

**North Pacific High (Hawaiian High)** A *subtropical anticyclone and *centre of action. Its average location is around 30° N, but in summer it expands and moves north to about 40° N. In the northern winter it shrinks and migrates south becoming centred near 20° N.

**North Wall (sometimes known as the cold wall)** The abrupt boundary separating the cold waters of the *Labrador Current from the warm waters of the *Gulf Stream and *North Atlantic Current. *See* ARCTIC POLAR FRONT.

**Norwegian Current** The extension of the *North Atlantic Current that carries warm water northeastwards into the Arctic Ocean, parallel with the Norwegian coastline.

**nor'wester** A *squall line that occurs in the area of the Ganges delta in late spring. Nor'westers are triggered when warm humid air from the ocean undercuts dry, cold air from the land.

**Novaya Zemlya effect** A form of abnormal refraction, where ducting of light from the Sun along an *inversion for extremely long distances enables an image of the body to be seen when several degrees below the geometrical horizon. First reported from Novaya Zemlya in 1596, it occurs in both Arctic and Antarctic regions. The term is sometimes mistakenly applied to any *mirage phenomenon affecting the image of the Sun.

**nowcasting** The preparation of a short-term forecast (generally for less than six hours). Such forecasts are often coupled with the description of current weather (or that prevailing a short time ago), based on observational data that are available in real time rather than at fixed, *synoptic times. Radar images of rainfall and *sferics fixes of *lightning discharges are two examples, which show *mesoscale systems that are not well defined by synoptic data. The term is also applied to forecasting conditions in the next 30 minutes for a specific site, such as the occurrence of *microbursts at an airport.

**nuclear winter** The term used for possible climatic change (*see feature*) following a nuclear war. It is suggested that the smoke caused by extensive wildfires and dust would blanket most of the Earth and prevent solar radiation from reaching the surface for a large number of years, leading to prolonged cooling and frigid conditions over most of the globe.

**nucleation** Any process that initiates a phase change to a more condensed state, such as *condensation (*water vapour to liquid droplets) and *freezing (liquid water to ice). These phase changes are normally initiated by a solid *nucleus, in a process known as **heterogeneous nucleation**. When the phase change occurs in the absence of nuclei, the process is known as **homogeneous** or **spontaneous nucleation**. The latter process does not occur for water vapour in the atmosphere, but can take place within small water droplets at low temperatures (*see* SUPERCOOLING).

**nucleus** A tiny solid or liquid particle that is suspended in the atmosphere, generally one that initiates a phase change between the various forms of water. The very smallest and most numerous particles (known as **Aitken nuclei**), with radius <0.01–0.1 μm, do not act as *condensation nuclei, which are found among the **large** (0.2–1 μm) and **giant nuclei** (>1 μm). *See also* FREEZING NUCLEUS; ICE NUCLEUS; SUBLIMATION NUCLEUS.

**numerical weather prediction (NWP)** The preparation of weather forecasts by the use of complex computer modelling of the behaviour of the atmosphere. The models, implemented on some of the most powerful supercomputers in existence, proceed by solving a set of equations that describe the changes that occur in various parameters (such as *pressure, *temperature, wind speed, and *humidity) in the *model atmosphere over a given span of time. (These time-scales may range from a few hours to several days.)

A typical model consists of a set of grid points covering the area of significance for a number of *isobaric levels in the atmosphere. *Synoptic observations are used to derive initial values for each grid point. Supercomputers are used to calculate the evolution of the system at appropriate intervals (often in 15-minute steps). To take the forecasts prepared by the *European Centre for Medium-Range Weather Forecasts (ECMWF) as examples, the atmospheric global model has a grid with a horizontal resolution of *c*.25 km, and with 62 levels. This model is used to prepare daily 3-day and 10-day global forecasts. The forecast to 21 days uses a grid of 80 km, as does the ECMWF *ensemble of 51 forecasts to 10 days. Global wave forecasts (10 days) are prepared with a 55 km grid, and those for European waters (5 days) with a 27 km grid. Similar procedures are used by national meteorological services to prepare forecasts for the next 24–72 hours.

Current models largely rely upon *radiosonde ascents and data reports from aircraft to model conditions at the various levels in the atmosphere. The latest generation of meteorological satellites, however, are able to provide data from *topside sounding that may be used as direct input to the atmospheric models. Instruments such as those carried by the *MetOp satellite are able to provide detailed, high-resolution profiles of temperature and humidity (and other parameters) throughout the depth of the atmosphere. *See also* GENERAL CIRCULATION MODELS.

# O

**oblique visibility** The maximum distance at which an object may be detected when viewed in neither the horizontal nor the vertical plane. The oblique visibility is extremely important in aviation, both for maintaining the vertical separation of aircraft when unaided by instruments, and for detecting objects on or near the ground, particularly during a descent. Also known as **slant visibility**.

**observation** The record of the measurement or assessment of a particular meteorological variable obtained at a specific time. A *weather report consists of the complete set of observations for a particular station.

**occluded depression** A late stage in a *depression's lifetime, when an *occlusion develops and the system generally begins to fill and decay. *See also* WAVE CYCLONE.

**occlusion (occluded front)** A *front that develops in the later stages of a *depression's lifetime, when the *cold front overtakes the *warm front, and the

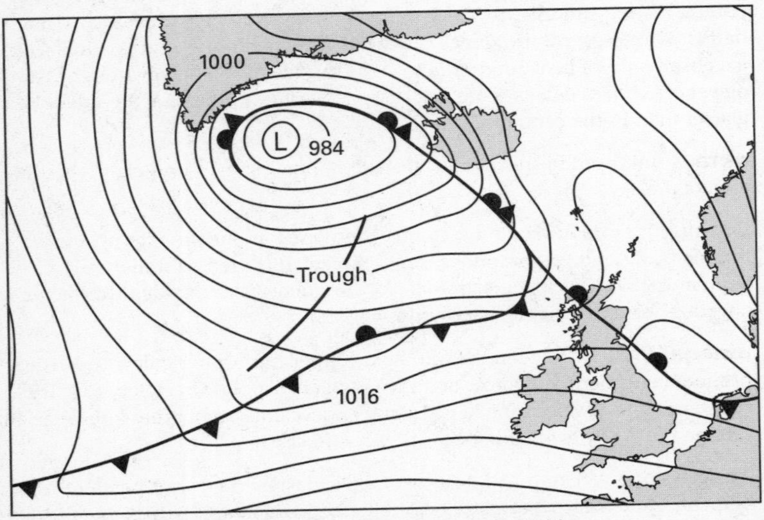


**Occlusion** An ageing *depression with a well-developed occluded front.

warm air is lifted away from the surface. The *warm sector is reduced to a *trough line, which is accompanied by cloud and *precipitation, and a *veer in the wind direction. Two types are distinguished, and the surface front has a different slope depending on whether the air originally ahead of the warm front or that behind the cold front is the cooler. The first case (cooler air in advance) is a **warm occlusion**, where the cloud and precipitation lies mainly ahead of the surface front. In the second case (a **cold occlusion**) with cooler air behind, the cloud and rain tend to follow the front. *See also* BACK-BENT OCCLUSION.

**ocean current** *See* CURRENT, OCEAN.

**oceanicity** The degree to which a specific region of the Earth is subject to the influence of the oceans; thus the opposite of *continentality. *See also* MARITIME CLIMATE.

**oceanography** The field of science devoted to the study of all aspects of the seas and oceans, including their physical, geological, chemical, and biological features. The interactions between the *atmosphere and the oceans are of fundamental importance in determining both global and local *climate, and are an essential part of *climate modelling.

**OCO** *See* ORBITING CARBON OBSERVATORY.

**ODP** *See* OZONE-DEPLETING POTENTIAL.

**ogive** Curved, shallow troughs and ridges running across a *glacier, appearing as alternating light and dark bands, and representing annual growth curves. During summer, the surface tends to melt and accumulate dirt and debris. Winter growth produces clean ice. When summer returns, the darker ice absorbs more heat and partially melts, producing a shallow trough. The ogives are convex downstream, because of the slower motion at the sides of the glacier than in the centre.

**okta** A unit used to specify the *cloud amount, equal to one-eighth of the area of the sky.

**Omega** A radio navigation system, widely used at one time for position finding by aircraft, *radiosondes, shipping, etc. It ceased operation in September 1997. By using extremely low frequencies it provided worldwide coverage from just eight base stations.

**omega high** A *ridge in the middle or upper tropospheric flow that has pronounced *meridional flow and thus appears like the Greek letter Ω. The situation often develops into a *blocking anticyclone (sometimes known as an **Omega block**) by extending down to the surface.

**opacus (op)** A *cloud variety in which a widespread layer or patch of *altocumulus, *altostratus, *stratocumulus, or *cumulus is so thick that the Sun or Moon is completely hidden.

**optical phenomena** Various phenomena caused by *diffraction, *reflection, *refraction, or *scattering by dry particles, water droplets, or ice crystals in the atmosphere or on the ground, or by the atmosphere itself. The phenomena include: *anthelic arcs, *anthelion, *arcs of contact, *blue sky, *Brocken spectre, *circumhorizontal arc, *circumzenithal arc, *corona (1), *dewbow, *fogbow, *glory, *green flash, *halo, *heiligenschein, *paraselene, *parhelic circle, *parhelion, rainbow (*see feature*), *shadow of the Earth, *subsun, and *sunrise/sunset colours.

**Orbiting Carbon Observatory (OCO)** A NASA satellite, scheduled for launch in 2008, that is specifically designed to obtain worldwide measurements of the daily and seasonal distribution of *carbon dioxide and the functioning of the *carbon cycle, particularly within the context of climatic change (*see feature*). The satellite's sole instrument consists of three high-resolution grating spectrometers, which will obtain measurements in three different modes: *nadir, from vertically beneath the satellite; *sun glint, for high-precision measurements; and targeted, for observations of a specific site from different orbital positions. It will operate in conjunction with a worldwide network of sites that will provide ground-truth observations. Once in orbit, OCO will become the lead satellite in the *Afternoon-Train, and its observations will complement those obtained by the *Aqua satellite.

**OrbView-2** *See* SEASTAR.

**orographic** Of, related to, or caused by surface relief such as a range of mountains.

**orographic cirrus** *Cirrus cloud that forms in the *lee wave created by hills or mountains, and which may produce a vast plume that extends hundreds of kilometres downwind.

**orographic cloud** Any cloud that is produced by the forced uplift that occurs when air passes over hills or mountains. The exact type of cloud depends on prevailing conditions, particularly the *stability, but because of the localized nature of the uplift the clouds tend to remain in one place. When conditions are unstable and *cumulonimbus clouds form, these may give rise to extremely long-lasting *showers with large amounts of *precipitation that may lead to *flash floods. Under stable conditions, *stratiform clouds often occur, frequently in the form of *wave clouds. *Banner clouds are also of orographic origin.

**orographic depression (sometimes known as a lee depression or an orographic low)** A low-pressure area that forms to leeward of a range of hills or mountains that act as a barrier to a broad airflow. It may take the form of a *trough or, when the mountains are high, become a closed circulation. In both cases the area is quasi-stationary while the airflow persists. Through descent and the *föhn effect the cloud cover may become broken or disappear completely, particularly in summer. In winter, it may create an area of cold air and persistent fog.

**orographic lifting** Forced uplift of air caused by an airflow over rising ground, which may, depending on the air's properties, give rise to *stratiform cloud or initiate convective activity.

**orographic precipitation (orographic rain)** *Precipitation that is caused by the uplift of humid air over an elevated barrier, such as a hill or mountain. When the *precipitation occurs mainly over the windward side of the hills, this may accentuate a *föhn effect to leeward. *See also* RAIN SHADOW.

**orphan anvil** A *cumulonimbus *anvil that persists as a dense mass of *cirrus (cirrus *spissatus) after the generating cells have died away, and which may last for many hours. There may be some precipitation from the anvil in the form of rain.

**oscillation** The variation of a specific parameter above or below a mean value. In meteorology, the term is normally applied to large-scale phenomena that exhibit *teleconnections and in which specific states succeed one another, but without a fixed periodicity. *See* ANTARCTIC CIRCUMPOLAR WAVE; ARCTIC OSCILLATION; EL NIÑO SOUTHERN OSCILLATION *feature*; NORTH ATLANTIC OSCILLATION; PACIFIC DECADAL OSCILLATION; SOUTHERN OSCILLATION.

**outflow boundary** The edge of a large pool of evaporatively cooled air created by the merging of *gust fronts from a number of active cells.

**overcast** **1.** The official designation for sky cover that exceeds 95 per cent when caused by clouds rather than surface obscuration by blowing snow, sand, or rain. **2.** A popular term for clouds that apparently cover the whole sky, particularly those associated with a *warm front.

**overseeding** In *cloud seeding, the undesirable production of too many individual ice crystals. The resulting crystals are too small to grow sufficiently large to fall out of the cloud and produce the desired *precipitation.

**overshooting top (penetrating top)** A dome of cloud that protrudes above the general level of the anvil top of a *cumulonimbus *incus, a *multicell storm, or, in particular, a *supercell thunderstorm. It is a sign of an extremely vigorous *updraught and thus of potentially severe weather beneath the storm. In the strongest storms updraughts may penetrate as much as 4 km above the level of the capping *inversion (which is frequently the *tropopause), but 1 km is a more common value.

**overturning** The rapid interchange and mixing of air or water from different levels, either through forced *convection or through changes in *buoyancy. In the atmosphere, such overturning often results from extreme heating during the day. Sudden overturning induced by landslides is thought to have been the basic cause for the release of *carbon dioxide gas from Lake

Monoun in 1984 and Lake Nyos in 1986 (both in Cameroon), which led to the deaths of 37 and 1 700 people, respectively.

**oxygen (O)** The second most abundant atmospheric gas after *nitrogen, amounting to 20.98 per cent by volume (*see* AIR) and normally present in molecular form ($O_2$). *Photodissociation by *ultraviolet radiation from the Sun is important in the *stratosphere, leading to the formation of *ozone. Absorption of energy at even shorter wavelengths is a significant factor in the formation of the *ionosphere. *See also* OZONE HOLE; OZONE LAYER.

**oxygen-isotope method** A method of determining the temperatures during a past geological epoch. The ratio between the oxygen *isotopes $^{16}O$ and $^{18}O$ that is found in marine carbonate shells, for example, is dependent upon prevailing temperatures. During an *ice age, the lighter $^{16}O$ is preferentially evaporated (and deposited as glacier ice), so oceanic water becomes enriched in the heavier $^{18}O$, this enrichment being reflected in the shells of marine organisms. Determination of oxygen-isotope ratios thus allows temperature variations during the various *glacial and *interglacial stages to be reconstructed.

**oxygen-isotope stage** A specific layer (often known as a **unit**) within deep-sea sediments that has been identified on the basis of the *oxygen-isotope method, and the age of which has been determined. Such layers then function as reference horizons in the study of other deep-sea cores.

**ozone ($O_3$)** Triatomic oxygen, a very reactive, irritant gas, mainly formed in the atmosphere by various photochemical reactions. Its concentration is measured in *Dobson units. It is found primarily at three distinct levels: in the *ozone layer at 10–50 km, where the major portion is created through the absorption by oxygen molecules of *ultraviolet radiation from the Sun; in the upper *troposphere (9–13 km) where *nitrogen oxides from *lightning and aircraft exhaust gases are significant sources; and at ground level, where it is produced by reactions involving various pollutants (particularly nitrogen oxides and hydrocarbons). Its lifetime at this level is *c.*25 days. In the upper troposphere it acts as a powerful *greenhouse gas, and its increasing concentration at this altitude is a cause for concern (*see also* SOLAR WIND).

**ozone-depleting potential (ODP)** A measure of the capacity of a substance to deplete atmospheric *ozone. It is measured relative to the *chlorofluorocarbon trichlorofluoromethane (CFC-11), which is assigned an arbitrary ODP value of 1.0.

**ozone hole** An area over the polar regions in which the *ozone layer is depleted or non-existent, particularly as a result of reactions involving chlorine (primarily derived from *chlorofluorocarbons), *bromine, and other chemicals. The reactions that destroy ozone are promoted by the presence of *polar stratospheric clouds, and increase rapidly with the return of sunlight during the polar spring, when the holes are most extensive. The Antarctic ozone hole is particularly strong because, in winter, the southern *polar vortex of high-speed winds isolates the polar region from the general atmospheric circulation and consequent mixing. Definitive evidence of Antarctic depletion was first published in 1985.

Destruction of the ozone layer is of great concern because it allows damaging solar *ultraviolet radiation to reach the surface. Although some ultraviolet wavelengths always penetrate to the ground, the range known as UV-B (315–280 nm) is normally absorbed. Increased exposure to UV-B is known to cause skin cancers in humans and genetic damage to most living organisms. In addition, the energy absorbed by ozone is the primary source of heating in the *stratosphere. Although the interactions between the stratosphere and underlying *troposphere are imperfectly understood, changes in heating at height exert an influence on weather and climate patterns at the surface.

Low altitude (1–2 km) ozone holes that last for a few days have been found in the Arctic. They may arise from the destruction of ozone by bromine, in the form of bromine oxide, possibly released from melting sea ice (*see* FROST FLOWERS).

**ozone layer** The layer between 10 and 50 km that contains most atmospheric *ozone, with the greatest concentration at altitudes of 20–25 km. It absorbs most of the energy from *ultraviolet radiation from the Sun, and thus acts as a protective shield for life on the Earth's surface. The energy absorbed causes the rise in temperature with height that occurs in the *stratosphere. It has recently been established that solar proton events (*see* SOLAR WIND) may cause transient decreases in ozone concentrations by *c.*10 per cent and a simultaneous increase in concentration in the upper *troposphere.

**ozonesonde** A *radiosonde that carries instrumentation to determine *ozone concentration (usually as a separate module linked to the basic telemetry package).

**ozonosphere** A term sometimes used to designate the *ozone layer.

# P

**Pacific Decadal Oscillation (PDO)** An *oscillation in *sea-surface temperatures in the central northern Pacific Ocean and those in a narrow belt along the western coasts of North America. The *anomalies are related to the strength of the *Aleutian low and appear to have a 20–30-year periodicity. When the Aleutian low is pronounced, with cold water in the central Pacific and warm water along the coast (positive PDO), strong westerlies bring warm, dry winters to the Pacific North-West. In the opposite phase (negative PDO), the temperature anomalies are reversed, the Aleutian low collapses, and winters are colder with higher *precipitation. It appears to have a positive correlation with the less-pronounced and less-well-known *Atlantic Multidecadal Oscillation.

**Pacific high (also known as the Pacific anticyclone)** The semi-permanent *subtropical anticyclone over the North Pacific Ocean, whose centre, on average, lies near 30–40° N, and 140–150° W. It is an important *centre of action in the northern hemisphere.

**pack ice** Any form of sea ice other than *fast ice; in particular, ice that covers more than 50 per cent of the surface. Three specific states are recognized: **open pack**, 50–80 per cent cover; **close pack**, over 80 per cent but with some *leads or *polynias; and **unbroken pack**, no open water.

**palaeoclimatology** The study of the climates that prevailed at earlier geological epochs, and the factors that have caused climatic change (*see feature*). The techniques employed for recent periods include *dendrochronology, the examination of ice cores, deep-sea sediment cores, *carbon-isotope determination, the *oxygen-isotope method, and *pollen analysis; and for more remote periods, the distribution of fossils and rock types, the chemical signatures of weathering, *palaeomagnetism, and various other studies.

**palaeomagnetism** The study of the changes in the direction, strength, and reversals of the Earth's magnetic field that have occurred over geological time. Such studies are of relevance to the understanding of both past and current climatic change (*see feature*). *See* GEOMAGNETISM.

**palaeotempestology** A rather ugly, recently introduced term for the study of prehistoric tropical cyclones (*see feature*) by the detection in sediment cores of specific layers that were deposited by such events.

**pampero** A severe *line squall that occurs over the pampas of Argentina and Uruguay. It is followed by a cold, blustery, southerly or south-westerly wind and often brings heavy rain and thunderstorms, together with a considerable drop in temperature.

**pancake ice** Approximately circular, floating pads of ice, often with raised rims. The pads form through the aggregation of *slush in the water, either derived from falling *snow or present as *frazil ice. Collisions with neighbouring pads are responsible for both the circular shape and the rims, where the slush deforms plastically on impact. Splashes of water assist freezing to consolidate both the disks of ice and the rims.

**pannus (pan)** [Latin: 'shred'] An *accessory cloud that consists of ragged shreds of cloud, which sometimes form a semi-continuous layer beneath the main cloud. The latter may be *altostratus, *cumulus, *cumulonimbus, or *nimbostratus.

**papagayo** A squally, northeasterly *fall wind on the Pacific coast of Nicaragua and Guatemala. It originates as a *norther that crosses the central mountain range and remains cold despite *adiabatic warming. *See also* TEHUANTEPECER.

**PAR** *See* PHOTOSYNTHETICALLY ACTIVE RADIATION.

**paranthelion** A *parhelion that appears at the same altitude as the true Sun, but differing by 120° in *azimuth.

**paraselene (pl. paraselenae; mock moon, moon dog)** A lunar counterpart to the *halo phenomenon known as a *parhelion, consisting of a bright spot of light at the same altitude as the Moon. Because of the lesser intensity of light from the Moon when compared with sunlight, spectral colours are less frequently seen.

**Parasol** A small satellite, developed by the French Centre National d'Études Spatiales (CNES) to study clouds and *aerosols through their microphysics and radiative properties. It carries a wide-field imaging *radiometer/polarimeter. It forms part of the *Afternoon-Train satellite constellation, flying in formation with several other satellites. *See also* AQUA; AURA; CALIPSO; CLOUDSAT; ORBITING CARBON OBSERVATORY.

**parhelic circle (also known as a mock sun ring)** A *halo phenomenon that takes the form of a white circle that passes through the Sun and lies parallel to the horizon. It arises through reflection from the faces of ice crystals with vertical axes, and is rarely visible though a full 360°, except in the polar regions.

**parhelion (pl. parhelia)** A relatively frequent *halo phenomenon (commonly known as a **mock sun** or *sun dog), consisting of a bright white or coloured spot of light at the same altitude as the true Sun. The effect arises from the refraction of light through hexagonal ice crystals and is common in *cirrostratus (where parhelia often occur in pairs, one on each side of the Sun) and in patches of *cirrus (Plate 6).

Depending on the Sun's altitude, a parhelion may lie at the same distance from the Sun as the 22° halo (which may not be visible), or as much as 14° farther away. When spectral colours are present, red is closest to the Sun, and a white tail frequently extends along the *parhelic circle.

On rare occasions, parhelia may also occur at other points on the parhelic circle, most commonly at 120° from the Sun (*paranthelion) and 180° (*anthelion); less frequently at 46°, 90°, and 140°. *See also* SUBSUN.

**Parry arc** A fairly rare *halo phenomenon, normally consisting of a principal arc, concave to the Sun, that lies above and just outside the 22° halo. The term has now been extended to other arcs, of varying curvature and complexity (depending on solar elevation), that are believed to arise from *refraction through the same type of ice crystal.

**partial pressure** The contribution that any single gas makes to the overall pressure exerted at a particular point. The gas is assumed to be the sole gas present, and to occupy the whole of the volume concerned. In the atmosphere, for example, the largest partial pressures are those exerted by *nitrogen and *oxygen, which are the dominant components (*see* AIR).

**Pascal, Blaise** (b. Clermont-Ferrand, France, 19 June 1623; d. Paris, 19 August 1662) French mathematician, physicist, and philosopher who carried out experiments to verify the results of *Galilei and *Torricelli. Although not the first person to suspect that pressure decreased with height, he obtained proof when, at his urging, his brother-in-law, Florin Perier, carried a barometer to the top of the Puy de Dôme on 19 September 1648. He subsequently developed the hydrostatic equation, known as **Pascal's law**, which states that two points at the same elevation in a continuous fluid or gas are at the same pressure. The SI unit of pressure is named in his honour.

**pascal (Pa)** The SI unit of pressure, defined as one newton per square metre, and named after Blaise *Pascal. In meteorology, the most commonly used multiple is the hectopascal (100 Pa = 1 hPa), which is equivalent to one *millibar.

**PDO** *See* PACIFIC DECADAL OSCILLATION.

**pearl-necklace lightning** *See* BEAD LIGHTNING.

**penetrating top** *See* OVERSHOOTING TOP.

**penitent** A free-standing column of ice, *firn, or compacted snow that has been formed by differential melting and *sublimation. Small, initial depressions collect dust, which absorbs radiation efficiently in summer, causing the ice at the bottom of the hollows to melt, whereas clean ice at the peaks sublimes. Melting requires far less energy (about one-eighth) of that required for sublimation, so the hollows deepen more rapidly than the peaks, eventually reaching the ground and producing separate columns of ice. Frequently known by the Spanish term *nieve penitente* ('penitent snow').

**pennant** *See* WIND ARROW.

**pentad** A period of five days. Frequently used in the analysis of meteorological data, because (unlike the week) it is an exact fraction of a normal year of 365 days.

**percolation** The flow of water through the soil under the influence of gravity.

**pereletok** *See* PERMAFROST.

**perfluorinated compounds (PFCs)** A family of persistent organic pollutants (POP) that appear to be ubiquitous in the environment, and which are causing considerable concern, because their origin and toxicity are unknown. They were thought to be relatively inert and unlikely to leach into the environment. One such substance, known to be toxic, perfluorooctane sulphonate (PFOS), has been found in animals and humans worldwide, and the major manufacturer announced a voluntary ban on its production in 2000. A similar group of compounds, perfluorocarboxylates, includes perfluorooctanoic acid (PFOA), which is a precursor of polytetrafluoroethylene (Teflon), with many commercial and industrial uses, such as non-stick cooking utensils, and stainproofing of carpets, fabrics, and furniture.

**perfluorocarbon (PFC)** A series of chemical compounds of carbon and fluorine that are significant *greenhouse gases. The most important are perfluoromethane ($CF_4$) and perfluoroethane ($C_2F_6$), with lifetimes of 50 000 and 10 000 years, respectively. Their use is being phased out under the terms of the *Kyoto Protocol.

**perigee** The point of its orbit at which the Moon or an artificial satellite is closest to the Earth. For the Moon, it is the point at which the tide-raising force is greatest.

**periglacial** Appertaining to the area bordering on a *glacier, *ice cap, or *ice sheet, particularly those regions that are affected during an *ice age.

**perihelion** The point of an astronomical body's orbit at which it is closest to the Sun. For the Earth, this occurs on about 1 January, when the Sun–Earth distance is *c.*147 099 600 km, or 98.33 per cent of its mean value. *Compare* APHELION.

**perlucidus (pe)** A *cloud variety in which there are broad layers or patches of cloud, with distinct gaps—even though some may be very narrow—through which blue sky, the Sun, or the Moon may be seen. Applies to *altocumulus and *stratocumulus.

**permafrost** A region of permanently frozen ground where summer heating is insufficient to counteract the low temperatures attained during winter. This occurs where the mean annual surface temperature is below approximately −5 °C, thus primarily in the polar regions, but also at high altitudes at lower latitudes. Permafrost is estimated to underlie approximately 20 per cent of the Earth's land surface.

Freezing may penetrate to considerable depths (reaching over 1 500 m in parts of Siberia). Ground ice is normally (but not always) present. The surface layer is usually subject to thawing in summer to a depth that depends upon prevailing conditions. A layer that may thaw in certain years is known as a **supra-permafrost layer** or **pereletok**. A layer of permanently unfrozen ground that lies above permafrost but beneath a seasonally frozen surface layer is known as **talik**, as are areas surrounded by permafrost that do not themselves freeze.

Permafrost regions exhibit characteristic surface features, including ice wedges (ice-filled cracks that taper downwards), polygonal ground (regular patterns of stones caused by localized *frost heave, *solifluction, and cracking), ice mounds, *pingoes, and frost blisters.

**Peru Current** A cold current that flows northwards along the western side of South America. Because of the *Ekman effect the flow is diverted towards the west, causing *upwelling of nutrient-rich *bottom water, which normally supports large fish stocks. The upwelling and the northern part of the current are particularly affected during an *El Niño event. Originally named the **Humboldt Current** after Alexander von *Humboldt, who first studied it.

**phenology** The study of the seasonal changes in the natural world, such as the times of first flowering of plants, the dates of the migration of birds, and similar phenomena, particularly as they are influenced by the weather and climate.

**photochemistry** The study of chemical reactions caused by visible light or *ultraviolet radiation. Significant photochemical processes occur both at high altitudes (in the *chemosphere) and at low levels, where one result is the production of photochemical *smog.

**photodissociation** The *dissociation of a molecule into two or more neutral molecules or atoms by the absorption of *ultraviolet radiation or photons at still shorter wavelengths. The production of *ozone in the atmosphere largely proceeds through this process.

**photoionization** The *ionization of an atom or molecule through the absorption of an energetic photon. An electron thus freed is known as a **photoelectron**. The process is particularly important in the formation of the *ionosphere.

**photosphere** The visible surface of the *Sun, from which most of its visible and infrared radiation is emitted. It is approximately 500 km deep and has an effective temperature of 5 800 K.

**photosynthetically active radiation (PAR)** The portion of the electromagnetic spectrum associated with photosynthesis, ranging from *c.*0.4 to 0.7 μm. Certain *pyranometers (PAR sensors) are specifically designed to measure this radiation.

**physical climatology** The principal field of *climatology, which aims to study and describe the *climate of a region in terms of the physical factors (such as *continentality, *oceanicity, topography, etc.) by which it is determined. *See also* SYNOPTIC CLIMATOLOGY.

**PIBAL** *See* PILOT BALLOON.

**piedmont glacier** The broad terminal region of a *glacier that has expanded over a lowland plain on leaving its parent valley. The term is particularly applied where several glaciers have combined, the most extreme example being the Malaspina Glacier in Alaska, where 25 glaciers combine to blanket an area of 3 000 $km^2$, which is roughly 40 km wide.

**pile d'assiettes** [French: 'pile of plates'] The usual term for a series of *lenticular clouds, stacked one above the other, caused by *wave motion in multiple humid layers of air.

**pileus (pil)** [Latin: 'cap'] An *accessory cloud, also known as a **cap cloud**, that appears as a smooth cap, or hood, of limited horizontal extent above a cumuliform cloud, usually *cumulus or *cumulonimbus. It forms when a humid layer is lifted to its *dew point above a rising *thermal. This may later penetrate the pileus, normally causing it to be absorbed into the main body of the cloud. Sometimes several pileus layers are observed above one another.

**pillow eddy (lee eddy)** An *eddy that forms to *leeward of an obstacle, and often associated with *turbulence and *gusts. When the leeward slope is particularly abrupt, there may be a wide zone where the surface wind blows towards the obstacle. *Compare* BOLSTER EDDY; *see also* ROTOR.

**pilot balloon** A small, free meteorological balloon, i.e. a balloon that does not carry instrumentation (such as a *radiosonde), and which ascends at a constant rate. Small balloons are used to determine the height of the cloud base, but larger versions are capable of reaching altitudes of 15 km or more. Tracking with a theodolite enables the latter to be used for a **pilot balloon observation** (PIBAL), i.e. the determination of direction and rate of upper-level winds.

**Pineapple Express** A period of extremely heavy rain (with potential flooding) in the Pacific Northwest of North America. So-called because the development of a southerly branch of the *jet stream and a low-pressure region in the northern Pacific draw a deep plume of tropical moisture (originating in enhanced *Madden-Julian Oscillation rainfall) northeast from the tropics over Hawaii and on to the Pacific Northwest.

**pingo** A hill that largely consists of an ice core with a relatively thin earth or gravel covering. Pingoes occur in swampy areas above a *permafrost layer through the progressive growth of an ice lens that gradually raises the overlying soil. They may reach heights of 65 m and diameters of more than 800 m.

**Pitot tube** An instrument for measuring the relative velocity of a fluid. It functions by comparing the **total pressure** in an open tube facing into the airflow with the **static pressure** in a similar tube with an opening perpendicular to the flow. The **dynamical pressure** thus determined may be converted into wind speed. Often used on aircraft and, suitably mounted, as an *anemometer or *pressure-tube anemograph.

**pixel** [contraction of 'picture element'] The smallest individual element in the image returned by a satellite sensor, and which is thus directly related to the image resolution.

**planetary albedo** The *albedo of a planet, considered as a whole, as measured from a point in space. For the Earth, the planetary albedo is thought to be approximately 40 per cent, signifying that, averaged over the whole range of incident radiation, 40 per cent is reflected back into space without change of wavelength.

**planetary boundary layer (also known as the friction layer)** The layer of the atmosphere closest to the Earth's surface, within which the air is subject to *turbulence, friction effects, and surface heating. Typical depths are 500 m over the oceans and 1 500 m over the land. The planetary boundary layer may be considered as consisting of two subsidiary layers: the *surface boundary layer, and the higher *Ekman layer, at the top of which the wind flow becomes *geostrophic.

**planetary scale** A term applied to the very largest atmospheric phenomena, which are of the same general order as the planetary radius. Examples are the general circulation (*see feature*) and *planetary waves. *See also* MESOSCALE; STORM SCALE; SYNOPTIC SCALE.

**planetary wave** A wave whose wavelength is a significant fraction of a planet's radius. In the case of the Earth, such waves have wavelengths of thousands of kilometres. The term is also used in a more general sense for a *Rossby wave.

**plasma** Material in a hot, almost completely ionized state, and thus consisting of atomic nuclei and electrons. Often described as the fourth state of matter (after solids, liquids, and gases), because of its highly specific properties. It occurs (inter alia) throughout the Sun, as the *solar wind, and in *lightning channels.

**plates, pile of** *See* PILE D'ASSIETTES.

**plate tectonics** The concept that the Earth's *lithosphere consists of a number of distinct oceanic and continental plates, whose relative positions alter over the course of geological time, driven by motions in the deep interior. Variations in the location of the plates have been a major factor in climatic change (*see feature*). *See also* CONTINENTAL DRIFT; POLAR WANDERING.

**pluvial** An interval during the Quaternary Period in which there was increased rainfall in tropical and subtropical regions, alternating with drier **interpluvials**. Although occurring within the same geological period as the last major *ice age (the Pleistocene glaciation), the pluvials and interpluvials do not correlate particularly well with the *glacial and *interglacial phases found at higher latitudes.

**PMC** *See* POLAR MESOSPHERIC CLOUDS.

**POES** *See* POLAR OPERATIONAL ENVIRONMENTAL SATELLITE.

**polar air** An *air mass that has originated at high latitudes, particularly as air flowing out of the *polar anticyclones. *Continental polar air is dry, cold, and stable in its lower layers, whereas *maritime polar air has a higher humidity and is unstable. *See also* POLAR FRONT; TROPICAL AIR.

**polar-air depression** *See* POLAR MESOCYCLONE.

**polar anticyclone** A high-pressure area over one of the polar regions. That over Antarctica is semi-permanent, but the weak anticyclone over the Arctic Ocean tends to be seasonal, generally appearing in late spring and persisting until early autumn. Also known as the **Arctic** (or **Antarctic**) **high** or **polar high**.

**polar barrens** Regions within the Arctic and Antarctic Circles, where the climate is too extreme for any vegetation to exist, except in small, isolated, protected spots. These regions are therefore distinct from the vegetated *tundra.

**polar cap absorption (PCA)** A weak *aurora that occurs inside (i.e. poleward of) the *auroral oval, when energetic protons arrive from the Sun. Generally too faint to be visible to the naked eye.

**polar climate** The climate of the polar regions, within the Arctic and Antarctic Circles, characterized by long periods of continuous sunshine or night, generally low *precipitation, and extremely low temperatures during the winter. Type E (or more specifically, EF and EW) in the *Köppen classification.

**polar continental air** *See* CONTINENTAL POLAR AIR.

**polar cusp aurora** A weak *aurora that arises when energetic particles from the *solar wind enter the Earth's atmosphere directly in the funnel-shaped regions of the geomagnetic field above the magnetic poles (the **polar cusps**). Because of the geometry of the situation, these aurorae occur around midday and are thus visible from high latitudes during the polar winter only.

**Polar Front** A semi-permanent and largely continuous *front that occurs at the boundary between the cold *polar air mass and the warmer *tropical air mass at middle latitudes in each hemisphere. Because of the strong temperature gradient across the front, it is the site at which the majority of *depressions originate. It is closely associated with the *Polar Front Jet Stream, and may often be traced for thousands of kilometres around the globe. Its latitude is highly variable and tends to move down towards the Equator as *secondary depressions form on the trailing *cold front of an earlier depression, until eventually the sequence breaks and the front reforms at a higher latitude.

**Polar Front Jet Stream (sometimes known as the Mid-Latitude Jet Stream)** A westerly *jet stream that is located above the *Polar Front, just below the mid-latitude *tropopause and within the *tropical air mass. It is extremely variable in latitude, in contrast to the *Subtropical Jet Stream.

**polar ice** An *ice sheet that occurs in the polar regions, but which is predominantly seasonal rather than permanent. The ice covering the Arctic Ocean is of this nature, and thus differs from the permanent sheets covering Greenland and Antarctica.

**polarization of the sky** The polarization of sunlight caused by *scattering (more specifically, *Rayleigh scattering) by molecules in the atmosphere. This polarization reaches a maximum 90° from the Sun. There are three **neutral points** at which no polarization is present. These are **Arago's point**, 20° above the *antisolar point; **Babinet's point**, 20° above the Sun; and **Brewster's point**, 20° below the Sun. *See also* HAIDINGER'S BRUSH.

**polar low** *See* POLAR-AIR DEPRESSION.

**polar maritime air** *See* MARITIME POLAR AIR.

**polar mesocyclone (formerly commonly called a polar-air depression)** A weak *secondary depression that forms, without associated fronts, in the unstable *air mass behind the *cold front of the primary depression. Its main feature is a line of *precipitation associated with a *trough that extends on the side away from the primary depression.

**polar mesospheric cloud (PMC)** Clouds that occur just below the *mesopause at *c.*85–88 km, most probably as polar caps that extend equatorward to latitudes of about 70° N and S. They appear to be closely related to *noctilucent clouds, but separated from the latter by a clear break of some 5–10° of latitude. They are visible at ultraviolet wavelengths from *polar-orbiting satellites and have also been detected in visible-light images obtained by *Meteosat around local midnight.

**Polar-Night Jet Stream** A *jet stream that develops at high latitudes (around 60°) from time to time during the winter. Its maximum lies within the *stratosphere at an altitude of *c.*25 km. It is the counterpart of the *Equatorial Jet Stream in the summer hemisphere.

**Polar Operational Environmental Satellite (POES)** Any of various *polar-orbiting meteorological satellites launched by the United States, in particular the current *NOAA series, and as distinct from the *geostationary satellites (*see* GEOSTATIONARY OPERATIONAL ENVIRONMENTAL SATELLITE).

**polar-orbiting satellite** A satellite that is placed in a highly inclined orbit (close to 90°) at altitudes of 800–1000 km, to obtain coverage of the polar regions. As the Earth rotates beneath the orbit, subsequent passes of the satellite (at intervals of *c.*90–100 minutes) cover adjacent swathes of the surface (Plate 13). Various Earth-resource and meteorological satellites, including the *Meteor, *NOAA, *Tiros, and *MetOp series, are in this category.

**polar stratospheric cloud (PSC)** Any of three types of cloud that occur in the polar stratosphere at altitudes of 15–30 km. All have sulphuric acid particles as primary *condensation nuclei. At temperatures below 195 K (−78 °C), *nitric acid trihydrate ($HNO_3{\cdot}3H_2O$) is deposited on these nuclei, producing cloud particles up to 1 μm in diameter. These nitric acid trihydrate clouds often occur as multiple, kilometre-thick layers that may extend for thousands of kilometres. They are tenuous and difficult to detect with the naked eye.

When temperatures drop below 190 K (−83 °C), water vapour is deposited as ice on these $HNO_3{\cdot}3H_2O$ particles to give the other two forms of cloud. If deposition is rapid, as in *lee waves, many small particles (<2 μm) are formed, giving rise to *nacreous clouds. When deposition is slow, as with gradually falling temperatures at the onset of winter, a smaller number of larger particles are created, with diameters up to 10 μm. The resulting clouds are more tenuous and less visible than nacreous clouds.

Polar stratospheric clouds are an essential factor in the destruction of *ozone, both because chlorine is activated on their surfaces and also because they remove nitrogen from the surrounding atmosphere, where it would otherwise trap chlorine as inert chlorine nitrate ($ClONO_2$).

**polar vortex** A large-scale *cyclonic circulation that exists in the middle and upper *troposphere, centred over the polar regions, and which is particularly strong in winter. The southern polar vortex isolates the polar region in winter, and is largely responsible for the severity of the southern *ozone hole. In the north, in winter, there tends to be one main low-pressure centre over the Canadian Arctic, and a subsidiary centre over eastern Siberia.

This pattern disappears with the weaker summer circulation. *See* Arctic Oscillation.

**polar wandering** The apparent motion of the Earth's magnetic or rotational poles as revealed by *palaeomagnetism and other geological techniques. Rather than a motion of the poles relative to the continents—as originally thought—it is now interpreted as a sign of *continental drift, as incorporated into the modern theory of *plate tectonics.

**Polar Year** *See* International Polar Year.

**pole of cold** *See* cold pole.

**pollen analysis** The determination of the plant species present at a particular past epoch from the examination of the pollen recovered from suitable deposits, such as peat. Because individual species have specific growing requirements, detailed information about past climate may be obtained. *See also* palaeoclimatology.

**pollen count** A measure of the quantity of pollen in the air, often given as an estimated value for the next 24 hours, to provide sufferers from hay fever with an indication of the likelihood that they may experience discomfort.

**pollution** *Aerosols, gases, or particulates that occur in the *atmosphere at a concentration that is a threat to the health of living things, or which change atmospheric conditions from those prevailing in pre-industrial times. (Similar criteria also apply to substances that occur in the *hydrosphere.) Some examples are: sulphuric acid ($H_2SO_4$) and nitric acid ($HN0_3$), which give rise to *acid rain; the particulates, *nitrogen oxides, and *ozone ($O_3$) responsible for the various forms of *smog; the *chlorofluorocarbons and other substances that create the *ozone holes; and, by extension, the *carbon dioxide and other *greenhouse gases responsible for global warming (*see feature*).

**polynia (polynya)** [from Russian *poly*: 'uncovered'] A large area of open water surrounded by *sea ice, as distinct from a narrow crack or *lead. The area of polynias may sometimes amount to thousands of square kilometres.

**ponente** A westerly wind along the coasts of the north-western Mediterranean. It tends to be a *land-breeze around the Golfe du Lion, and a weakened flow from the *mistral on the Côte d'Azur and over the Ligurian Sea, where it often precedes the *tramontana.

**poniente** A westerly wind in the Straits of Gibraltar.

**POPS** *See* Protocol on Persistent Organic Pollutants.

**potential energy** In meteorology, the energy that an object possesses as a result of its position in a gravitational field. Specifically, the energy required to lift a parcel of air to a given height above sea level.

**potential instability** *See* CONDITIONAL INSTABILITY.

**potential temperature** The temperature that a parcel of air would assume if it were brought at the *dry adiabatic lapse rate to the standard pressure (1000 hPa). It may be determined easily by means of an *aerological diagram. *See also* WET-BULB POTENTIAL TEMPERATURE.

**praecipitatio (pra)** [Latin: 'fall'] A *supplementary cloud feature: any form of *precipitation from a cloud (*drizzle, *hail, *ice pellets, *rain, or *snow) which is reaching the surface; this excludes *virga.

**precipitable water** The amount of water vapour in a column of air (of unit cross-section) between two specified levels. The quantity may be expressed in terms of either the depth of water that would be obtained if all of the vapour could be condensed, or its mass. Although a theoretical concept, it is a useful measure of the amount of water present.

**precipitation** Water in either liquid or solid form that is derived from the atmosphere and falls to the surface. It thus includes *drizzle, *rain, *freezing rain, *hail, *ice pellets, *ice crystals, *snow, and other forms. It specifically excludes *clouds, *dew, *fog, *frost, *mist, and *rime (which are either suspended in the atmosphere or deposited directly on to the surface), together with *virga (which do not reach the ground).

**precision aneroid barometer (also known as a digital barometer)** An *aneroid barometer in which the measurement of pressure is not through a mechanical linkage, but which makes use of electrical contacts. These may be adjusted by a micrometer screw to give an extremely sensitive reading.

**prefrontal squall line** A *squall line that develops within the warm *air mass, ahead of an advancing *cold front. Such squall lines are often far more severe than the front itself. The precise mechanism by which they form is uncertain, but may be through a *solitary wave propagating ahead of the front.

**pressure** The force exerted on unit area by a mass subjected to some acceleration. In meteorology, generally taken to be the force exerted by the (hypothetical) column of air extending from the surface to the outer limit of the atmosphere and subject to the Earth's gravitational attraction. Measured by a *barometer or *barograph and commonly quoted in *hectopascals, a multiple of the *pascal. *See figure overleaf.*

**pressure altitude** The altitude above sea level that corresponds to a particular pressure reckoned from the *ICAO standard atmosphere, i.e. with a fixed sea-level pressure and *lapse rate. It is the value indicated by an altimeter with an *altimeter setting of 1 013.25 hPa.

**pressure gradient** The difference in pressure between two points, measured in hPa per *nautical mile. In general, the **pressure gradient force** is taken to be the force acting in a horizontal plane at right angles to the *isobars.

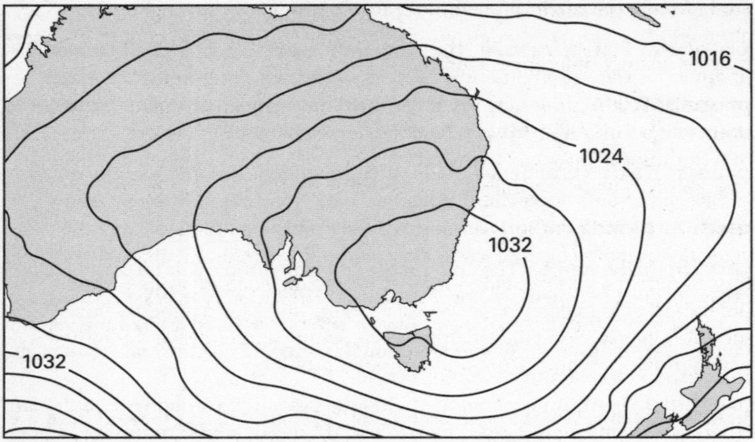


**Pressure** The simplest form of *isobaric chart, showing just the pressure distribution. Compare with the charts for *geopotential height and *thickness for the same area and time.

**pressure jump** A sudden increase in atmospheric *pressure, usually amounting to several hPa, which may accompany a *line squall, *morning glory, or similar phenomenon.

**pressure-pattern flying** The planning and execution of a flight routing in such a way as to minimize flight time for any given aircraft and load. Named from the fact that the principal aids to planning are *isobaric (constant-pressure) *charts and the pattern of *contours that they show.

**pressure-plate anemometer** An *anemometer that determines wind speed from the force exerted by the wind on a solid plate. The basic principle is ancient, a swinging-plate anemometer having been described by Leonardo da Vinci. A later form maintained the plate perpendicular to the wind. *See* DRAG-FORCE ANEMOMETER.

**pressure tendency** In *synoptic meteorology, the amount and nature of the change in atmospheric *pressure over the past three hours (or some other specified period). Also known as the **barometric tendency**. The amount (in 0.1 hPa) and direction of change (i.e. to higher or lower values) are reported using a specific code.

**pressure-tube anemograph** An obsolescent form of *anemograph that records atmospheric *pressure by means of a *Pitot tube, usually arranged so that pressure changes alter the vertical position of a recording drum.

**prevailing wind** The wind direction that occurs most frequently at a specific site. For most locations there is a definite tendency for the wind to come from a particular direction or range of directions throughout the year, although at certain sites, such as those affected by *monsoon conditions, there may be distinct differences with the seasons.

**probability** A numerical value given to the expectation that a particular event will occur. Although in scientific usage it is normally reckoned on a scale of 0 (for impossibility) to 1 (certainty), in forecasts to the general public, probability is usually specified as a percentage, 100 per cent being certainty.

**profiler (wind profiler)** A *Doppler radar designed to determine the velocities and directions of winds throughout the air column above its site, and thus to derive a vertical wind profile. In addition, such instruments are potentially able to determine vertical motion, *turbulence, and the location of the *tropopause. Some installations also include a microwave *radiometer, which provides temperature and humidity profiles.

**profiling float** An oceanic float, designed to sink to a depth of *c.*1.5–2 km, where it remains for a period (typically 10–14 days), before returning to the surface. Temperature and salinity are measured on the ascent, this data (and the float's position) being transmitted to the operations centre by satellite. The float then descends again to repeat the cycle for its lifetime of 2–3 years. *See also* ARGO.

**prognostic chart** A *synoptic chart produced by a forecaster for a specific future time, showing the expected distribution of *isobars (or *contours) and *fronts.

**propeller anemometer** An *anemometer that consists of a horizontal shaft rotated by propeller-like vanes. The whole housing pivots to point into the wind—and thus may also function as a *wind vane—and the rotating shaft generates an electrical current proportional to the wind speed.

**Protocol on Persistent Organic Pollutants (POPS)** An international treaty, sponsored by the United Nations, which came into force on 23 October 2003, governing the reduction or complete elimination of harmful chemicals with long lifetimes. Substances covered include: dioxins, furans, polychlorinated biphenyls (PCBs), various pesticides (including DDT), polycyclic-aromatic hydrocarbons (PAHs), and a flame retardant. The United States is not a signatory.

**PSC** *See* POLAR STRATOSPHERIC CLOUD.

**pseudo-adiabatic lapse rate (also known as the pseudo-saturated lapse rate)** The *lapse rate of a saturated air parcel from which some condensation products are removed as *precipitation. The air parcel will cool at a slightly greater rate than the *saturated adiabatic lapse rate.

**psychrometer** A form of *hygrometer that uses comparison of the readings from dry-bulb and *wet-bulb thermometers to determine the humidity of the air. In its simplest form, consisting of two such thermometers, it is also known as a **dry- and wet-bulb hygrometer** or **Mason's hygrometer**. Although readings may be obtained from such thermometers in a standard *thermometer screen, an *Assmann or *whirling psychrometer provides consistent ventilation and greater accuracy. A recording form is known as a **psychrograph**.

**puelche** An easterly *föhn wind encountered on the western side of the Andes. The term is also sometimes applied to the *land-breeze along the Pacific coast. *See also* TERRAL; VIRAZON.

**pumping** An obsolescent term for fluctuations in the height of a marine mercury-column barometer that are caused by gusts of wind or the rolling of a ship.

**purga** *See* BURAN.

**purple light** The highest segment of the *twilight arch, which appears a purple shade. It may become extremely strong with a vibrant colour when large quantities of material have been injected into the *stratosphere by volcanic eruptions. *See also* ALPINE GLOW.

**pyranometer** An instrument for measuring (depending on design) either the total amount of radiation received on a horizontal surface, or else the diffuse radiation from the sky, excluding that from the Sun. Various wavelength ranges may be measured, including *photosynthetically active radiation. A shielded, downward-facing version may be used to determine *albedo.

**pyrgeometer** An instrument for measuring the amount of *nocturnal radiation, i.e. the amount of heat being radiated away from the Earth's surface into space.

**pyrheliometer** An instrument for measuring the amount of direct solar radiation (i.e. excluding diffuse radiation from the sky) at normal incidence on a surface. Measurements are obtained at noon and at times when the path-length through the atmosphere is an integer multiple of the vertical thickness of the atmosphere. Such sensors may also be operated continuously, being driven to track the Sun.

**pyrocumulus** [Greek: 'fire' + Latin: 'heap'] An unofficial term for a *cumulus cloud that has formed over a heat source such as a forest fire.

# Q

**QBO** *See* QUASI-BIENNIAL OSCILLATION.

**Q-burst** An abnormal, abrupt spike in the low-frequency radio noise generated by *thunderstorms. For decades their source has been unknown, but they have recently been associated with the exceptionally strong *lightning strokes that generate *sprites.

**quasi-biennial oscillation (QBO; also known as the stratospheric oscillation)** An *oscillation in the mean *zonal wind in the tropical *stratosphere which switches between westerly and easterly phases on an approximately 27-month time-scale. The oscillation is at a maximum at an altitude of *c.*25 km, and also involves fluctuations in temperature and *ozone levels.

**quasi-stationary front (commonly known as a stationary front)** A *front that shows little or no horizontal motion. By convention, the term is applied to any front that is moving at less than 5 *knots. In *synoptic meteorology, a front that has not changed its position appreciably since the last *synoptic chart was prepared, three or six hours previously. The slow horizontal motion, combined with considerable vertical motion of the warm air, often gives rise to persistent and sometimes extreme amounts of *precipitation, leading to flooding or major snow or *ice storms.

# R

**radar meteorology** The use of radar for determining *precipitation in connection with short-term forecasts, and also for the study of specific meteorological phenomena such as the structure of *supercell thunderstorms and tornadoes (*see feature*). The latter aspect often involves the use of specialized *Doppler radar.

**radar wind** The upper-atmosphere wind as determined from radar tracking of a meteorological balloon.

**radiation** The transfer of any form of energy by means of electromagnetic waves. Such transfer may occur without any intervening medium, as in the radiation of heat and light from the *Sun to the Earth through essentially empty interplanetary space. The Earth itself radiates energy to space at infrared wavelengths. Radiation is a primary means of energy transport, together with *conduction and *convection. In meteorology, radiation processes are of particular importance at the surface, as in the formation of *radiation fog and a *radiation inversion, but radiation is also significant in other processes. The loss of energy through radiation from the top of a sheet of *stratus, for example, may be sufficient to initiate cooling, producing descending air currents, and to cause it to break up into *stratocumulus.

**radiation balance** The equilibrium between the overall annual amount of incident solar radiation received by the Earth, and the energy radiated into space.

**radiation fog (often called a ground fog)** A common type of *fog that forms in association with a *radiation inversion when the temperature of the air is reduced below its *dew point.

**radiation inversion** An *inversion that develops at night in the lowest layer of the atmosphere above a land surface as the ground loses heat through radiation to space. The depth of the inversion layer increases throughout the night as heat is transferred downwards from higher levels by *turbulence.

**radiation night** A night on which there is significant cooling of the ground by radiation to space. Such conditions occur when both clouds and wind are mainly absent, and are amplified by the presence of dry air, long hours of darkness, and low soil conductivity.

**radiation pressure** The extremely small pressure exerted on a body that is exposed to photons of light or other electromagnetic radiation. The force is so low that it is of significance only in interplanetary space, and for the smallest particles ($<1\,\mu m$), such as some of those shed by comets or that create the *zodiacal light.

**radiative forcing** The *forcing of the climate system by various atmospheric processes that involve radiation. This includes the influence of *clouds, *greenhouse gases, and particulates such as *aerosols.

**radiatus (ra)** A *cloud variety. Broad parallel bands of cloud that appear to converge because of perspective. May apply to *cumulus, *stratocumulus, *altocumulus, *altostratus, or *cirrus.

**radioactive carbon** The radioactive *isotope of carbon, $^{14}C$ (carbon-14), with a half-life of *c.*5 500 years. Its decay is the basis of *carbon dating.

**radioactive fallout** The deposition of radioactive particles on the Earth's surface following a nuclear explosion or an accident involving nuclear materials. The pattern of deposition depends strongly on the nature of the original event (an explosion may be expected to inject material high into the atmosphere), surface and upper-atmosphere winds, and the removal of radioactive material by *precipitation.

**radioactivity** The spontaneous fission of the nuclei of certain unstable (radioactive) elements or *isotopes into more stable configurations, which gives rise to daughter elements or isotopes. In certain cases, these products may themselves be radioactive, and subsequently decay, the process eventually ending with a stable element or isotope.

**radio duct** *See* ANOMALOUS PROPAGATION.

**radiometer** An instrument for measuring the amount of energy radiated by an object. Radiometers may be designed to cover both broad and narrow ranges of the electromagnetic spectrum, and some are known by specific names, including *pyranometer, *pyrheliometer, *pyrgeometer, etc. The general term is often used for the primary sensing instrumentation carried by Earth-resource and *meteorological satellites, which, by application of the radiation laws, may very accurately determine the temperature of the object being imaged. From the absorption in specific narrow wavebands the abundance and distribution of particular atmospheric components may also be obtained to a high degree of accuracy.

**radiosonde** An instrumented package, carried aloft by a balloon, that returns observations to a ground station by radio telemetry. In the most common form, *pressure, *temperature, and *humidity (PTU) readings are obtained. Other versions may report pressure alone (*windsonde), *ozone concentrations (*ozonesonde), or *radioactivity.

To determine wind vectors, radiosondes may be tracked from the ground by radar or *radio-theodolite, or (in the form sometimes known as a *rawinsonde) may be equipped to determine their positions from *LORAN-C or *Global Positioning System signals and to radio this information to the ground. *See also* DROPSONDE.

**radio-theodolite** An electronic system for tracking a *radiosonde, providing *azimuth and *altitude (elevation) readings. These may be used in conjunction with the height (derived from the telemetry) to estimate wind vectors (the *radio wind), although generally to a lesser degree of accuracy than those obtained by radar methods (*see* RADAR WIND).

**radio wind** The upper-atmosphere wind as determined by tracking a *radiosonde with a *radio-theodolite.

**radon (Rn)** A radioactive *inert gas, atomic number 86, the decay product of radium. The most stable *isotope, radon-222, has a half-life of 3.825 days. Relatively high concentrations of radon may occur in mines and inside buildings built above certain rocks, such as granite. This is regarded as a health hazard requiring remedial measures, although currently unconfirmed research suggests that the danger in homes may have been overestimated. *See also* THORON.

**RAFOS float** *See* SOFAR FLOAT.

**rain** Liquid *precipitation in the form of drops that are larger than 500 μm (0.5 mm) in diameter, smaller drops being classed as *drizzle. For *synoptic purposes rain may be classified as 'slight' ($<0.5\,mm\,h^{-1}$), 'moderate' ($0.5$–$4\,mm\,h^{-1}$), and 'heavy' ($>4\,mm\,h^{-1}$). *See also* RAINDROPS; RAINFALL.

**rainband** **1.** A band of *precipitation or increased precipitation found in a *depression. The bands may be several hundred kilometres long, with widths of 5–50 km, and lie approximately parallel to one of the fronts. Broad bands tend to be found ahead of the *warm front, ahead of the *cold front (in the warm sector), and also behind the cold front, which may itself be a much narrower rainband. Rainbands are *mesoscale phenomena, poorly revealed by the *synoptic network. **2.** An obsolete term for absorption bands in the solar spectrum that are caused by atmospheric water.

**rain day** An obsolete term, once used in *synoptic reports to indicate any 24-hour day (beginning at 09:00 GMT) on which more than 0.2 mm of rain had been recorded.

**raindrops** Drops of liquid water, normally taken to be all those larger than 500 μm (0.5 mm), but excluding the smaller *drizzle droplets (200–500 μm). Most raindrops lie between 1 and 2 mm in diameter. The largest drops are distorted, being flattened at the bottom by aerodynamic forces, and have spherical equivalent diameters of 5–8 mm. These large drops break up into

## Rainbow

An optical phenomenon that appears in the form of one or more arcs of spectral colours when sunlight is both reflected and refracted by falling raindrops. The most common bow (the **primary bow**) is a circular arc, with a radius of approximately 42°, centred on the *antisolar point. This bow is caused by refraction when the light enters and leaves each raindrop, together with a single reflection from the rear of the drop. The spectral colours run from red on the outside of the bow to violet on the inside, with radii of approximately 42° and 40°, respectively. In a **secondary bow**, which is also fairly common, the light undergoes two reflections within each raindrop and the sequence of colours is reversed. This bow has a radius of approximately 51° about the antisolar point.

Theoretical calculations and laboratory experiments show that numerous higher-order bows exist but are unlikely to be seen in nature. The faint tertiary bow, for example, has a radius of about 40° but is centred on the Sun.

The region between the primary and secondary bows appears darker than the background sky and is known as **Alexander's dark band**. Light from raindrops in this region does not reach the observer's eye.

Interference between light that has taken slightly different paths may give rise to a series of arcs within the primary bow. These **supernumerary bows** (or **interference bows**) generally appear pale violet and green to the naked eye.

Additional bows may be seen if a reflecting surface, such as a sheet of fairly calm water, lies behind the observer. Reflected sunlight acts as a secondary source, giving rise to primary or secondary **reflected-light rainbows**, centred between the antisolar point and the *zenith.

Large raindrops (>1 mm) produce brilliant bows in which red is clearly visible, whereas red is less prominent with smaller drops (*c*.0.3 mm). The smaller the droplet size, the greater the spacing between any supernumerary bows. With very small drops, *diffraction becomes dominant and the bow appears white, when it is known as a *fogbow. Such a fogbow may show a white supernumerary bow.

The Moon also gives rise to rainbows, but the colours are normally too faint to be readily detectable. All-red bows occur when the Sun is low on the horizon and other colours have been removed by atmospheric *scattering.

smaller droplets when they reach terminal velocity with respect to the surrounding air.

**rainfall** For *synoptic purposes, rainfall is defined as the total amount of liquid-water equivalent of all forms of *precipitation. It thus includes *hail, *sleet, and *snow, together with the forms that may be deposited directly from the air, i.e. *dew, *frost, and *rime.

**rainfall station** A station at which the principal measurement is of rainfall. Such stations are often maintained by water authorities for their own purposes, although the measurements are generally passed on to the main meteorological network.

**rain-free base** A dark, horizontal cloud base with no visible *precipitation. It often indicates the base of the main *updraught in a *thunderstorm system. Tornadoes (*see feature*) may develop either directly from a rain-free base or from a lower *wall cloud.

**rain gauge** An instrument for determining the depth of *precipitation that occurs over unit area. It may be designed as a simple passive collector—when the amount of precipitation is determined by use of a graduated measuring glass—or as a recording device, generally a *tipping-bucket rain gauge. More sophisticated forms enable rainfall intensity to be determined.

**rain shadow** An area that has lower average *rainfall than nearby regions because it is sheltered by a range of hills or mountains, lying to windward with respect to the predominant rain-bearing winds.

**rain splash** *Raindrops that hit a surface (such as the ground) and break up into smaller droplets. Such rain splash is a significant factor in the transport of many bacteria, viruses, and fungal spores and may greatly influence the spread of plant pathogens.

**rainy season** The period of the year in certain tropical and subtropical climatic regions during which precipitation is a maximum; distinct from a *dry season. In a *Mediterranean climate, the rainy season occurs in winter, but in *savanna and *monsoon climates it comes in the summer. Some equatorial regions experience two rainy seasons during the year.

**RAPID** A programme to monitor the strength and nature of the *Atlantic Meridional Overturning Circulation, primarily to assess its significance in determining the climate of northern Europe. Because of the importance of oceanic heat transport between latitudes 25–35° N, a major feature of the programme is the establishment of a series of moorings across the Atlantic at 26.5° N to determine factors such as temperature, salinity, pressure, and currents. These provide continuous monitoring of the transatlantic pressure gradient. In addition, the voltages induced on submarine cables across the Florida Strait by charged particles within the moving sea-water enable the strength of the *Florida Current to be determined on a continuous basis.

**rare gas** *See* INERT GAS.

**rasputitsa** [Russian: 'season of bad roads'] The brief period in spring when there is a rapid thaw of lying snow and frozen ground.

**ravine wind** A strong wind that flows through a valley or pass in response to a pressure gradient between the two sides of a range of hills or mountains.

The force of the wind may be greatly increased by the funnelling effect of the valley sides. The *kosava, which occurs on the Danube near the Yugoslav–Romanian border, is one example.

**rawinsonde** An obsolescent term for a *radiosonde with intrinsic wind-finding capability, through reception of broadcast signals such as those from the *LORAN or (now terminated) *Omega systems. Modern units, including those using the *GPS system, are generally termed 'radiosondes'.

**ray, auroral** An auroral form (*see* AURORA) consisting of narrow lines of light that extend up into the sky, approximately vertical to the horizon. They follow the lines of force of the Earth's magnetic field, and their appearance is normally a sign that the display is becoming more active. They may be seen in isolation or, more commonly, associated with auroral *arcs and *bands. When folded, rayed bands pass overhead they may produce a spectacular *corona (3).

**Rayleigh scattering** The *scattering of electromagnetic radiation by spherical particles whose radius is below approximately one-tenth of the wavelength of the incident radiation. The frequency and thus the wavelength of the radiation remain unaltered, but changes in phase and polarization occur. Rayleigh scattering by air molecules is responsible for *blue sky and *polarization of the sky. It is also significant in the scattering of radar waves by raindrops, although the latter is more appropriately treated by the more complex theory of *Mie scattering.

**rear-flank downdraught** A flow of subsiding dry air that partially surrounds the rear of a *mesocyclone, producing a *clear slot around the *wall cloud, and often indicated by a strong *hook radar echo.

**Réaumur scale** An obsolete temperature scale, in which the freezing point of water was 0° and its boiling point 80°. It was introduced in 1731 by the French physicist René Réaumur (1683–1757) after experiments with a thermometer that used a mixture of alcohol and water.

**recombination** Any process by which positive and negative ions, or positive ions and electrons, combine to give a neutral element or compound. In the atmosphere, such processes are significant in, for example, the night-time *ionosphere.

**recurvature** A change in the direction of the path of a tropical cyclone (*see feature*) towards the north-east (in the northern hemisphere) or south-east (in the southern). Following recurvature, which often occurs at latitudes 20–30° and carries the systems over cooler waters, tropical cyclones frequently degenerate into deep mid-latitude *depressions.

**reflection** The process by which all or a portion of incident radiation is returned in the general direction of the source, the exact amount being dependent on the wavelength of the radiation, the nature of the reflecting surface, and the angle of incidence. The majority of natural surfaces will give rise to **diffuse reflection**, where the radiation is sent out in a wide range of directions, but a few surfaces, such as calm water, will produce **specular reflection** (as in a mirror), with the radiation restricted to a narrow range or single direction, as occurs in *sun glint. Reflection is important in many atmospheric optical phenomena, including *heiligenschein, rainbows (*see feature*), and *sun pillars.

**refraction** The deviation of light or other electromagnetic radiation from its original path on passing from one medium to another. It is usually accompanied by dispersion, which, in the case of light, gives rise to the colours of the spectrum. Refraction is a significant factor in atmospheric optics, particularly in the various *halo phenomena, including *circumhorizontal and *circumzenithal arcs and *parhelion, and in *dewbows and rainbows (*see feature*).

Refraction also occurs in the lowest layers of the atmosphere, even if the surface and the adjacent air are at the same temperature, because of the decrease in density with height. The effect is to raise objects on the horizon (such as the setting Sun) by 34′. When there are distinct temperature (and hence density) contrasts, refraction produces *mirages and associated phenomena such as *looming, *sinking, *stooping, and *towering.

**refractometer** An airborne instrument for determining the refractive index of the atmosphere by means of microwaves.

**reg** [Arabic] One of the three main desert landforms, where the ground surface is covered in loose rock fragments or gravel. *See also* ERG; HAMMADA.

**regelation** The refreezing of water that has been melted by pressure. At temperatures close to 0 °C, ice turns into liquid water with the application of pressure, such as beneath the blade of an ice skate, or when ice crystals are squeezed together to make a snowball. Regelation takes place when the pressure is removed.

**relative humidity** The ratio of the actual *vapour pressure to the *saturation vapour pressure with respect to water at the same temperature. This value is generally expressed as a percentage.

**remote sensing** The measurement of a property or a phenomenon by instrumentation that is situated at a distance and not in direct physical contact with the object of study. In meteorology, the instrumentation may be carried by an aircraft or *meteorological satellite, or else be ground based, as in the case of *skywave radar sounding and wind *profilers. Both active and passive forms are used, the former employing *lidar, *radar, or *sonar techniques, for example, and the latter involving devices such as *radiometers.

**remote sounding** *See* TOPSIDE SOUNDING.

**resistance thermometer** A *thermometer in which the sensing element consists of a substance whose electrical resistance varies with temperature. Such sensors are both sensitive and rugged, and are commonly employed in *automatic weather stations, *dropsondes, *radiosondes, and similar devices.

**retrograde** A term applied to motion in the opposite direction to normal. Used of a pressure system (*depression, tropical cyclone (*see feature*), etc.) that reverses its general direction of motion, or of a wave that moves in the opposite direction to the flow of which it is part.

**return flow** The winds that occur on the rear flank of an *anticyclone that is moving eastwards, and which are southerly in the northern hemisphere and northerly in the southern. These usually bring warm (and often moist) air in behind the system.

**returning air** In one (older) classification scheme, an *air mass that has been drawn towards the Equator and then returned towards the pole. Indicated by a prefix (r). For example, returning *maritime polar air (rPm) that was initially brought south behind a *depression and then returned polewards. Having first become warmer and more humid, it subsequently travels over progressively colder water, so the lower layers cool and thus become more stable.

**return period (also known as the recurrence interval)** A statistical measure of the average interval between occurrences of a particular phenomenon, such as a flood of a given magnitude, a devastating storm, an exceptionally heavy snowfall, etc.

**return stroke (also known as the main stroke or return streamer)** The principal discharge in a *lightning stroke, occurring after the *stepped leader reaches the ground (or other object). The return stroke carries the main current, which flows upwards from the ground to the cloud.

**reversal** The change in the dominant flow in the *stratosphere at high latitudes which occurs in spring and autumn. In winter, the flow is characterized by very strong westerlies, whereas in summer there are much weaker easterlies. The reversals generally occur in late April and early September.

**revolving storm** *See feature* TROPICAL CYCLONE.

**ribbon lightning** A rarely seen form of lightning in which a stroke appears to have a considerable width or even to consist of individual parallel channels. It occurs when a very strong wind is blowing at right angles to the observer's line of sight. The multiple strokes that form a single discharge are thus displaced horizontally from one another. A similar appearance may occur when the observer (or camera) is moving approximately perpendicular to the line of sight.

**Richardson, Lewis Fry** (b. Newcastle upon Tyne, 11 October 1881; d. Kilmun, Scotland, 30 September 1953) British meteorologist and physicist who developed numerical techniques for forecasting the state of the atmosphere from a knowledge of the situation at some specific instant. Because of the time required for computation, his methods remained impractical until the development of fast electronic computers capable of handling the immense number of parameters and calculations that are needed. His work is the basis of all modern *numerical weather prediction.

**ridge** An elongated region in which atmospheric *pressure is higher than that at the same altitude in the immediately adjoining areas, and thus usually an extension of an *anticyclone or high-pressure area. The *anticyclonic curvature is a maximum along the axis of the ridge. It is normally accompanied by relatively settled weather. The counterpart of a low-pressure *trough.

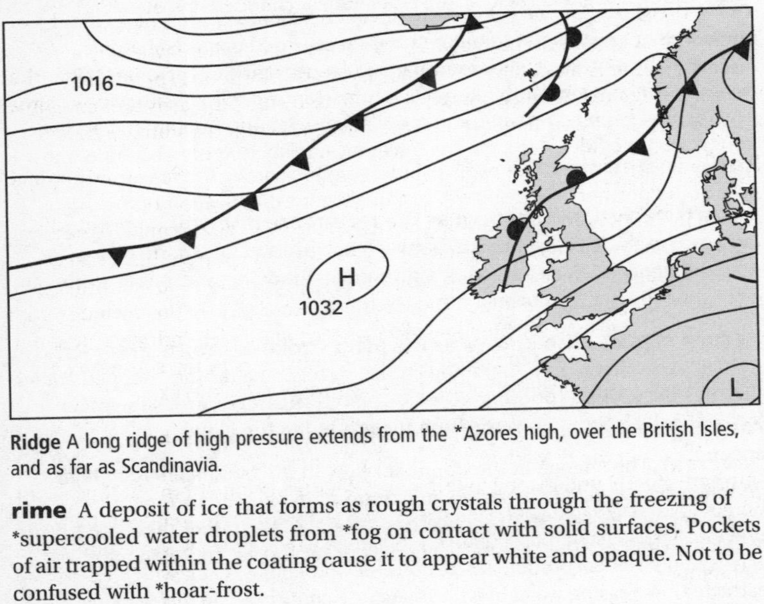


**Ridge** A long ridge of high pressure extends from the *Azores high, over the British Isles, and as far as Scandinavia.

**rime** A deposit of ice that forms as rough crystals through the freezing of *supercooled water droplets from *fog on contact with solid surfaces. Pockets of air trapped within the coating cause it to appear white and opaque. Not to be confused with *hoar-frost.

**rime ice** Ice that forms (sometimes in large masses several metres across) by direct freezing of water vapour and droplets on rocks or other exposed surfaces, particularly in humid conditions, such as those found on the Antarctic Peninsula and the sub-Antarctic islands. Rime ice may break away and add substantial amounts of ice to valley *glaciers.

**roaring forties** The band of strong westerlies that encircles the southern hemisphere at approximately latitude 40° S. Originally a nautical term, but now in common usage.

**rocket lightning** An extremely rare form of *lightning, in which the discharge propagates so slowly that it may be followed by eye. There appears to be only anecdotal evidence, so there is no adequate scientific explanation, and many experts doubt the reality of the phenomenon.

**rocketsonde** An instrumented package, ejected from a *meteorological rocket, normally equipped with a temperature sensor, but frequently also carrying pressure, humidity, or ozone sensors. It descends by parachute and, like a *radiosonde, may be tracked by radar to determine upper-atmospheric winds.

**rocket sounding** The investigation of the Earth's upper atmosphere to altitudes of several hundred kilometres by means of instrumented rockets. Typical parameters measured include: temperature, pressure, density, composition, magnetic-field strength, electrical conductivity, etc.

**rockoon** A system for investigating the upper atmosphere whereby a research rocket is lifted into the *stratosphere by a balloon. The rocket is subsequently fired at an appropriate moment to study the arrival of energetic particles from a *solar flare or *aurora or to obtain other scientific data.

**Rocky Mountains** The principal topographic barrier to the westerly *zonal flow in the northern hemisphere, responsible for the location of the major trough in the *Rossby waves, which lies at about 70° W. For complex dynamical reasons, the flow turns northwards as it rises over the barrier, but because *vorticity must be conserved, turns southwards as it descends to leeward. From this initial disturbance, additional oscillations occur further downwind in the zonal flow. The Tibetan Plateau exerts a similar, but lesser, influence at about 150° E, but there are no comparable mountain barriers in the southern hemisphere. *See also* OROGRAPHIC DEPRESSION; POLAR VORTEX.

**rogue wave** An exceptionally large wave that far exceeds those encountered in the prevailing *sea state, with heights reaching 25–30 metres. Currently poorly understood, but initial studies using data from the *European Remote Sensing Satellites reveal that such waves are more frequent than previously suspected, particularly at locations where normal wind-induced waves and swell encounter strong currents and eddies, from which energy is extracted, increasing wave height. The study confirmed the occurrence of rogue waves on the notoriously dangerous *Agulhas Current off southern Africa, but also revealed severe conditions where the *Gulf Stream interacted with waves from the Labrador Sea.

**roll cloud** **1.** A general term for the *arcus type of *supplementary cloud feature. **2.** More specifically, a relatively rare, low, horizontal roll of cloud that accompanies a *cold front, a *line squall, or a *thunderstorm *gust front. It is completely detached from the main cloud mass, unlike the more common *shelf cloud, and normally appears to be rotating about a horizontal axis.

**roll cumulus** An obsolete term for *stratocumulus in the form of extended individual rolls.

**rope cloud** A narrow line of *cumuliform clouds that forms along a shallow *cold front, and is normally particularly well developed over the sea. Such a cloud line may arise when a front weakens, its slope decreases, and there is a cold air outflow at the surface, creating increased convection ahead of the front. There is often a clear band behind the rope cloud, where air is subsiding between it and the residual cloud along the original cold front.

**Rossby, Carl-Gustaf Arvid** (b. Stockholm, 28 December 1898; d. Stockholm, 19 August 1957) Swedish-American meteorologist, who worked under Vilhelm *Bjerknes before emigrating to the United States in 1926. He discovered and studied the atmospheric waves subsequently named after him, and also identified the *jet stream.

**Rossby wave (long wave)** A long-wavelength oscillation in the *westerlies that predominantly occurs in the middle and upper *troposphere, and is particularly marked in the northern hemisphere. Occasionally referred to as a *planetary wave. (*See also* POLAR VORTEX.) The Rossby waves are quasi-stationary, with between three and six oscillations being encountered around the Earth. In the northern hemisphere, major semi-permanent *troughs at about 70°W and 150° E lie to leeward of the major topographic barriers formed by the *Rocky Mountains and the Tibetan Plateau, and are just to the west of the wintertime location of the Atlantic and Pacific zones of the *Polar Front. Rossby waves play a large part in the formation and decay of *extratropical cyclones and *anticyclones, and in their motion. The longest waves occasionally move westwards against the flow, but waves with a shorter wavelength occur in the overall *zonal flow, and these generally move eastwards, being steered by the longer waves.

Rossby waves may be explained in simplified terms as arising because of the variation in the *Coriolis effect with latitude. When a zonal flow is deflected towards either the pole or the Equator, conservation of absolute *vorticity causes the flow to return towards its original latitude. This initiates a north–south oscillation with a typical wavelength of *c.*2 000 km.

The term is also applied to all similar waves in the atmosphere and oceans. Westward-propagating Rossby waves in the equatorial Pacific Ocean are an important factor in the El Niño Southern Oscillation (*see feature*), and eastward waves form the *Antarctic Circumpolar Wave. *See also* KELVIN WAVE.

**rotor** A closed, vertical circulation that develops in the lee of high mountain barriers, or in the valley between two mountain ranges, when conditions are appropriate. A form of *lee eddy. At the surface the wind is in the opposite direction to the gradient wind. Rotors are often the site of extreme *turbulence and the top of the circulation may be marked by a **rotor cloud** (such as *altocumulus *lenticularis).

**run-off** The percentage of *precipitation that falls on a specific area of the Earth's surface (such as a catchment basin) which is drained away by streams or rivers, rather than being absorbed to become groundwater.

**runway visual range** The maximum horizontal distance down a specified runway at an airfield at which an observer can detect (and identify) a standard light. It is normally measured by a *transmissometer, and is an all-important factor in determining whether the airfield may be declared open for traffic.

# S

**SAF** *See* SATELLITE APPLICATION FACILITY.

**Saffir-Simpson scale (Saffir-Simpson damage potential scale)** A scale originally developed for describing the potential severity of a *hurricane, but subsequently applied to all tropical cyclones (*see feature*). The assessment is in terms of both the wind speed and possible storm-surge height. Note that the scale is defined in non-metric units.

| CATEGORY | | CENTRAL PRESSURE | | WIND SPEED | | STORM SURGE | |
|---|---|---|---|---|---|---|---|
| | | in | hPa | mph | $km\,h^{-1}$ | ft | m |
| 1 | weak | >28.94 | >980 | 74–95 | 119–153 | 4–5 | 1.2–1.5 |
| 2 | moderate | 28.50–28.91 | 965–979 | 96–110 | 154–177 | 6–8 | 1.8–2.4 |
| 3 | strong | 27.91–28.47 | 945–964 | 111–130 | 178–209 | 9–12 | 2.7–3.7 |
| 4 | very strong | 27.17–27.88 | 920–944 | 131–155 | 210–249 | 13–18 | 4.0–5.5 |
| 5 | devastating | <27.17 | <920 | >155 | >249 | >18 | >5.5 |

**St Elmo's fire (corposant)** The luminous glow that often accompanies a *corona discharge, the term being originally applied to the glow at the tips of the masts and spars of ships. An Italian corruption of the name of the patron saint of Mediterranean sailors, Sant Ermo (St Erasmus), who was believed to indicate his protection in this manner. The alternative name, *corposant,* comes from the Portuguese *corpo-santo* ('holy body').

**St Luke's summer** The term for a period of fine weather that (in popular folklore) supposedly occurs around St Luke's Day, 18 October.

**St Martin's summer** The term for a period of particularly fine weather that (in popular folklore) supposedly occurs around St Martin's Day, 11 November.

**St Swithin's Day** According to folklore, if rain falls on St Swithin's Day (15 July), rain will fall at the same location on each of the following 40 days. There is no foundation in fact for this superstition.

**salinity** The amount of inorganic salts dissolved in water, i.e. the number of grams per kilogram of water. Generally given as parts per thousand (per mille, ‰). Although it varies to a certain extent, the average salinity of seawater is 34.7 parts per thousand. *See also* THERMOHALINE CIRCULATION.

**sand haze** A *haze consisting of fine sand particles raised by a recent *sandstorm.

**sand pillar** *See* DUST DEVIL.

**sandstorm** A strong, turbulent wind that lifts sand particles from the surface, but rarely beyond a height of *c.*15 m; in contrast to a *dust storm. In *synoptic reports, a sandstorm is reported if the *horizontal visibility is reduced below 1000 m, and a severe sandstorm if visibility is below 500 m.

**Santa Ana** A hot, dry, easterly wind that affects the Los Angeles region of southern California, named after the valley through which the main flow occurs, and which acts to intensify it as a *mountain-gap wind. The wind originates over the dry interior plateau, and its temperature and desiccating effects are increased by *adiabatic heating during its descent, as with a *föhn wind.

**SAR** *See* STABLE AURORAL RED ARC; SYNTHETIC APERTURE RADAR.

**sastruga (zastruga, pl. sastrugi/zastrugi)** [Russian: 'groove'] A wave-like irregularity in hard snow, normally perpendicular to the wind direction, sloping gently on the windward side and steeply to leeward. Sastrugi arise by the coalescence of snow dunes. The term is also applied to ridges parallel to the wind direction produced by mechanical abrasion by wind-borne ice particles, rather than simple *deflation. The size of sastrugi depends on the state of the surface and the strength and duration of the wind. They may be up to 2 m high and often pose a considerable hazard for travellers over *glaciers, *ice caps, or the polar *ice sheets.

**satellite, meteorological** An artificial satellite specifically designed for meteorological purposes, and whose main instrumentation is some form of scanning *radiometer. There are two main variants: *geostationary satellites and *polar orbiting satellites.

**Satellite Application Facility (SAF)** A specialized development and processing centre set up by *Eumetsat to exploit data from meteorological satellites, especially the *Meteosat Second Generation and *MetOp satellites. Each facility (a national meteorological organization) handles one particular aspect (or group of aspects) of data application, such as the development of specific software and data-handling techniques. Example SAFs are those for *nowcasting (Spain), ocean and sea ice (France), *numerical weather prediction (UK), and climate monitoring (Germany).

**saturated adiabatic lapse rate (SLR; also known as saturation adiabatic lapse rate or moist adiabatic lapse rate)** The *lapse rate that prevails when a parcel of air reaches *saturation. For a rising air parcel, because of the *latent heat released by *condensation, this lapse rate is less than the *dry adiabatic lapse rate (*c.*10 deg C km$^{-1}$) and is normally between 4 and 7 deg C km$^{-1}$. If all condensation products (cloud droplets, *rain, *snow, and *hail) remain within the parcel and are available for melting and evaporation, the air will warm at the same rate on descent. In many cases, however, some condensation products will fall out as *precipitation, removing heat from the rising parcel, which will cool at a somewhat higher *pseudo-adiabatic lapse rate. The loss of precipitation may result in significantly different behaviour on subsequent descent and warming.

**saturation** The state in which the *partial pressure of any component equals the maximum possible under the prevailing conditions. A parcel of air, for example, is saturated when it is in neutral equilibrium with a flat surface of water or ice at the same temperature and pressure. In the absence of *condensation nuclei, a higher partial pressure may be achieved, giving rise to *supersaturation, but such conditions are not thought to exist in the atmosphere.

**saturation fog** A form of *fog that occurs when water vapour evaporates from the ground, saturated after heavy rain, into relatively cold air and immediately condenses to give a shallow layer of fog.

**saturation vapour pressure (SVP; also known as the equilibrium vapour pressure)** The *vapour pressure at which an equilibrium exists between *water vapour and an ice or water surface. The rates of evaporation and condensation are equal, i.e. identical numbers of molecules are entering and leaving each phase. This vapour pressure is dependent only on the substances involved and the temperature.

**savanna** A subtropical or tropical region consisting of open grassland with drought-resistant trees and shrubs. The climatic regime is type A in the *Köppen classification, with a long dry season (usually in winter—Köppen subtype Aw) and an intense rainy season.

**scale height** The vertical distance within an atmospheric column in which the pressure falls by a factor of *e* (2.718). In the Earth's *troposphere it is *c.*8.5 km.

**Scanning Imaging Absorption Spectrometer for Atmospheric Chartography (SCHIAMACHY)** An instrument carried by *Envisat, which covers an extremely wide range of wavelengths (240–2380 nm), and is able to detect trace gases, *nitric oxide (NO), *ozone, and cloud and dust particles, providing both total amounts and profiles.

**scattering** The process by which radiation that is incident on small particles suspended in a medium with a different refractive index is dispersed in various directions. The intensity of the scattered beam in any particular direction is strongly dependent on the wavelength of the incident radiation, and on the size of the particles. Polarization may also be introduced. In the atmosphere, scattering by gaseous molecules gives rise to the *blue sky, and by *aerosols and larger particles to the *blue moon, *haze, and other phenomena. *See* MIE SCATTERING; POLARIZATION OF THE SKY; RAYLEIGH SCATTERING.

**scatterometer** A radar instrument used in *remote sensing from aircraft or satellites. It is particularly applied to the determination of wave height and wind direction, which may be measured from the properties of the reflected beam.

**SCHIAMACHY** *See* SCANNING IMAGING ABSORPTION SPECTROMETER FOR ATMOSPHERIC CHARTOGRAPHY.

**scintillation** The rapid changes in apparent brightness, colour, or position exhibited by stars, particularly those that are close to the horizon. These fluctuations arise from slight differences in density (and thus of refractive index) of small portions of the lower atmosphere, which are generally just a few centimetres across. A similar phenomenon (*shimmer) is often seen during the daytime in strongly heated air.

**scirocco** *See* SIROCCO.

**Scotch mist** A combination of *drizzle and thick *mist or *fog that often occurs in upland areas (such as the Scottish Highlands), particularly associated with a moist *tropical maritime airstream.

**scud** **1.** Tattered fragments of low cloud (usually *stratus *fractus) that are often observed in windy conditions below rain clouds, particularly *nimbostratus. Originally a nautical term, it now has a more general usage. **2.** A term used in parts of Canada for a sudden gust of wind.

**SEA** *See* SUDDEN ENHANCEMENT OF ATMOSPHERICS.

**sea area** A division of inshore or offshore waters into specific regions, primarily for ready reference in the preparation and broadcast of weather forecasts. The individual areas are defined such that conditions are likely to be consistent over their full extent. Forecasts are broadcast, area by area, in a fixed pattern for intelligibility and to lessen confusion.

**sea-breeze** A local, onshore wind that arises from the differential heating of land and sea during the day and the consequent difference in pressure over the two surfaces. Its nocturnal counterpart is the *land-breeze. A sea-breeze normally commences before noon, reaches a maximum in the afternoon, and dies away in the evening. It may extend tens of kilometres inland and often gives rise to a region of *convergence, known as a **sea-breeze front**, with enhanced formation of *cumuliform cloud. Sea-breezes propagating from both sides of a peninsula may give strong convergence in the centre. In certain parts of the world, such vigorous interaction may produce a *morning glory.

**sea fret** *See* HAAR.

**sea ice** **1.** Ice that has formed through the freezing of sea-water. Under most circumstances, the ice crystals are intermixed with brine, but separation begins to occur even at the *pancake ice stage. Under pressure, the ice may form pressure ridges and hummocks, from which the brine drains away, leaving essentially pure ice. **2.** Any ice that is floating on the surface of the sea.

**sea-level pressure** The pressure at sea level, calculated from the observed pressure at the station. The value plotted on the most commonly available *isobaric charts.

**sea smoke** *See* ARCTIC FROST SMOKE.

**season** A division of the year for climatological purposes. Although varying considerably with latitude and by geographical region, the division is commonly regarded in terms of the conventional four seasons that apply to most northern temperate regions: winter—December to February; spring—March to May; summer—June to August; and autumn—September to November. Even here, however, a more rigorous treatment finds discrepancies. In Britain, for example, five periods of distinctive weather may be found: early winter (late November to late January); late winter (late January to end of March); spring and early summer (April to mid-June); high summer (mid-June to early September); and autumn (early September to late November).

In the *tropics, where temperatures remain generally high throughout the year, the division is often into dry and rainy seasons, or else referred to in terms of the prevailing *monsoon. In subtropical regions, where greater temperature variations are found, the terms 'hot season' and 'cold (or cool) season' may be employed. At high latitudes, there is often an abrupt change between winter and summer (or summer and winter) conditions, so spring and autumn may be extremely short or essentially non-existent.

**SeaStar** A small satellite, air-launched by a Pegasus rocket in August 1997, later renamed OrbView-2. It carries a single instrument, the *SeaWiFS sensor. A highly successful, relatively cheap mission.

**sea state (state of sea)** A system for the classification of average wave height, with a scale that ranges from 0 to 9.

| WAVE HEIGHT (m) | STATE | DEFINITION |
|---|---|---|
| 0 | 0 | calm, glassy |
| 0–0.1 | 1 | calm, rippled |
| 0.1–0.5 | 2 | smooth |
| 0.5–1.25 | 3 | slight |
| 1.25–2.5 | 4 | moderate |
| 2.5–4 | 5 | rough |
| 4–6 | 6 | very rough |
| 6–9 | 7 | high |
| 9–14 | 8 | very high |
| >14 | 9 | phenomenal |

**sea-surface temperature (SST)** The temperature of the surface layer of sea or oceanic water. The SST is an important factor in many *teleconnections and is then often assessed as an *anomaly.

The average range of SST over both polar and equatorial regions is *c.*6 deg C. At middle latitudes it may reach 10 deg C, but even larger ranges (up to 30 deg C or more) may occur in small areas of the north-western Pacific and north-western Atlantic Oceans.

**SeaWiFS (Sea-viewing Wide Field-of-view Sensor)** The sole instrument on the *SeaStar satellite, monitoring the oceans in eight spectral bands. When it ceases to operate, its functions will be covered by the more versatile *MODIS instrument on the *Terra satellite.

**Secchi disk** A black-and-white disk that is lowered into the sea or a lake to determine the transparency of the water by noting the depth at which the disk disappears or reappears.

**seclusion** A term occasionally used for a very late stage in the lifetime of a *depression, following the *back-bent occlusion stage, when the cold air has virtually encircled the centre of the low and cut off (secluded) a pool of warm air above the centre.

**secondary depression (wave depression)** A low-pressure centre that forms within the overall circulation of another, pre-existing *depression. The secondary has its own closed circulation, and initially has a higher central pressure than the primary. In general it deepens at the expense of the primary and, particularly in the case of a secondary that has formed on a *cold front, may eventually assimilate the primary. The majority of secondary depressions form on fronts (*see* COLD-FRONT WAVE; WARM-FRONT WAVE) or at the point of *occlusion (*see* COLD-OCCLUSION DEPRESSION; WARM-OCCLUSION DEPRESSION), but they may also arise in an unstable *air mass (*see* POLAR-AIR DEPRESSION). *See figure overleaf.*

**secondary front** A *front that arises between two *air masses that, despite nominally originating from the same *source region, have different thermal characteristics, because they have followed different paths or are of differing ages.

**secondary organic aerosols (SOA)** Organic compounds that form in the atmosphere through chemical reactions between various primary pollutants, probably through the oxidation of long-chain hydrocarbons, and often promoted by sunlight. Recent research (particularly in connection with the *Megacity Initiative: Local and Global Research Observations) has revealed that SOAs are far more significant pollutants, especially downwind of major pollution sources, than previously appreciated. They may comprise the majority of pollutants to which people are exposed. Long-chain hydrocarbons had previously been thought to be relatively stable, and confirmation of their role will require extensive changes to pollution models.

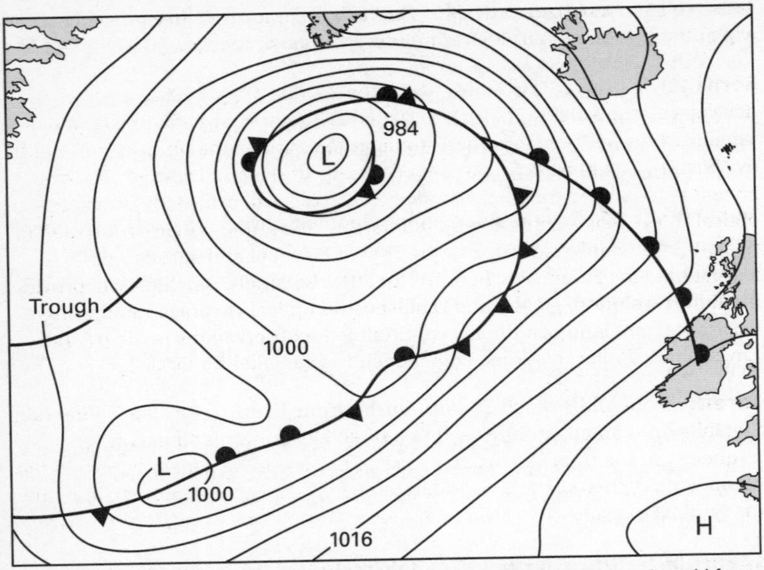


**Secondary depression** Yet another depression may develop at the wave on the cold front behind the closed circulation in mid-Atlantic.

**secular trend** A persistent increase or decrease in a particular measured variable (such as temperature) when the effects of short-term fluctuations have been removed. The increase in carbon dioxide concentration in the atmosphere is one example of such a trend.

**seiche** A regular oscillation of the water level within a closed or semi-enclosed basin, such as a lake or bay. Seiches may be generated by changes in atmospheric *pressure, the wind, earthquakes, landslides, or *tsunamis. They are *gravity waves and their frequency of oscillation and amplitude are determined by the physical size, shape, and depth of the basin. Repeated reflection from the ends of the body of water produces a pattern of standing waves, with points (nodes) or nodal lines where there is no vertical motion. The lifetime of seiches ranges from a few minutes to several days.

**seistan (also known as bad-i-sad-o-bist-roz)** [Persian: 'wind of 120 days'] A strong north-westerly over eastern Iran and Afghanistan. It is associated with the summer *monsoon low-pressure region over north-western India and begins in late May or early June, blowing almost continually until the end of September. It transports considerable quantities of sand and dust, which it deposits as large dunes. *See also* SHAMAL.

**sensible heat** The energy absorbed or emitted by a substance during a change in *temperature that is not accompanied by a phase change; *compare*

LATENT HEAT. In meteorology, it is generally taken to apply to a parcel of air in which the *water vapour does not undergo a phase change.

**Sentinel** A programme under which the European Space Agency is developing three different forms of satellites for Earth observation. Sentinel-1 satellites will use *synthetic aperture radar to continue the observations made by *Envisat, including day and night coverage of sea and ice state. The first launch is scheduled for 2011. Sentinel-2 satellites will provide high-resolution optical images for environmental monitoring, including the provision of data related to the implementation of the *Kyoto Protocol and subsequent legislation. First launch is scheduled for 2012. Sentinel-3 satellites will provide medium-resolution, global, and multispectral optical coverage monitoring ocean, ice, and land conditions, continuing the observations made by *ERS, Envisat and similar missions. First launch is scheduled for 2011–12.

**serein** [from Old French *serain*: 'dusk', from Latin *serus*: 'late'] Fine rain that falls from an apparently clear sky, often after sunset and occurring frequently in the tropics. This may arise either when the generating cloud has dissipated or when there is considerable *wind shear between the cloud and the ground, causing the precipitation to be carried well away from its origin.

**severe weather threat index (SWEAT index)** A numerical index used to indicate the probability of extreme *convection and the accompanying *thunderstorms and tornadoes (*see feature*). It takes into account the *dew point at 850 hPa, the wind speeds and difference in wind directions (i.e. the *wind shear) at 850 and 500 hPa, and a temperature index.

**SEVIRI** *See* SPINNING ENHANCED VISIBLE AND INFRARED IMAGER.

**sferics** The study of *atmospherics, particularly with a view to locating and tracking the *thunderstorms that are creating the radio disturbances.

**sferics fix** The determination of the location of the *lightning discharge producing specific *atmospherics radiation. There are two basic forms of system. The first uses standard radio direction-finding techniques. In the second, the arrival time difference (ATD) system employed by the British Meteorological Office, the use of several (currently seven) stations enables the location to be determined accurately even at distances of thousands of kilometres. This system is likely to be enhanced to provide details of individual discharges such as polarity, amplitude, and type (cloud-to-cloud, cloud-to-ground, or multiple).

**shade temperature** The temperature of the air as recorded by a *thermometer that is protected from direct sunlight, from heat radiated by nearby surfaces, and from any form of *precipitation. *See* THERMOMETER SCREEN.

**shadow bands** An optical phenomenon seen fleetingly during the course of a total solar eclipse, shortly before and after totality, when just a small portion of the Sun is uncovered. They consist of low-contrast bands of alternating light and shade, normally a few centimetres wide, that are visible against light-coloured objects or the ground. They arise through *refraction of the light from the narrow segment of the Sun by small regions of differing density in the atmosphere. Two types are known: type 1 have separations of *c.*200 cm and are produced below altitudes of 3 500 m, whereas type 2, with separations of *c.*8 cm, arise below 150 m. *See also* SCINTILLATION.

**shadow of the Earth** The shadow of the solid Earth cast on to distant objects (such as mountain-tops) or the atmosphere itself, and visible at sunrise or sunset. It takes the form of a blue-grey segment, normally with a red upper border (known as the *counterglow), that rises in the east as the Sun sets in the west. It gradually darkens as it rises until it merges with the dark sky overhead. The reversed sequence is seen at sunrise. *See also* ALPINE GLOW.

**shallow moist zone** *See* SPLIT FRONT.

**shamal** A hot, dry, north-westerly wind that occurs over Iraq and the region around the Persian Gulf. It is particularly strong and persistent during the daytime in summer, but tends to weaken at night. Like the *seistan, it is associated with the summer *monsoon low over northern India.

**sharav** [from Hebrew 'heat of the land'] The name given in Israel to the southeasterly *khamsin that originates over the Arabian desert.

**Shaw, Sir William Napier** (b. Birmingham, 4 March 1854; d. London, 23 March 1945) British meteorologist who was largely responsible for turning the Meteorological Office from a primarily clerical organization into a centre for scientific meteorology, becoming its first Director in 1905. He strongly supported various scientific investigations, particularly those into the nature of the upper atmosphere, and Lewis Fry *Richardson's work on *numerical weather prediction. He also introduced the *tephigram as a meteorological tool.

Shaw actively promoted international cooperation, and was President of the *International Meteorological Organization 1907–23. He secured agreement on the free exchange of meteorological data around the world and the methods by which this was to be achieved. With Vilhelm *Bjerknes, he promoted the adoption of the *millibar as the international unit of pressure.

**shear** *See* WIND SHEAR.

**sheet clouds** *See* LAYER CLOUDS.

**sheet lightning** A popular term for a cloud-to-cloud *lightning discharge that illuminates the clouds but where the actual channel is not seen. It is commonly, but erroneously, thought to be fundamentally different from *fork lightning.

**shelf cloud** A specific type of *arcus *supplementary cloud feature in the form of a wedge of cloud on the leading edge of a *thunderstorm *gust front or along a *cold front. It is an extension of the main cloud mass, unlike a *roll cloud. The top of the wedge is normally smooth, occasionally appearing as multiple layers, but the lower surface is often turbulent.

**shimmer** Rapid fluctuations in the images of objects that are viewed (essentially horizontally) through a layer of heated air, and caused by the changes in refraction produced by small parcels of air at different temperatures, and hence densities. *Compare* SCINTILLATION.

**short-wave radiation** In meteorology, electromagnetic radiation (between *c.*0.3 and *c.*4.0 μm) that penetrates down to the lower *troposphere and the Earth's surface. This may be compared with the *long-wave radiation that is emitted by the Earth.

**shower** In meteorology, liquid or solid *precipitation from *convective clouds (*cumulonimbus or *cumulus *congestus) as distinct from *stratiform clouds (*nimbostratus, *stratus). Showers generally consist of groups of cumulonimbus cells. They are usually of relatively short duration and may show rapid changes in intensity (Plate 8). In *synoptic reports, they are classified by their rate of precipitation: slight, 0–2 $mm h^{-1}$; moderate, 2–10 $mm h^{-1}$; heavy, 10–50 $mm h^{-1}$; and violent, over 50 $mm h^{-1}$. *See also* THUNDERSTORM.

**shred cloud** *See* FLOCCUS.

**Siberian anticyclone (Siberian high)** A shallow, cold *anticyclone (2–3 km deep) that forms over Siberia in winter (November to March), with its centre located in the general region of Lake Baikal. It is a *centre of action and has a sea-level pressure of at least 1030 hPa.

**Siberian express** An informal North American term for a surge of extremely cold air from the *Siberian anticyclone that may occur in winter, after the surface of the Arctic Ocean has completely frozen. It crosses the pole (at speeds of as much as 100 $km\ hr^{-1}$) and may cause temperatures in Canada and the northern United States to drop by 6–12 deg C in just a few minutes.

**SID** *See* SUDDEN IONOSPHERIC DISTURBANCE.

**sidereal day** The time required for a planet or other astronomical body to complete one rotation on its axis relative to the distant stars. In the case of the Earth the mean sidereal day is currently 23 hours 56 minutes 4 seconds. Primarily because of the action of the Moon, the length of day is slowly increasing, having been considerably shorter in the distant past, with correspondingly more days in the year.

**sidereal period** The time required for a planet or other astronomical object to complete one orbit around a central body (e.g. the Sun), relative to the distant stars. In the case of the Earth, the sidereal period is 365.25636 days.

**silver frost** *Dew that was initially deposited as droplets but which has subsequently frozen with the decline in temperature overnight. Initially, the dew exists in a *supercooled state, but usually freezes when temperatures fall below –3 to –5 °C.

**silver iodide (AgI)** A compound that is commonly used to provide *freezing nuclei in *cloud-seeding operations. The particles may be created by burning a mixture of silver iodide and acetone in ground-based or aircraft-mounted generators or else released from a pyrotechnic mixture fired into suitable *supercooled clouds by rockets or artillery shells.

**silver thaw** A North American term for a thaw that occurs when a warm, humid wind suddenly follows a period of intense cold. In the initial stages objects remain so cold that moisture from the air is deposited upon them and they become covered in a layer of *glaze.

**simoom** [Arabic: 'poisonous'] A hot, dry wind or *whirlwind that occurs, particularly in summer, in North Africa and Arabia. It often carries large quantities of suffocating sand and dust, but normally persists for just 10–20 minutes.

**singularity** The annual recurrence of a specific meteorological phenomenon on or close to a particular date. Examples are to be found in the onset of the monsoon at certain places, and the January thaw that may occur in New England. *See also* BUCHAN SPELL; ST SWITHIN'S DAY.

**sinking** A *refraction effect in which an object normally visible on the geographical horizon apparently sinks beneath it. It occurs when air density decreases with height at a lower rate than normal, or actually increases. Rays of light from the object are refracted upwards to pass above the observer's head. The opposite effect is known as *looming. *See also* MIRAGE.

**sirocco** [from Italian *scirocco,* from Arabic *sharq*: 'east wind'] A warm, dust-laden, southerly wind in the Mediterranean region. Variously known as the *chili in Morocco, Algeria, and Tunisia, *ghibli in Libya, *khamsin in Egypt, and the south-easterly *leveche in Spain. It is particularly prevalent in spring and autumn and is normally drawn northwards ahead of the *cold front of a *depression. With a long *fetch across the Mediterranean it may become very humid and produce extensive *stratus cloud. The leveche has a short fetch and remains hot and dry.

**site** The location of an observing *station, which must be chosen with due regard to the necessary *exposure for the various instruments. With the exception of *rain gauges, which require specific precautions to prevent measurement errors, in general the more exposed and open the site, the better.

**Six's thermometer** A U-shaped thermometer, designed by J. W. Six, with a bulb that contains creosote. Mercury occupies the bend in the tube and a spring-loaded index rests on the mercury's *meniscus in each leg of the tube. These indexes are displaced by the expansion and contraction of the mercury, remaining at their extreme positions to give readings of the maximum and minimum temperatures. They are reset by the use of a magnet.

**skip distance** The distance along the Earth's surface at which a radio transmission is undetectable. It is the distance between the limit at which the ground wave is detected and that at which the first reflection from the *ionosphere is received. It corresponds to the *zone of silence for sound waves.

**sky, state of** In *synoptic reports, the *cloud amount measured in *oktas (or tenths, in the United States), or the description in *Beaufort letters. *See also* CLOUDINESS.

**skywave radar sounding** The use of radar for the determination of oceanic surface wind and wave fields at considerable distances (1000–3000km). The outgoing radar signal is reflected down on to the ocean surface by the *ionosphere. The weak scattered echo returns along a similar path to the receiver.

**sleet** In Britain, a mixture of rain (or *drizzle) and melting snow. In the United States (not Canada), the term is more frequently employed for *ice pellets, although the British usage is occasionally encountered.

**sling psychrometer** *See* WHIRLING PSYCHROMETER.

**sling thermometer** A thermometer, mounted in a frame with a rotating handle (somewhat like a football rattle), that may be whirled in the air to obtain a rapid reading of air temperature. *See also* WHIRLING PSYCHROMETER.

**SLR** *See* SATURATED ADIABATIC LAPSE RATE.

**slush** A viscous mixture of *snow and water and, in particular, a stage in the formation of sea or lake ice, where the solid crystals may have originated as snow or *frazil ice.

**small hail** Frozen *precipitation in the form of *snow pellets that have acquired a translucent coating of ice, formed either by collision with water droplets and subsequent freezing, or by partial melting and refreezing. Generally spherical, but occasionally conical in shape and with diameters that sometimes exceed 5mm.

**smog** ['smoke' + 'fog'] A *fog that contains a high concentration of smoke or other pollutants, in particular when accompanied by unpleasant or injurious physiological effects. The term **photochemical smog** is applied to the air pollution and accompanying haze that occurs when stagnant, polluted air is subject to *photochemical reactions through the action of sunlight. *Ozone and other irritants are produced from nitrogen oxides and various hydrocarbons, primarily, but not exclusively, emitted from vehicle exhausts.

**smoke** A suspension in the atmosphere of small particles produced by combustion, typically consisting of carbon or various hydrocarbons. These particles generally have diameters of *c.*0.1 μm and normally remain suspended for some 1–2 days.

**snow** Solid *precipitation in the form of individual tiny *ice crystals when temperatures are low, or larger *snowflakes when the air temperature is near 0 °C. Snow *showers are classified according to their rate of accumulation (with no drifting or melting): slight, less than 0.5 $cm h^{-1}$; moderate, 0.5–4 $cm h^{-1}$; and heavy, more than 4 $cm h^{-1}$. *See also* SNOWFALL; SNOW GRAINS.

**snow, day of** A period of 24 hours, ending at midnight, during which snow is observed to fall.

**Snowball Earth** The controversial hypothesis that the Earth underwent a worldwide glaciation, when even tropical regions became ice covered, perhaps as many as four times between 750 and 580 $\times 10^6$ BP (during the Precambrian). The trigger for these events appears to have been a concentration of land masses in the equatorial region. (This is in contrast to other *ice ages, when some continental areas lay at high latitudes, but tropical seas remained unfrozen.) In addition to the normal polar *ice sheets, snow and ice were deposited on high ground near the Equator, giving a dramatic increase in the planetary *albedo and a runaway decrease in temperature, which may have reached as low as –50 °C. All the oceans froze, and glacial conditions persisted for 10 million years or more. Because of the global lack of *precipitation in the form of rain, the *carbon cycle was interrupted, but continuing volcanic activity caused a gradual increase in the *carbon dioxide content of the atmosphere, eventually causing abrupt global warming (*see feature*), bringing the glaciation to an end.

Evidence for these glaciations is found in massive carbonate rocks (which form under tropical conditions) immediately above major glacial deposits that *geomagnetism reveals were laid down near the Equator. The various rocks show carbon *isotope ratios consistent with a drastic, worldwide decline in biological activity during the glaciation, and a gradual recovery thereafter.

Recent research has, however, thrown doubt on some details of this hypothesis, by determining that the glaciation (although extremely extensive) began and finished very rapidly. The time-scale was too short for the tropical oceans to respond, so these therefore remained ice-free.

**snow blink** Bright, white illumination of the underside of a low cloud layer, caused by the reflection of light from a snow covered surface. It is brighter than *ice blink and readily differentiated from the darker appearance of clouds over ice-free water (*water sky) or land.

**snowdrift** Snow that has been blown by the wind from exposed areas of ground and has accumulated in the lee of natural or artificial obstructions.

**snowfall** The rate at which *snow falls, measured either directly using a graduated scale and given as a specific depth, or else obtained by melting the snow and obtaining the rainfall equivalent: 30 cm of fresh snow is approximately equal to 25 mm of rain.

**snow fence** A barrier placed some 15–20 m to windward of a road or railway line. It is generally semi-permeable to the wind and designed to create an eddy immediately behind it, where a *snowdrift will accumulate, leaving a parallel strip farther downwind relatively clear of snow.

**snowflake** An aggregate of *ice crystals, and the most common form in which *snow is deposited. Under cold conditions, the snow may consist of tiny, individual crystals, but at higher temperatures, particularly those close to freezing, numerous crystals may clump and freeze together to produce snowflakes that are several centimetres across.

**snow gauge** An instrument for determining snowfall, either by retaining the snow, which is subsequently melted to provide a suitable reading, or by weighing. The latter form of instrument is capable of providing a continuous record of the rate of fall.

**snow grains** *Precipitation in the form of small, opaque white grains of ice, normally flat or slightly elongated, with diameters of less than 1 mm. *See also* SNOW PELLETS.

**snowline** A useful but inexact term for the lower limit of perpetual snow. Although the snowline has an overall trend towards lower elevations as one moves towards the poles—it is at sea level in the polar regions—it varies considerably because of many other factors, including the actual *snowfall during the previous winter, summer temperatures, and general exposure.

**snow lying** The description used when half or more of the ground around an observing *station is covered with *snow. The land should be flat and easily visible from the station, and not differ in altitude by more than 30 m.

**snow pellets** *Precipitation in the form of opaque white particles of ice, which are approximately spherical or occasionally subconical, with diameters of 2–5 mm. *See also* SNOW GRAINS.

**snow rollers** Approximately cylindrical, natural accumulations of snow, with diameters that sometimes reach 300 mm, and a similar length, which lie at the end of meandering tracks. They arise when an old, moderately cohesive snow layer is covered by a layer of new, uncompacted snow, followed by a rise in temperature and a moderately strong wind. Small, initial clumps of snow are blown downwind, growing as they travel. A slight gradient appears to assist in the rollers' formation.

**SOA** *See* SECONDARY ORGANIC AEROSOLS.

**sodium (Na)** A metal (atomic number 11) that exists in trace amounts in the atmosphere, primarily concentrated at an altitude of approximately 85km. It has a characteristic pair of emission lines at 589.3nm that are seen in the *nightglow and *twilight glow.

**SOFAR float** A neutrally buoyant float designed to study the subsurface flow in the oceans and making use of the SOFAR (SOund Fixing and Ranging) acoustic channel, present at about 1000–1300 metres in tropical and subtropical oceans. It was originally named the Swallow float after the inventor. Floats were fitted with a piezoelectric transducer, and the sounds, which were found to propagate over hundreds of kilometres, were recorded by a moored, autonomous listening system (ALS). The system was used successfully over the whole of the Atlantic basin.

Subsequently, the system was greatly improved by introducing floats that recorded signals from fixed sources. These RAFOS floats (SOFAR spelled backwards to indicate that the direction of the signals was reversed) are much simpler, lighter, and cheaper. Three or four moored SOFAR floats are used as sources and the positions of RAFOS floats are obtained by triangulation, with an accuracy of a few kilometres. At the end of the mission, the floats drop ballast and return to the surface, and transmit their data by satellite link. Further developments have produced isopycnal floats, which remain within a specific parcel of water, regardless of its temperature or salinity; floats that follow bottom water that spills from one ocean basin to another; and floats that may be used in areas of strong *upwelling.

**soft hail** *See* GRAUPEL.

**SOI** *See* SOUTHERN OSCILLATION INDEX.

**soil moisture** The moisture content of soil that lies above the water table, including any water vapour that may be present in pores.

**soil temperature** The temperature as measured at specific depths, for example 50, 100, 200, 300, and 1000mm.

**soil thermometer** A thermometer for determining *soil temperature. Two forms are in use. One consists of a glass tube bent at right angles, so that the bulb is held at a depth of 50, 100, or 200mm, with the graduated portion of the tube remaining above and parallel with the ground surface. In the second type, for depths of 300 and 1000mm, the thermometer bulb is encased in wax within an outer tube. The whole may be withdrawn from the ground, with the wax acting to prevent any change in the reading while the observation is being made.

**solano** An easterly or south-easterly wind on the south-eastern coast of Spain, generally hot and humid, but which may be hot, dry, and dust laden. It normally originates as a *sirocco.

**solar activity** The general term for all forms of active phenomena on the Sun, including *solar flares and *sunspots. All solar activity is linked with the Sun's magnetic fields and with increases and decreases in the *solar cycle.

**solar constant** The total amount of solar radiation (over the whole electromagnetic spectrum) incident on a given area, outside the atmosphere, at the Earth's average distance from the Sun. Observations by satellites indicate that it is currently 1 368 $Wm^{-2}$, but that there is a slight variation over the *solar cycle of about 0.1–0.3 per cent. It is believed that there may be greater fluctuations over longer periods of time.

**solar cycle** A quasi-periodic variation in the number of *sunspots and other *solar activity, with an average interval between maxima of *c.*11 years. Because there is a reversal of magnetic polarity with each cycle, the true length of the underlying physical cycle is *c.*22 years. Sunspot numbers themselves are one index of variations in solar activity and energy output. Recent research has established that there is a small (*c.*0.16 deg C) variation in Earth's atmospheric temperature between maximum and minimum phases of the eleven-year cycle. This finding confirms that short-term solar fluctuations have no part to play in global warming (*see feature*) and that current models are essentially correct.

**solar day** The time required for the Earth to complete one rotation on its axis with respect to the Sun. Because of various factors, including the eccentricity of the Earth's orbit (and thus the Earth's velocity in its orbit and varying distance from the Sun) and the effect of the Sun's changing altitude, the interval between two successive transits of the Sun across the meridian varies considerably throughout the year. Civil time is therefore based on a mean solar day of 24 hours. *See* SIDEREAL DAY.

**solar flare** A sudden release of energy in the *Sun's *corona (3) that emits radiation over the whole electromagnetic spectrum. Flares also eject high-speed charged particles (mainly electrons and protons, but also some atomic nuclei), with velocities that may reach as much as 70 per cent of the speed of light, reaching Earth within 15–20 minutes. The radiation and particles interact in various ways with the Earth's *geomagnetic field, giving rise to *aurorae, *geomagnetic storms, *sudden enhancement of atmospherics, and *sudden ionospheric disturbances.

**solarimeter** A form of *pyranometer designed to measure the amount of radiation from the Sun or, with a shielded design, the diffuse radiation from the sky. There are thermal and photoelectric forms.

**Solar Radiation and Climate Experiment (SORCE)** A NASA satellite measuring X-ray, ultraviolet, visible, near-infrared, and total solar *radiation.

The observations are specifically designed to monitor short-term variability, long-term changes (particularly those related to climate), atmospheric *ozone concentrations, and levels of UV-B (*see* ULTRAVIOLET RADIATION). It continues measurements carried out by the *Upper Atmosphere Research Satellite.

**solar radiation thermometer** *See* BLACK-BULB THERMOMETER.

**solar-terrestrial relationships** The Sun exerts an influence on the Earth not only through the light and heat that it emits, but also through the *solar wind, which controls the size and shape of the *magnetosphere and the flux of *cosmic rays received by the Earth. There are thought to be secular changes in the *solar constant, and these may have long-term implications for climatic change (*see feature*). More immediate fluctuations in the *geomagnetic field are caused by variations in *solar activity, particularly *solar flares, which fluctuate in numbers and strength with the *solar cycle. Energetic particles ejected by solar flares cause *aurorae, *geomagnetic storms, and alterations in the *ionosphere.

Although numerous attempts have been made to find correlations between solar activity (as shown by *sunspot numbers, for example) and weather, no links have been reliably established until recently. It has been found that there may be a mechanism that could link solar activity and weather processes. When the Sun is active, heating causes the upper atmosphere to expand farther into space (incidentally causing the orbits of artificial satellites to decay more rapidly). The increased depth of the atmosphere alters the way in which vertical *planetary waves propagate around the globe, and may affect a coupling between the upper and lower atmosphere and thus influence weather processes. A direct link has been found between solar activity and a small variation in atmospheric temperature (*see* SOLAR CYCLE), and between *ozone levels and solar proton events (*see* SOLAR WIND).

**solar wind** The stream of charged particles from the *Sun's *corona (3), predominantly consisting of electrons and protons, with some atomic nuclei. This flow carries the Sun's magnetic field into space, and at the distance of the Earth's orbit generally has a minimum velocity of 350–400 $km\ s^{-1}$. At minima of the *solar cycle, high-speed streams are observed, which may reach 750 $km\ s^{-1}$. The interaction of the solar wind with the Earth's *geomagnetic field gives rise to the *magnetosphere. When the solar wind is highly active, the flux of *cosmic rays is reduced, leading to a drop in the production of radioactive carbon ($^{14}C$) in the atmosphere (*see* CARBON DATING). It has recently been determined that events with greatly enhanced proton fluxes (known as solar proton events, SPEs) trigger a decrease in *ozone in the *ozone layer and an increase in the upper *troposphere.

**solifluction** In *tundra and *periglacial regions, the slow downslope migration of thawed ground that overlies a *permafrost layer.

**solitary wave** A wave-like disturbance that propagates across the surface and which may intensify *convection and *precipitation as it passes. Of uncertain origin, solitary waves may be generated ahead of an advancing cold front (*see* PREFRONTAL SQUALL LINE) or other major atmospheric disturbance.

**solstice** **1.** The two points at which, in the course of a year, the Sun reaches its greatest or least *declination (i.e. is farthest above or below the *celestial equator), and thus attains its greatest or least angular distance above the horizon. **2.** The dates at which this occurs: 20 or 21 June and 21 or 22 December.

**Somali Current** A current in the northern Indian Ocean that reverses direction during the year. In the northern winter season, it is a warm current that is fed by an anticlockwise circulation in the Arabian Gulf and flows south-westwards along the coast of Somalia. With the onset of the summer *monsoon, the circulation reverses. The current flows north-eastwards from approximately 10° S, becoming a cold current with *upwelling, and the source of an important fishery.

**sonar (sound navigation and ranging)** A system for determining the distance or other properties of a remote object by the emission of pulses of sound waves and the detection of any reflected signal. Apart from well-known applications such as echo sounding, similar techniques have been applied to determining the refractive index of layers in the atmosphere, measuring ice thickness (from moored submerged buoys), and long-range thermometry of the oceans (*see* ACOUSTIC THERMOMETRY OF OCEAN CLIMATE).

**sonic anemometer** *See* ULTRASONIC ANEMOMETER.

**SORCE** *See* SOLAR RADIATION AND CLIMATE EXPERIMENT.

**sounding** A measurement of the vertical distribution of the chemical or physical properties of the atmosphere. Observations may be made by a direct method, such as *balloon or *rocket sounding, or indirectly by *topside sounding from satellites. *See also* PROFILER.

**source region** The area of the globe over which an *air mass originates. Source regions are extensive regions of relatively uniform surface conditions, which are so placed in the general circulation (*see feature*) that air may stagnate over them for a sufficiently long period to acquire specific characteristics.

**South Atlantic Current** A current, formed by the merger of the *Brazil Current and the *Falkland Current, that flows eastwards across the South Atlantic at about latitude 35–40° S. To the south it is bordered by the *Antarctic Circumpolar Current.

**South Equatorial Current** Any of three westward-flowing currents in the southern Atlantic, Indian, or Pacific Oceans that are driven by the south-easterly *trade winds. In the Atlantic, the current flows between the Equator and *c.*10° S and, on reaching South America, divides, part crossing the Equator to become the *Guiana Current, and the other portion flowing southwards as the *Brazil Current. In the Indian Ocean the flow is well south of the Equator (*c.*20° S) and becomes the *Agulhas Current. In the Pacific, the main flow is concentrated between 3° and 10° S, and a large portion forms a major *gyre in mid-ocean. In the west, some of the flow turns north along the coast of New Guinea, but most turns south to become the East Australian Current.

**southerly buster** An extremely abrupt cold front experienced on the coast of New South Wales, often accompanied by a dramatic *roll cloud (*see* ARCUS). A warm northerly flow suddenly veers to become a much colder, gale- or near-gale-force southerly. The southern end of the Great Dividing Range blocks the progress of a normal cold front from the south-west. Various factors then combine to create an intensified flow of cold air up the east coast. Often mistakenly called a southerly 'burster'.

**Southern Oscillation** An oscillation, discovered by *Walker in 1923, between the pressure in the central Pacific and that over the Indian Ocean. Its state is currently measured by the *Southern Oscillation Index: when pressure rises at Tahiti, it falls at Darwin in Australia, and vice versa. It is now known to be closely linked with *El Niño events, and the two are now discussed in terms of a single phenomenon, the El Niño Southern Oscillation (ENSO—*see feature*). *See also* TELECONNECTION.

**Southern Oscillation Index (SOI)** A measure of the state of the *Southern Oscillation. It is defined as the sea-level pressure at Tahiti minus that at Darwin in northern Australia. It is anti-correlated with the *Cold Tongue Index. *See also feature* EL NIÑO SOUTHERN OSCILLATION.

**South Pacific Convergence Zone (SPCZ)** An important zone of *convergence that exists in the southern winter in the western Pacific, extending from the area of the Torres Strait south-eastwards to approximately 30° S, 120° W. It is the zone where north-easterlies circulating around the *subtropical anticyclone over the South Pacific converge with south-easterlies. It may be regarded as a major discontinuity of the *Intertropical Convergence Zone, which at that time of year is generally located at *c.*10° N over the central and eastern Pacific.

**space charge** The electrical potential gradient that exists between the surface and higher levels of the atmosphere. This initially increases rapidly with height, but then climbs more slowly, reaching a positive potential of approximately $4 \times 10^5$ V at an altitude of *c.*15 km. There is no increase at higher levels because of their high conductivity. This large positive potential is maintained by various mechanisms associated with *thunderstorms and only partially offset by the positive charge that is carried to the ground by rain. *See also* ATMOSPHERIC ELECTRICITY.

**Spanish plume** A layer of potentially warm air (*see* POTENTIAL TEMPERATURE) that originates over the Iberian Peninsula and flows north towards France. It forms a capping *inversion that restricts small-scale convection in the lowest layers, which become extremely humid. The arrival of an eastward-moving upper *trough leads to the breakdown of the inversion and the development of *multicell or *supercell thunderstorms over France, which often cross the English Channel into southern England. Similar to the *Mexican plume in the southern United States.

**SPARC** *See* STRATOSPHERIC PROCESSES AND THEIR ROLE IN CLIMATE.

**SPCZ** *See* SOUTH PACIFIC CONVERGENCE ZONE.

**specific heat (heat capacity)** The heat required to raise the temperature of unit mass of a particular substance by one *kelvin. The specific heat of water (for example) is $4\,187\,\mathrm{J\,kg^{-1}\,K^{-1}}$.

**specific humidity** For a system of humid air, the ratio of the mass of *water vapour to the total mass of the system. Normally expressed as the number of grams of water vapour per kilogram of air.

**spectrobolometer** A *bolometer that is specifically designed to determine the intensity of incident radiation at different wavelengths, particularly in the infrared region of the spectrum.

**spectrophotometer** An instrument for determining the intensity of radiation at specific wavelengths. Both ground-based and satellite-borne instruments are used for the determination of the abundance of particular components in an atmospheric column, most notably *ozone with instruments such as the *Dobson spectrophotometer and the *Global Ozone Monitoring by Occultation of Stars (GOMOS) equipment carried by *Envisat.

**spindrift** In meteorology, dry snow and ice crystals transported by the wind. To sailors and in oceanography, persistent foam generated by breaking waves that may be blown some distance inland by the wind.

**Spinning Enhanced Visible and Infrared Imager (SEVIRI)** The advanced *radiometer carried by *Meteosat Second Generation satellites, with twelve spectral channels and giving a complete full-disk scan of the Earth every 15 minutes. The amount of data returned by SEVIRI is ten times that given by earlier radiometers.

**spissatus** A *cloud species: dense *cirrus that appears dark grey when viewed against the Sun, and which may be thick enough to hide it completely.

**splashing cirrus** A term applied to an *overshooting top when it loses its firm outline, becomes fibrous, and drops down to the layer of the main *cumulonimbus *anvil.

**splintering** The fragmentation of ice crystals during cooling, or the ejection of small crystals during the sudden freezing of a *supercooled water droplet. The process leads to a rapid increase in ice crystals at the freezing level. In addition, individual fragments acquire opposite electrical charges, and this may be part of the mechanism that creates a charge separation in *cumulonimbus clouds that then become *thunderstorms.

**split front** A form of *cold front in which air from the *cold conveyor belt descends forming the *dry intrusion. This overruns the *warm conveyor belt, creating two distinct boundary surfaces: an *upper cold front, which is primarily a moisture boundary, rather than a thermal one, and a lower 'conventional' surface cold front. Between the two fronts, the dry air overlies a **shallow moist zone** with reduced *precipitation. The upper front exhibits enhanced *convection and precipitation (often accompanied by *thunderstorms). The following surface front has less extreme activity. On satellite images there is often a clear distinction between the accompanying cloud bands. The position of the upper front is marked by a cellular structure and much higher (colder) cloud tops—appearing white on visible images—whereas the clouds on the surface front are lower and warmer—grey on visible images. Split fronts are now recognized to be relatively common.

**Spörer Minimum** A period of reduced *solar activity from about AD 1450 to 1550, which shows a decrease in the number of recorded *sunspots and, as shown by *dendrochronology, an increase in $^{14}C$ production (*see* CARBON DATING). It was similar to the better-defined *Maunder Minimum and coincided approximately with the onset of the *Little Ice Age.

**spring** *See* SEASON.

**sprite** A weakly luminous, but extensive optical phenomenon that occurs in the *mesosphere, high above the decaying region of a *thunderstorm. Lasting just a few milliseconds, sprites are barely detectable by the naked eye. Anecdotally reported for many years, they were first confirmed by low-light video techniques in 1989. There are at least three common forms, the most frequent shaped like a jellyfish, with the brightest blood-red region at about 65–70 km, a fainter red glow or wisps that may reach 90 km, and blue tendrils that may extend down to 40 km. The other types resemble a carrot, and also a simple column of light.

Sprites frequently occur in clusters up to 50 km across, and often appear at the same time as the higher *elves. They are associated with positive cloud-to-ground *lightning flashes, not the more frequent negative type. After powerful positive discharges there is an abrupt change in potential between the cloud top and the *ionosphere. This accelerates electrons upwards to levels at which they interact with air molecules, which then radiate the energy as visible light. *See also* BLUE JET; GAMMA-RAY AND X-RAY EVENTS.

**squall** An abrupt, severe increase in wind speed that lasts for some minutes and then dies away. Defined technically by the *World Meteorological Organization as a rise of at least $8\,m\,s^{-1}$ (16 knots) to a minimum of $11\,m\,s^{-1}$ (22 knots) and lasting in excess of one minute. On the *Beaufort scale, a rise of at least 3 in *force to force 6 or more. The term is often used for the *cumulonimbus cell or *thunderstorm that causes the wind, together with the accompanying heavy *precipitation, *thunder, and *lightning.

**squall line** **1.** An obsolete term for a *cold front, originally introduced by Vilhelm *Bjerknes. **2.** A non-frontal line or belt of violent convective activity, often with vigorous *thunderstorms, i.e. a particularly pronounced form of *instability line. Often precedes a cold front (*see* PREFRONTAL SQUALL LINE) and is followed by extensive *stratiform clouds and moderately heavy *precipitation. *See also* LINE SQUALL.

**squamish** A strong, often violent north-easterly or easterly wind that blows through certain fjords in British Columbia. It is a *katabatic wind, whose strength is increased by the funnelling effect of the fjords' sides.

**SSI** *See* STORM SEVERITY INDEX.

**SST** *See* SEA-SURFACE TEMPERATURE.

**stability** The property of a system such that any perturbation to the steady state will decrease in amplitude (*compare* INSTABILITY). In meteorology, it is the capability of an air parcel to return to its original level after having been forced to ascend or descend, rather than continuing its motion. A parcel will be stable if, on ascending, the *environmental lapse rate is less than the *dry adiabatic lapse rate or the *saturated adiabatic lapse rate (as appropriate), so that the parcel becomes cooler than its surroundings. If, after ascending or descending, the parcel has precisely the same temperature as the environment, it is said to have **neutral stability**. *See also* ABSOLUTE STABILITY.

**stable auroral red arc (SAR)** Relatively common, weak auroral emission from oxygen that occurs at middle latitudes and altitudes of between 300 and 700 km. SARs strengthen during geomagnetic storms, but are not detectable visually.

**stadial** A period within an *ice age when there is a major advance of the edge of the *ice sheet. *See also* GLACIAL; INTERGLACIAL; INTERSTADIAL.

**standard atmosphere** A hypothetical atmosphere, based upon the assumption of a fixed surface pressure and layers with fixed *lapse rates, so chosen that the model corresponds reasonably well with the average state of the actual atmosphere. It defines the pressure and temperature at specific heights. Various standards are used, two of the most important of which are the *ICAO standard atmosphere (used for aviation purposes), up to 20 km, and the US standard atmosphere, which is essentially identical to the ICAO standard but extends it to an altitude of 32 km. This second standard is widely used for scientific purposes to standardize instrument calibration and measurements. *See figure overleaf.*

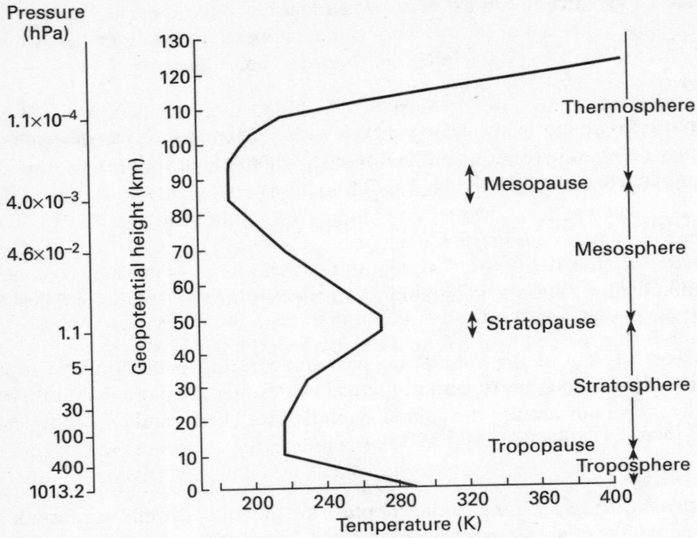


**US Standard Atmosphere** representing average, worldwide conditions.

**standard atmospheric pressure** A pressure of 1013.25hPa, used in certain physical definitions (*see* BOILING POINT; DOBSON UNIT; ICAO STANDARD ATMOSPHERE). It is the pressure exerted by 760mm of mercury of *standard density and subject to one *standard gravity.

**standard density** The density of mercury, defined as 13595.1 kg $m^{-3}$ at 0 °C, used for calibration purposes and for conversion between pressure and altitude.

**standard gravity** In meteorology, a gravitational acceleration of 9.80665 m $s^{-2}$, defined as a standard by the *World Meteorological Organization to ensure uniformity. It differs slightly from the value used in other branches of physics.

**standard temperature** The temperature at which, under one *standard gravity, the reading of a mercury barometer is correct, and defined for the purposes of calibration by the *World Meteorological Organization as 0 °C.

**standard temperature and pressure (STP)** A temperature of 0 °C and a pressure equal to 1013.25hPa (*standard atmospheric pressure).

**standing wave** A wave in which the nodes, crests, and troughs remain fixed relative to a reference point, with the fluid flowing through the wave, which does not itself migrate. Many *planetary waves are of this nature, because the topographical features that create them are stationary. *Lee waves are also often standing waves while the wind speed remains constant.

**state of ground** *See* GROUND, STATE OF.

**state of sea** *See* SEA STATE.

**state of sky** *See* SKY, STATE OF.

**station** In meteorology, a site at which weather observations are obtained at regular intervals, and including *automatic, *climatological, *rainfall, and *synoptic weather stations.

**stationary front** *See* QUASI-STATIONARY FRONT.

**station index number** A code number that identifies the meteorological *station within a specific geographical area (often identical with a political unit) for reporting purposes.

**station plot** A method of plotting past and current weather on a *synoptic chart. The symbols, codes, and numerical values that are plotted, and their relative positions around the station symbol, have been standardized by the *World Meteorological Organization, the only exception to a fixed position being the combined symbols for wind speed and direction (*see* WIND ARROW).

**steam devil** A term sometimes applied to *steam fog (Arctic sea smoke) when the columns of humid air rising from the surface of the sea exhibit distinct rotation. *See* DEVIL.

**steam fog** A low, usually fairly thin *fog that arises when water evaporates into cold air from relatively warm water and almost immediately condenses. At high latitudes, when such fog forms over areas of open water in *pack ice it is often known as **Arctic sea smoke** or, if the air is below freezing, **Arctic frost smoke.**

**steering** The influence exerted by some aspect of the state of the atmosphere (such as the pressure distribution) on the motion of some specific disturbance.

**steering flow** A fundamental flow (e.g. the *westerly *zonal flow) that exerts a major influence on the motion of some specific smaller disturbances that are embedded within it. In the case of tropical cyclones (*see feature*), the steering flow is largely responsible for their actual track.

**steering line** A term applied by Vilhelm *Bjerknes to the inner edge of the area of precipitation at a *warm front, because it apparently determined the direction of motion of the front.

**steppe** [Old Russian: 'lowland'] The treeless grassland plains of south-eastern Europe, Central Asia, and Siberia. The term is sometimes applied to somewhat similar, semi-arid regions in other continents.

**stepped leader** A faintly luminous streamer that creates the first path in a *lightning stroke. It propagates downwards from the cloud in stages, each some 50 m long, with slight pauses between each step, sending out branches as it does so. When one of the branches nears an object on the ground, a *return stroke carrying the main discharge follows the path of that branch, but in the reverse direction. *See also* DART LEADER.

**Stern Review** A report, entitled 'The economics of climate change', prepared for the British Government in late 2006 by a team under the chairmanship of Sir Nicholas Stern. Its principal finding is that in terms of both national and global economies, early action to mitigate the effects of climatic change (*see feature*) minimizes the overall economic repercussions. The cost of early action would be *c.*1 per cent of GDP, whereas delayed action could result in the loss of 10 per cent (or more) of global output. The report has been welcomed by various governments as an extremely valuable contribution for determining policies and actions.

**Stevenson screen** An instrument shelter with double-louvred sides that allow a free flow of air while protecting the instruments from direct sunlight and *precipitation. It was devised by the Scottish civil engineer, Thomas Stevenson. Although often used generically, the term should, strictly speaking, be restricted to the small screen that houses a set of thermometers, not the larger instrument screens. *See also* THERMOMETER SCREEN.

**sting jet** A zone of exceptionally strong, surface winds found in systems undergoing extreme *cyclogenesis (*see* BOMB). The high winds originate in the *cloud head of the *depression and descend towards the surface behind the depression centre in the general region of the *dry intrusion. In satellite images, the cloud head has a characteristic hooked shape (resembling a scorpion's tail and sting), and the sting jet descends where evaporation is occurring at its very tip, giving rise to the highest (potentially extremely destructive) winds in the system.

**stooping** A *refraction effect that causes distant objects to appear compressed vertically. It occurs when the lowermost layer of air has a markedly lower density than higher layers, i.e. when it is considerably warmer. Rays of light from the bottom of an object are subject to much greater refraction than normal and appear to come from higher in the sky. Higher rays are less affected. The opposite effect is known as *towering. *See also* MIRAGE.

**storm** **1.** A general term for any violent atmospheric disturbance, such as a *thunderstorm, strong *gale, *dust storm, *hail storm, *blizzard, etc. **2.** A term for an active *depression, commonly used by both the general public and forecasters in North America, especially the United States. **3.** A specific force (10) on the *Beaufort scale of wind speeds.

**storm scale** A term applied to weather systems that are of the same general order of size as an individual *thunderstorm or large *cumulonimbus, i.e. a few kilometres across. *See also* MESOSCALE; SYNOPTIC SCALE.

**Storm Severity Index (SSI)** A measure of the severity of *storms. It is calculated from a formula of the type: $V^3 \times A \times D$, where $V$ is the maximum surface wind speed (*knots), $A$ is the greatest area with damaging winds ($10^5$ km$^3$), and $D$ is the duration (hours). A system of six classes may be applied, giving (in decreasing order of severity): class I, SSI > 5000; class II, *c.*4000–1800; class III, *c.*1600–700; class IV, *c.*600–300; class V, *c.*250–150; class VI, <100.

**storm speed** The forward velocity, relative to the surface, of the centre of a tropical cyclone (*see feature*) or similar system.

**storm surge** A deviation (either positive or negative) in the height of the sea surface from normal tidal levels, sometimes known as a **storm tide** or **storm swell**. The principal cause is normally *wind stress, but low atmospheric *pressure becomes significant in raising sea levels in very deep depressions and tropical cyclones (*see feature*). Largely unnoticeable on oceanic islands surrounded by deep water, tropical-cyclone-related storm surges may be many metres high on gently shelving coasts, where they frequently overwhelm sea defences and cause widespread flooding. The effects are accentuated where geography or the accompanying winds cause water to be funnelled into a confined area (such as an estuary). The greatest recorded surge was 12.2 m, produced by a cyclone at Hatia Island, Bangladesh, on 12 November 1970.

Storm surges frequently cause great loss of life. On 1 February 1953, a depression travelled south in the North Sea, producing a surge estimated at 3.3 #x2009;m in East Anglia, where the death toll was 307, and in the Netherlands, where 1 800 lives were lost. Far greater cyclone disasters have hit the low-lying land at the head of the Bay of Bengal. Over 15 000 people died in Andhra Pradesh on 18 November 1977 when a surge 4–6 m high travelled 15 km inland, and it is estimated that about 300 000 died in the area around Bakarganj (now in Bangladesh) on the Ganges delta in 1970.

**storm swell** *See* STORM SURGE.

**STP** *See* STANDARD TEMPERATURE AND PRESSURE.

**stratiform** Clouds that occur as layers or sheets, specifically: *stratus, *nimbostratus, *altostratus, and *cirrostratus. *Stratocumulus, *altocumulus, and *cirrocumulus are sometimes regarded as having both stratiform and *cumuliform characteristics.

**stratiformis (str)** A *cloud species. An extremely extensive horizontal sheet or layer of *stratocumulus, *altocumulus, or occasionally *cirrocumulus.

**stratiform-precipitation area** A broad area to the rear of a *squall line in which there is moderately heavy *precipitation. It is separated from the leading edge of the squall line (with its very heavy precipitation) by a transition zone of lesser rainfall.

**stratocumulus (Sc)** One of the main *cloud genera. A distinct layer of low-level (*see* ÉTAGE) cloud, in the form of regular clumps or rolls, with dark shading (Plate 11). An extremely common cloud type, especially over the oceans. It may occur as the *cloud species *castellatus, *lenticularis, and *stratiformis, and in the *cloud varieties *duplicatus, *opacus, *perlucidus, *translucidus, *radiatus, and *undulatus.

**stratopause** The boundary between the *stratosphere and the overlying *mesosphere, an *inversion, at which the temperature again begins to decrease with height. Its average altitude is *c.*50 km.

**stratosphere** The atmospheric layer above the *troposphere and below the *mesosphere, in which the temperature does not decrease with height, i.e. the stratospheric *lapse rate is zero or negative. In the lower stratosphere, between the *tropopause and an altitude of *c.*20 km, there is little temperature change, but above this there is an overall increase to the *stratopause at *c.*50 km. The heating is largely the result of the absorption of solar *ultraviolet radiation by the relatively abundant *ozone. Strong stratospheric winds are often closely associated with tropospheric weather systems.

Circulation within the stratosphere occurs as a *meridional *Hadley cell with rising air over the tropics, forced by the underlying tropospheric circulation, and descending air at high latitudes in the winter hemisphere. In addition, a high-latitude westerly *Polar-Night Jet Stream may form in the winter hemisphere, as well as the easterly *Equatorial Jet Stream.

**stratospheric clouds** *See* NACREOUS CLOUDS; POLAR STRATOSPHERIC CLOUDS.

**stratospheric oscillation** *See* QUASI-BIENNIAL OSCILLATION.

**Stratospheric Processes and Their Role in Climate (SPARC)** A *World Climate Research Programme project to coordinate international studies of the *stratosphere. There are four main areas of research: influence of the stratosphere on the *troposphere and *climate; physical and chemical processes involved in the changes in stratospheric *ozone; stratospheric variability; changes in the ultraviolet flux at the Earth's surface.

**stratospheric warming** *See* SUDDEN WARMING.

**stratus (St)** One of the main *cloud genera. A predominantly grey, low-level layer cloud that is relatively featureless and has a fairly uniform base (Plate 12). If visible, the Sun or Moon appear sharply defined and do not show any optical phenomena. Stratus may produce small amounts of *precipitation, in the form of *drizzle, *snow, or even grains of ice.

Stratus is created by *orographic or general uplift (as in a *depression), but not by *convection. *Fog may be regarded as stratus at ground level, and fog that has formed overnight will often lift with daytime heating to produce a layer of low stratus. Stratus may occur as the *cloud species *fractus or *nebulosus, and the *cloud variety *opacus.

**streak lightning** *Lightning in which there is a single main discharge channel (often very tortuous) and minimal subsidiary branches. *Compare* FORK LIGHTNING.

**streamline** A line representing the path of individual particles in a horizontal flow. On a chart, a line drawn parallel to the wind direction. Only in the case of a strictly *geostrophic wind do streamlines correspond to *isobars.

**striation** A fine groove or channel in a cloud formation, formed parallel to the airflow, which thus gives an indication of the flow relative to the parent cloud. Often observed in the rotating *updraught of strong *thunderstorms, where striations may lead to a 'barber's pole' effect. *See also* CORRUGATIONS.

**sublimate** A solid that has been produced by the process of *deposition. In meteorology, the material is taken to be ice, and the process is often referred to as *sublimation.

**sublimation** The process by which a substance changes phase from a solid to a gas, without passing through a liquid phase; the opposite of *deposition. The removal of *latent heat by the vapour reduces the temperature of the solid phase. In meteorology, the term is also frequently applied to the reverse process, such as the direct formation of *hoar-frost from *water vapour.

**sublimation nucleus** A solid particle (*nucleus) that promotes the formation of an ice crystal by *sublimation from water vapour (*compare* FREEZING NUCLEUS). It is uncertain whether true sublimation nuclei actually exist in the atmosphere.

**sublime** To turn from a solid to a gas without passing through a liquid phase.

**subsidence** The slow descent of a mass of air, accompanied by *adiabatic warming and an increase in pressure. It is normally accompanied by *divergence at low levels and an increase in *stability.

**subsun** A *halo phenomenon that appears as a white, vertically elongated spot of light as far below the horizon as the true Sun is above. It is thus visible only from an aircraft or elevated position. It arises through reflection by horizontal faces of ice crystals and is, in fact, a specific form of *sun pillar. Often also known as an **undersun**. Coloured subsuns (**subparhelia, subsun dogs**) may sometimes appear 22° away from a subsun, in the same relative position as *parhelia bear to the true Sun. They probably form through internal reflection from horizontal crystal faces.

**subtropical anticyclone (subtropical high)** One of the semi-permanent regions of high pressure centred approximately at latitudes 30° N and S, but which migrate north and south slightly with the seasons in response to changes in the strength of the *polar vortices. They are extremely important *centres of action, and the oceanic cells (the *Azores high, *Bermuda anticyclone, *Pacific high, and the South Atlantic high) are strongly persistent. Over the continents, the high-pressure cells tend to be weak in summer, but are intensified in winter. They lie over the Sahara, the southern and south-western United States (known as the Great Basin or Sonoran high), and (in winter) southern Australia.

**Subtropical Jet Stream** A westerly *jet stream, located at the point where there is a steep temperature gradient in the upper *troposphere and a major break in the height of the *tropopause over the *tropics and the middle latitudes. It is very persistent and usually lies approximately above latitudes 30° N and S, i.e. above the *subtropical anticyclones. Unlike the *Polar Front Jet Stream, it is not associated with an atmospheric *front.

**subtropics** The poorly defined, transitional climatic zones that lie between the *torrid and *temperate zones. The equatorial boundaries may be taken as the Tropics of Cancer and Capricorn, and the poleward boundaries as *c.*35° N and S.

**suction vortex** A small, extremely intense, secondary vortex that occurs within a tornado (*see feature*). Multiple vortices are known to exist around the circumference of many funnels found in the most violent tornadoes, and they are thought to be responsible for the most intense damage in such events. Also known as a **suction spot**, although this term may also be applied to the distinctive trail of damage a suction vortex leaves on the ground.

**sudden enhancement of atmospherics (SEA)** A sudden increase in the number of *atmospherics detected, caused by enhanced reflectivity of *ionospheric layers, allowing the radiation from more distant *lightning discharges to be received. The increased reflectivity arises when a *solar flare ejects charged particles that subsequently produce a sudden rise in ionization in the Earth's upper atmosphere. *See also* SFERICS.

**sudden ionospheric disturbance (SID)** An abrupt increase in the absorption of radio waves in the D-region of the *ionosphere. This arises from additional ionization caused by the intense X-ray radiation emitted by *solar flares.

**sudden warming (also known as stratospheric warming)** A relatively rapid increase in temperature at an altitude of 25 km or more in the *stratosphere, associated with subsidence at those levels and occurring primarily in the northern hemisphere in winter. The rise of some 50 K normally takes place over a period of *c.*10 days (but may sometimes occur in just a few days). The effects are also felt, to a lesser degree, at lower atmospheric levels. *See also* ARCTIC OSCILLATION.

**Suess effect** The change in the ratios of carbon *isotopes (specifically, the $^{14}C : ^{12}C$ and $^{13}C : ^{12}C$ ratios) caused by release of carbon from fossil fuels.

Such materials essentially lack any $^{14}C$, because, during their burial underground, $^{14}C$ decays radioactively to $^{14}N$ (nitrogen-14). They are also strongly depleted in $^{13}C$ because of the preferential uptake of $^{12}C$, relative to $^{13}C$, by the original bacteria and plants from which fossil fuels later formed. Burning the fuels produces *carbon dioxide, whose carbon consists almost entirely of the $^{12}C$ isotope, and thus subsequently dilutes the ratios in all carbon reservoirs. *See also* CARBON CYCLE; CARBON DATING.

**sukhovei** [Russian: 'dry wind'] An extremely hot, dessicating, dusty, and predominantly easterly wind that affects the steppes of southern Russia, particularly in the region south and west of the Ural Mountains. It occurs most frequently in spring and summer and may cause extensive crop damage.

**sulphur dioxide ($SO_2$)** A colourless and very pungent gas, produced in large quantities by volcanoes. Violent eruptions may inject considerable quantities high into the *stratosphere, where it may act to cause cooling of the surface by trapping incident radiation. It may also produce the rare optical phenomenon known as *Bishop's ring. In a *photochemical reaction it is oxidized to sulphur trioxide ($SO_3$), which readily combines with water to give sulphuric acid ($H_2SO_4$). In the *troposphere the latter is an important component of *acid rain, and in the stratosphere forms the initial nuclei for *polar stratospheric clouds.

**sulphur hexafluoride ($SF_6$)** An inert compound of sulphur and fluorine, which was employed as a useful dielectric, i.e. insulator. In gaseous form it is an exceptionally powerful *greenhouse gas (24 000 times as effective at absorbing infrared radiation as *carbon dioxide), with a lifetime of 3 200 years, and as such is being phased out under the *Kyoto Protocol.

**sultriness** The combination of high temperature and high humidity.

**sumatra** A form of *squall line that crosses the Malayan Peninsula from the west during the period of the south-west *monsoon and produces heavy rain and some thunder. Sumatras appear to originate in the *convergence of *land-breezes from both sides of the Malacca Straits.

**summer** *See* SEASON.

**Sun** The principal body in the solar system, defined technically as a main-sequence yellow dwarf star of spectral type G2 and luminosity class V. It has a diameter of 1 392 530 km, an effective surface temperature of 5 770 K, and an estimated central temperature of $15.6 \times 10^6$ K. Its energy arises from the conversion of hydrogen to helium in its core, and is eventually radiated into space from the visible surface, the *photosphere. The Sun's outer atmosphere, the *corona (3), is the source of the *solar wind, particularly at the sites known as *coronal holes. *See* SOLAR–TERRESTRIAL RELATIONSHIPS.

**sun dog** A popular term, of obscure origin, for a *parhelion (mock sun).

**sun drawing water** *See* CREPUSCULAR RAYS (1).

**sun glint (sunglint)** A bright area of water in both visible and near-infrared images of the Earth, caused by the reflection of sunlight from the surface, and particularly strong when wave-heights are low. Observations of this area may reveal minor changes in reflectivity, such as those caused by oil-slicks. The intense illumination also allows suitable instruments to obtain high-precision measurements of atmospheric components, such as carbon-dioxide determinations by the *Orbiting Carbon Observatory.

**sun pillar** A *halo phenomenon consisting of a vertical streak of light passing through the true Sun. It is caused by reflection from horizontal faces of ice crystals, i.e. faces that are parallel to the ground. On rare occasions, regarded with awe in more superstitious times, it may occur with part of the *parhelic circle to give the appearance of a cross centred on the Sun. *See also* SUBSUN.

**sunrise/sunset colours** The colours seen at sunrise and sunset arise from the same effect that produces *blue sky, namely *scattering. In their path through the atmosphere the shorter wavelengths are preferentially scattered out of the line of sight. If the air is clean and dry, the rising (or setting) Sun appears yellow, but if there are numerous dust particles or tiny water droplets suspended in the air, it will appear orange or even red. Varying mixtures of scattered light at shorter wavelengths and the light illuminating them from the Sun cause the varied cloud colours that are visible. Similar considerations apply to the colours seen in the sky itself (*see* TWILIGHT ARCH). *See also* ALPINE GLOW; GREEN FLASH; PURPLE LIGHT; SHADOW OF THE EARTH.

**sunshine** Direct solar radiation, as distinct from radiation that has been diffused by clouds or the atmosphere.

**sunshine recorder** An instrument for recording the duration of sunshine. The most common form is that of a *Campbell–Stokes recorder, but electronic devices are increasingly being introduced.

**sunspot** A relatively dark area on the Sun's surface. The centre of a sunspot is at a temperature of *c.*4 500 K, and therefore appears darker than the surrounding *photosphere, which has a temperature of approximately 5 800 K. Sunspot numbers vary with the *solar cycle and act as one indicator of fluctuations in *solar activity.

**superadiabatic lapse rate** A *lapse rate that exceeds the *dry adiabatic lapse rate (*c.*10 deg C per 1 000 m). Such a lapse rate may occur immediately above a strongly heated surface, and is often encountered in the *surface boundary layer. Any air parcel with a superadiabatic lapse rate has *absolute instability.

**superballoon** *See* ULTRA-LONG DURATION BALLOON.

**supercell thunderstorm** An extremely violent, persistent *thunderstorm, characterized by the formation of an extremely large, rotating *updraught (or *mesocyclone) that extends high into the top of the *cumulonimbus cloud, which may lie at *c*.8–15 km. Cool, dry air enters the system at middle levels and is cooled still further by the *evaporation of cloud and rain droplets, causing it to descend, partly as the *rear-flank downdraught, and partly as an outflow ahead of the storm system. Where the latter encounters the ground it produces a *gust front, which acts to feed warm, humid air into the storm. An additional downdraught (the **front-flank downdraught**) often brings air from the very highest levels down to the ground in advance of the storm. Heavy rain and *hail occur in both downdraughts. The horizontal separation of the main updraught from the downdraughts is the main reason for a *supercell thunderstorm's long lifetime (6 hours or more) when compared with smaller thunderstorms or even *multicell storms.

Supercell thunderstorms are associated with extremely heavy rain, frequent cloud-to-ground *lightning, and damaging hail. The intense *precipitation often produces *flash floods, and the downdraughts themselves may be damaging outflow winds. Supercells are also the source of the most violent tornadoes (*see feature*), which tend to occur where cool, downdraught air bleeds into the mesocyclone.

**supercooling** The state in which a substance (particularly water) exists as a liquid at a temperature below its normal freezing point. This often occurs in clouds that extend above the 0 °C *isotherm, being particularly common in *altocumulus. In the absence of *freezing nuclei, supercooled droplets may exist at temperatures as low as –40 °C.

**superior air** Air at upper levels that has become extremely dry as a result of *subsidence.

**superior mirage** *See* MIRAGE.

**supernumerary bow** *See feature* RAINBOW.

**super-refraction** An abnormally rapid decrease in the refractive index with height in the atmosphere, leading to anomalous propagation of radio waves, generally marked by an increase in their range.

**super-rotation** The condition in which a planet's atmosphere rotates independently of (and much faster than) the surface. It is primarily observed on Venus, where the upper atmosphere (but not the surface layers) has a retrograde rotation with a period of *c*.4 days, compared with the surface's retrograde rotation of 243 days. The mechanism by which this occurs is not fully understood.

**supersaturation** The state in which the *relative humidity of a parcel of air exceeds 100 per cent. Because of the presence of plentiful *condensation nuclei, such conditions rarely exist in the atmosphere, although they may do so with respect to ice.

**supplementary cloud feature** A specific form that is sometimes exhibited by particular *cloud genera or *cloud species. Such features are: *arcus, *incus, *mamma, *praecipitatio, *tuba, and *virga.

**surface boundary layer** The thin, lowermost layer of the atmosphere in immediate contact with the ground, and loosely defined as the layer below *anemometer level, i.e. below 10 m. Throughout the depth of this layer, friction effects are largely constant, and the wind direction and velocity are primarily determined by the roughness and contours of the ground, and by the vertical temperature distribution. Over a hot surface, the temperature profile often shows a *superadiabatic lapse rate. *See also* PLANETARY BOUNDARY LAYER.

**surface chart** A *synoptic chart that covers a specific geographical area and presents conditions at the surface. It generally shows the pressure distribution (reduced to sea level), by the use of *isobars, with the corresponding *anticyclones, *depressions, *ridges, and *troughs, together with *air masses and *fronts. Depending on the intended use, other information may be plotted, including individual station plots, *isallobars (showing *pressure tendency), wind directions and speeds, etc. *See also* UPPER-LEVEL CHART.

**surface inversion** An *inversion that occurs at ground level, i.e. where the lowest temperature is at the surface. Such inversions are produced when warm air is *advected over a cold surface, and are also commonly encountered as *radiation inversions, caused by nocturnal cooling.

**surface temperature** The air temperature as measured inside a standard *thermometer screen or, in the case of *sea-surface temperature, measured either from a sample taken from the surface layer with an appropriate bucket, through an inlet below the waterline, or by use of a specialized radiation thermometer.

**surface wind** The wind as measured by an *anemometer or wind vane at the standard height of 10 m. The distance between the anemometer and any obstruction should be at least ten times the height of the obstruction to avoid the effect of eddies.

**surge** **1.** An abrupt increase in pressure in an area, greater than any change that may be linked to a nearby *anticyclone or *depression. Such an increase may be produced by a *squall line or *mesoscale storm system. **2.** Water transported up a beach by breaking waves. *See also* STORM SURGE.

**surge band (prefrontal cold-surge band)** A belt of thick bands of cloud, with heavy *precipitation, which may form ahead of the surface location of either a *cold or *occluded front.

**surge line** A line along which there is a sudden change in wind speed, and sometimes also of wind direction. The wind is normally, but not always, stronger on the upwind side of the surge line.

**Sutcliffe, Reginald Cockroft** (b. Wrexham, North Wales, 16 November 1904; d. High Wycombe, Buckinghamshire, 28 May 1991) British meteorologist. The author of a standard work on aviation meteorology, a field in which he was active during World War II, his research specifically concerned atmospheric dynamics. He introduced the concept that pressure, rather than altitude, was the significant vertical parameter in the study of atmospheric dynamics and the development of weather systems. He became one of the foremost dynamical meteorologists, and first Director of Research in the UK Meteorological Office. President of the Royal Meteorological Society and president and chairman of various commissions and advisory committees of the *World Meteorological Organization. Elected a Fellow of the Royal Society in 1957.

**SWEAT index** *See* SEVERE WEATHER THREAT INDEX.

**swell** Sea waves that have been generated by a disturbance at some considerable distance, and which often persist long after the original storm has died away. As they radiate away from their origin, shorter wavelengths decline, as does the amplitude of the longer waves, leaving swell's characteristic low-amplitude long waves. The direction from which they propagate is often a useful indication of the position of the parent storm. *See also* DOUGLAS SEA AND SWELL STATE.

**symbols, international meteorological** A standardized set of symbols, published by the *World Meteorological Organization, for plotting weather phenomena (including some optical and electrical effects) on *synoptic charts.

**synoptic** Affording an overall view; used in meteorology in the sense that the data employed were obtained simultaneously and thus represent the conditions that apply over a wide area at a specific point in time.

**synoptic chart** Any chart or map that presents data and analyses of the state of the atmosphere at a specific point in time, generally for a large region of the Earth. Also frequently called a **weather map**. *See also* ISOBARIC CHART; SURFACE CHART; UPPER-LEVEL CHART.

**synoptic climatology** The study of the general weather over a particular region in terms of the various synoptic situations that may occur. This may be contrasted with the more general information about *climate represented by such classifications as the *Köppen classification, which averages conditions over all synoptic situations.

**synoptic meteorology** The division of meteorology that studies current weather across the globe by means of observations made at *synoptic stations, supplemented by observations made from aircraft, ships, satellites, and *radiosondes (*upper-air observations). These synoptic observations are plotted on *synoptic charts and form the basis of all forecasting, including *numerical weather prediction. The density of the network of observing stations is such that the phenomena detected are primarily on the *planetary and *synoptic scales.

**synoptic scale** Weather phenomena that are *c*.250–2 000 km across, on the scale of continents or oceans, and thus intermediate between *planetary-scale and *mesoscale phenomena. Such phenomena include *anticyclones, *depressions (*extratropical cyclones), and tropical cyclones (*see feature*), as well as *air masses and *fronts. These features may be studied and followed reasonably well by the current network of *synoptic stations.

**synoptic station** A meteorological station at which observations are made for *synoptic meteorology and at the standard synoptic hours of 00:00, 06:00, 12:00, and 18:00 UT (*Universal Time, also known as GMT and *Zulu). Most stations also make observations at the intermediate synoptic times of 03:00, 09:00, 15:00, and 21:00 UT. These observations are reported to the national meteorological service and plotted on the overall *synoptic charts.

**synthetic aperture radar (SAR)** A form of *remote sensing used as an aircraft- or satellite-borne system. Motion of the transmitting and receiving aerial along the aircraft's path or satellite's orbit, and recording of the received radar echoes, are used to synthesize a larger apparent aerial size and thus dramatically increase surface resolution. There are many earth-sensing applications, but the main meteorological uses are the direct determination of sea-surface height relative to datum, and (in *scatterometer mode) the roughness or wave height, and hence the surface wind field.

# T

**tablecloth** A popular term for *orographic *stratus cloud that forms on the windward side of a flat-topped mountain. Originally named (most appropriately) after Table Mountain behind Cape Town, South Africa, but now often applied to other locations.

**taiga** The sub-Arctic climatic zone (the boreal forest zone) that lies in an immense swathe across the northern hemisphere, between the treeless *tundra and the boreal grasslands of Eurasia and the Canadian prairies. Overall, the zone experiences moderate amounts of rain, with extremely severe winters and peak summer temperatures of 15 °C or less. As with the tundra, there is no counterpart in the southern hemisphere, where the corresponding latitudes are occupied by the Southern Ocean. It mainly consists of vast coniferous forests with a few very hardy deciduous species, notably birch. There is a distinct *treeline on the mountains, above which trees are unable to grow, generally because of the decrease in temperature with altitude.

**tail wind** In aviation, any wind that increases an aircraft's speed over the ground. *See* HEAD WIND; EQUIVALENT HEAD WIND.

**talik** *See* PERMAFROST.

**TAO** *See* TROPICAL ATMOSPHERE–OCEAN ARRAY.

**Tehuantepecer** A violent, squally, northerly or north-easterly wind in the Gulf of Tehuantepec on the Pacific coast of southern Mexico. It originates as a *norther in the Gulf of Mexico and crosses the isthmus in the gap between the Sierra Madre del Sur and the Guatemalan mountains. *See also* PAPAGAYO.

**Teisserenc de Bort, Léon Philippe** (b. Paris, 5 November 1855; d. Cannes, 2 January 1913) French meteorologist who discovered the *stratosphere and was a pioneer in the use of instrumented meteorological balloons. He established that at an altitude of *c.*11 km the temperature ceased to fall and then began to rise. He believed that at this altitude gases would lie in distinct layers in the absence of convective mixing, and thus named the region the 'stratosphere'. He introduced the term *troposphere ('sphere of change') for the lowermost region of the atmosphere, in which major weather systems occur.

**teleconnection** A correlation between atmospheric or oceanic variables or *anomalies that occur in widely separated regions of the world. Many examples are now known, but the earliest to be discovered (in 1897) was the strong correlation between periods of drought in Australia and India. Multiple teleconnections exist related to such phenomena as the El Niño Southern Oscillation (*see feature*) and the *North Atlantic Oscillation.

**temperate zone** One of the two theoretical climatic zones that lie between the Tropic of Cancer and the Arctic Circle (23° 27′ and 66° 33′ N, respectively) and between the Tropic of Capricorn and the Antarctic Circle (23° 27′ and 66° 33′ S, respectively). *See also* FRIGID ZONE; TORRID ZONE.

**temperature** The degree of heat of a solid, liquid, or gaseous material, which is actually determined by the overall *kinetic energy of the atoms within the substance. In meteorology, it is measured by means of a *thermometer, which is mounted in a *thermometer screen to ensure consistent measurements. The *temperature scales that are employed are based upon the phase transitions of water, specifically its *triple point and *boiling point.

**temperature scales** The principal temperature scales are the *Celsius and the *Kelvin scales, which are rigorously defined and thus recommended for scientific work. The *Fahrenheit scale is utilized extensively in the United States for all forms of measurement, and remains in popular use in the United Kingdom. Various obsolete scales were once employed, the most important of which was the *Réaumur scale.

**tendency** The rate of change of a particular meteorological element, such as pressure or temperature, over a specific period of time. The term is used (in particular) in *synoptic meteorology. *See also* PRESSURE TENDENCY.

**tephigram (T–ϕ–gram)** An *aerological diagram in which the abscissa is temperature ($T$) and the ordinate is *entropy ($S$, previously symbolized by $\phi$). Introduced by William Napier *Shaw in about 1915.

**Terra** A *polar-orbiting satellite launched on 19 December 1999 by NASA and part of the Earth Observing System. The advanced instrumentation is specifically designed to obtain comprehensive data for the study of climatic change (*see feature*). It carries a series of passive sensors. *See* ASTER; CERES; MOPITT; MISR; MODIS.

**terral** The relatively weak land-breeze that occurs along the Pacific coast, west of the Andes. *See* VIRAZON.

**terrestrial gamma flashes (TGF)** *See* GAMMA-RAY AND X-RAY EVENTS.

**terrestrial magnetism** *See* GEOMAGNETISM.

**tetroon** A superpressure, *constant-level balloon in the form of a tetrahedron that is used to obtain observations in the lower atmosphere.

**thaw** A period of warmer weather that melts lying snow and ice, often associated with the transition from a *continental *air mass to a *maritime one.

**thermal** A bubble of air that is rising through the atmosphere as a result of being warmer, less dense than its surroundings, and hence buoyant. Thermals arise when localized heating of the surface produces a *superadiabatic lapse rate. They initially take the approximate form of a toroidal vortex, with air rising in a central core and descending on the outside. *Entrainment of surrounding air weakens the circulation, especially when thermals merge within a cloud, when specific *updraughts and *downdraughts tend to be established.

**thermal anemometer** A form of *anemometer in which wind speed is determined from the heat loss from a wire heated by an electrical current. The measurement may be direct, by monitoring the variation in resistance of the wire itself, as in **hot-wire** and **hot-film anemometers**. In **heat-transport** and **heat-pulse anemometers**, measurement is indirect, using separate sensors to determine the temperature of the stream of warm air carried downwind.

**thermal equator** The latitude of the highest mean annual temperature, which, because of the unequal distribution of land and sea, does not correspond to the geographical Equator. It approximates latitude 10° N, the highest temperatures being found around 15° N in the northern summer and near 5° S during the southern summer.

**thermal high** An area of high pressure, i.e. an *anticyclone, that forms when air remains stationary above a cold surface, which cools the air from below. Surface pressure rises as the air density increases during the cooling process. As with *thermal lows, the circulation is generally weak and shallow.

**thermal low** A region of low pressure, i.e. a *cyclone, that forms through strong heating of the surface. The *cyclonic circulation in thermal lows is generally weak and relatively shallow, and *fronts are absent. Thermal lows are most common in the subtropical regions in the summer season. Unlike *depressions (travelling lows), they remain over the region in which they form.

**thermal wind** The theoretical wind that blows parallel to the *thickness lines in the layer between two specific *isobaric surfaces. The name derives from the fact that low thickness values correspond to cold air, and high values to warm air. The thickness lines lie perpendicular to the temperature gradient. The *geostrophic wind at the higher level is the sum of the vectors of the lower geostrophic wind and the thermal wind.

**thermocline** A layer, in a body of water, in which the temperature gradient is greater than in the overlying or deeper layers. A thermocline is *stable and prevents vertical mixing of the water. Permanent thermoclines exist in the oceans at low and middle latitudes at depths of *c.*100–1 000 m. Changes in the depth of the thermocline in the Pacific are an important factor in El Niño Southern Oscillation events (*see feature*). Seasonal thermoclines may occur when solar heating produces a shallow surface layer of warm water.

**thermocouple** A device for determining temperature that consists of a probe containing a junction of two dissimilar metals or semiconductors, in circuit with a second similar junction that is maintained at a reference temperature. The voltage generated is proportional to the difference in temperature between the junctions.

**thermodynamic diagram** *See* AEROLOGICAL DIAGRAM.

**thermodynamics** The domain within the science of heat that deals with the transformation of energy into heat or of heat into energy.

**thermogram** **1.** A continuous record of temperature as recorded by a *thermograph. **2.** An image, as recorded by an infrared sensor, in which differences in temperature are indicated by differences in colour.

**thermograph** An instrument that provides a continuous record of ambient temperature, usually in the form of a paper chart.

**thermohaline circulation** An oceanic circulation that is driven by changes in water density (and thus *buoyancy), caused by variations in temperature and *salinity. This circulation is not completely independent of the wind-driven, surface currents, because cooling and evaporation are proportional to wind speed. Although various individual flows are of this nature, the prime example is the *Great Ocean Conveyor Belt, which, from its source in the waters around Greenland, forms a slow circulation through all the major oceans of the world.

**thermohygrograph** *See* HYGROTHERMOGRAPH.

**thermometer** An instrument for the measurement of *temperature: of two basic forms, in which either the expansion and contraction of a suitable fluid is read against a graduated scale, or the reading is obtained from the variation of electrical resistance with temperature. Mercury-in-glass thermometers may be used down to −38.87 °C (at which mercury freezes), and alcohol-in-glass instruments for still lower (minimum) temperatures. Electrical-resistance thermometers are increasingly used, and are the only form suitable for *automatic weather stations, *radiosondes, and other similar devices. *See also* BLACK-BULB THERMOMETER; MAXIMUM THERMOMETER; MINIMUM THERMOMETER; SIX'S THERMOMETER; SLING THERMOMETER; SOIL THERMOMETER; WET-BULB THERMOMETER.

**thermometer screen** The standard enclosure for meteorological thermometers, designed to protect them from direct radiation while allowing a free flow of air and thus ensuring accurate readings. The thermometers are mounted so that the bulbs are at the standard height of 1.25 m above the ground. For manned observing *stations the form known as a *Stevenson screen is normally employed, but automatic stations use a different type of enclosure, originally developed for marine use.

**thermopause** The transitional layer between the *thermosphere and the overlying *exosphere. The transition is poorly defined, but may be taken to lie at between *c.*200 and *c.*700 km, depending on *solar activity.

**thermopile** In meteorology, an instrument to measure the amount of radiation received from the Sun. It normally consists of a series of *thermocouples, arranged so that one junction receives the radiation while the other is shielded from it. Depending on the arrangement, either the voltage generated or the amount of current flowing is used to determine the amount of incident radiation.

**thermoscope** The forerunner of the *thermometer, generally held to have been invented at the beginning of the 17th century. It normally consisted of a small flask with an extremely elongated neck. The (partial) vacuum, obtained by warming the flask before inverting the neck into a reservoir, drew liquid (usually water) into the tube. The level then fluctuated with variations in temperature (and pressure). Although the height of the column may have been marked by threads or similar devices, a thermoscope differed from a thermometer in having neither a fixed, graduated, and calibrated scale nor a truly evacuated, sealed tube.

**thermosphere** The tenuous outermost layer of the Earth's atmosphere, in which the temperature increases continuously with height. It lies above the *mesopause (at *c.*86–100 km), and extends into interplanetary space. The region beyond 700 km is sometimes defined as the *exosphere.

**thickness** The difference in *geopotential height between specific pressure levels (*see* ISOBARIC SURFACE). Information is drawn from *radiosonde data and plotted on **thickness charts**, with the appropriate contours, known as **thickness lines**, joining points of equal thickness. The thickness of a given layer (1 000–500 hPa, for example) is proportional to the mean temperature, i.e. high values correspond to warm air, and low values to cold air. The theoretical wind blowing parallel to the thickness lines is known as the *thermal wind. Analysis of thickness charts is also of great assistance in determining the way weather systems are likely to develop. *See figure overleaf.*

**thoron** The name given to a specific radioactive *isotope of *radon, $^{220}Rn$, which has a half-life of 51.5 seconds. Its name derives from the fact that it is a decay product of the radioactive element thorium.

**three-front model** A model used by Canadian meteorologists to represent the situation when *maritime Arctic and *maritime polar *air masses interact along their boundary, giving rise to three distinct frontal zones across the country.

**thunder** The audible sound produced by a *lightning discharge, arising from the intense heating and expansion of the air by the electrical current. Although lightning is essentially instantaneous, and a nearby stroke will produce a short,

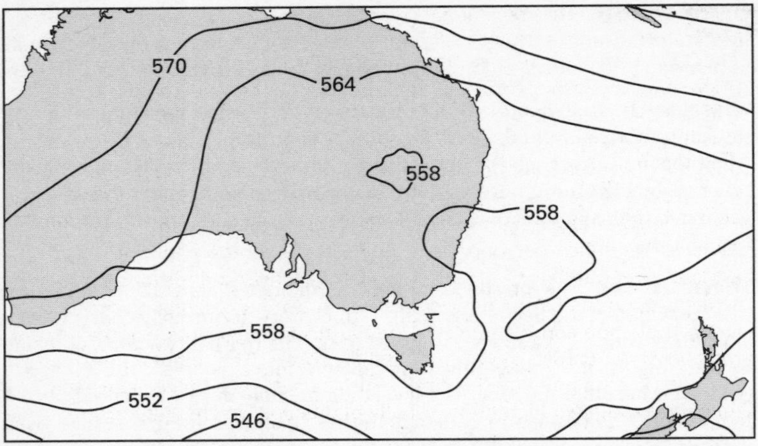


**Thickness** A chart of the thickness between 1 000 and 500 hPA. This may be compared with the charts for *geopotential height and *pressure for the same area and time.

intense crack, the sound is normally prolonged into a rumble because different parts of the discharge column are at various distances from the observer. The distance to the nearest part of the discharge may be estimated from the interval between seeing the flash and hearing the thunder: three seconds is approximately equivalent to 1 km. (For greater accuracy, multiply the interval in seconds by 330 m $s^{-1}$, the approximate speed of sound at sea level.)

Thunder is not normally heard at distances of more than 20 km, although under exceptional circumstances it has been recorded over more than 60 km. Lighting seen without audible thunder is sometimes popularly referred to as *heat lightning.

**thunderbolt** An obsolete, but persistent, popular term for a *lightning discharge that causes damage on the ground. It originates from antiquity, when it was believed that lightning strikes were material objects hurled from the clouds by mythological beings or gods.

**thundercloud** A popular name for a *cumulonimbus that contains a cell or cells producing *lightning and the associated *thunder.

**thunderstorm** A localized storm originating in a *cumulonimbus cloud, with heavy *precipitation (including *hail), *lightning, and the associated *thunder. As with *showers, thunderstorms usually consist of groups of cells, in which lightning develops at the mature stage (*see* CUMULONIMBUS). The inflow and outflow from individual cells may create an organized *multicell storm, in which two or more cells are active at any time. Under appropriate conditions, a *mesocyclone may develop, resulting in a *supercell

**Tide**

The rise and fall of sea level, normally occurring twice a day, predominantly through the action of the Moon. The value of the gravitational acceleration of the Moon (i.e. the force that acts on all portions of the Earth) varies from place to place because the points lie at differing distances from the Moon. These differential forces may be resolved into a force acting directly towards the Moon (which has little effect) and a force acting around the Earth (i.e. parallel to the surface). Although the solid Earth does flex slightly, in this context it may be treated as rigid. The water in the oceans, however, readily responds to the forces acting around the Earth and accumulates in two bulges, one on the side facing the Moon and the other on the Earth's far side. Note that the rotation of the Earth plays no part in this process. The rotation does, however, carry a point on the Earth's surface past each bulge once a day, thus producing two high tides (usually of different heights) at most locations.

The Sun exerts a similar, but lesser, influence. When the lunar and solar forces are aligned (i.e. around new and full Moon) the tide-raising forces are at a maximum, producing **spring tides**, with a maximum range between low and high water. When the forces act at right angles to one another (around first and last quarter), they partially cancel out and give rise to **neap tides**, with a reduced tidal range.

Friction between the *hydrosphere and the underlying solid surface causes the tidal bulges to feel the effect of the Earth's rotation. They therefore lie slightly in advance of (**lead**) the Earth–Moon line. For the same reason, however, as the Earth's rotation carries a point on the surface towards a tidal bulge, the Moon crosses the *meridian first, so high tide **lags** behind the Moon.

This is a simplified description of the forces at work in what are known as **equilibrium tides**. In reality, the timing of the tides is greatly complicated by the shape and size of the seas and oceans. In addition, there are various astronomical factors involved, such as the changing distance of the Moon from *perigee to *apogee, which affect the amplitude of the tidal forces. *See also* ATMOSPHERIC TIDES.

thunderstorm, the most violent and longest-lasting type of thunderstorm, associated with the production of tornadoes (*see feature*).

**thunderstorm day** A (calendar) day on which thunderstorm activity has taken place, the criterion being whether thunder has been heard from that particular site.

**tidal wave** *See* TSUNAMI.

**tilting-siphon rain recorder** A recording rain gauge, designed to provide a daily graph of rainfall versus time. A pen, linked to a float, records a trace on a continuously driven chart. When the chamber is full it overbalances, lifting the pen from the chart. The chamber drains rapidly, resetting the pen to the bottom of the chart.

**tilt of a trough** The angular departure of the axis of a *trough from a north–south line, anticlockwise rotation being defined as positive.

**timberline** The limit in latitude or altitude for the normal growth of trees. Beyond the timberline and as far as the *treeline, stunted or dwarf forms may occur.

**TIPP** *See* TRANS-IONOSPHERIC PULSE PAIR.

**tipping-bucket rain gauge** A *rain gauge that automatically determines the amount of *precipitation by means of a bucket consisting of a pair of small compartments constructed to hold a precise quantity of water. Rainfall is directed first to one compartment, which, when full (with the equivalent of 0.2mm of rain), overbalances, emptying itself of water and simultaneously bringing the other compartment into use. The rocking action actuates an electromagnetic counter to indicate overall rainfall.

**TIROS (Television and Infra-Red Observation Satellite)** An early series of *polar-orbiting satellites, launched by the United States, carrying a television camera and infrared detector. The first global cloud mosaic was constructed from TIROS data for 13 February 1965. Subsequently replaced by the more advanced *Nimbus series.

**TLE** *See* TRANSIENT LUMINOUS EVENT.

**TOGA** *See* TROPICAL OCEAN GLOBAL ATMOSPHERE.

**TOPEX-Poseidon** A satellite (a collaboration between the United States and France) launched in 1992 to study sea-surface levels by means of radar. Capable of detecting low-amplitude waves that slowly migrate across the oceans on a time-scale of months, the year-to-year variations associated with phenomena such as the El Niño Southern Oscillation (*see feature*), and the slow sea-level rise that may accompany global warming (*see feature*).

**topside sounding** The determination of various properties of the atmosphere from above by the use of instrumentation carried by artificial satellites. The term is used, in particular, for observations carried out vertically beneath the satellite, and thus normal to the surface. Also known as **remote sounding**, although this term is also applied to measurements made at angles other than normal to the surface.

## Tornado

**1.** A form of thundery *squall encountered in West Africa, particularly around the Gulf of Guinea. It often occurs as a *line squall, and is especially frequent at the beginning and end of the rainy season.
**2.** A violent, rapidly rotating column of air that extends downwards from the base of a *cumulonimbus cloud and reaches the ground. Popularly known as a **twister**. Even if a *funnel cloud (*tuba) is not visible, the presence of a *debris cloud classifies the *vortex as a tornado. Tornadoes may develop directly from a *rain-free base or from a rotating *wall cloud. The most violent tornadoes are produced by *supercell thunderstorms, which are the site of strong *mesocyclones and the resulting intense *updraughts and *downdraughts. Extremely strong non-supercell convection may give rise to weaker tornadoes, more properly described as *landspouts or *waterspouts.

The intensity of tornadoes is generally specified on the Enhanced Fujita Scale (*see* FUJITA SCALE), based on the damage caused, although in some respects the *TORRO scale has advantages, because it is based on wind speed. The record speed for a tornado ($512 km h^{-1}$) was determined by *Doppler radar in the extremely violent tornado that hit the outskirts of Oklahoma City on 3 May 1999. The pressure drop in the centre is estimated to reach 200–250 hPa, which causes the air to reach *saturation, giving rise to the visible vortex. Tornadoes have typical durations of 15 minutes, but occasionally much more; diameters of 100–2000 m; and path-lengths of 10–100km. Multiple outbreaks frequently occur, one extreme case being on 3–4 April 1974, when 148 were recorded. Some record path-lengths may actually represent a series of tornadoes forming and decaying one after another. *See also* DRY LINE; FEEDER BAND; GUSTNADO; HOOK ECHO; SUCTION VORTEX; UNDULAR BORE.

**Torricelli, Evangelista** (b. near Faenza, Italy, 15 October 1608; d. Florence, 25 October 1647) Italian natural philosopher and noted mathematician, who invented the mercury-in-glass *barometer, the crucial experiments for which were described in a letter of 11 June 1644.

**torrid zone** One of the three theoretical climatic zones, lying between the Tropic of Cancer and the Tropic of Capricorn, otherwise known as the *tropics. *See also* FRIGID ZONE; TEMPERATE ZONE.

**TORRO scale** A scale of tornado intensity (*see feature*), developed by the Tornado and Storm Research Organisation (TORRO) in the United Kingdom. Although similar in some respects to the *Fujita scale, it is defined on the basis of wind speed, rather than on the intensity of damage. It may thus be considered more appropriate when there are accurate measurements of wind speed by *Doppler radar or other methods. (Note that neither scale has been officially adopted by the *World Meteorological Organization.) *See table overleaf.*

| SCALE NUMBER | WIND SPEED | | NAME |
|---|---|---|---|
| | $m s^{-1}$ | $km h^{-1}$ | |
| T0 | 17–24 | 61–86 | light |
| T1 | 25–32 | 90–115 | mild |
| T2 | 33–41 | 119–148 | moderate |
| T3 | 42–51 | 151–184 | strong |
| T4 | 52–61 | 187–220 | severe |
| T5 | 62–72 | 223–259 | intense |
| T6 | 73–83 | 263–299 | moderately devastating |
| T7 | 84–95 | 302–342 | strongly devastating |
| T8 | 96–107 | 346–385 | severely devastating |
| T9 | 108–120 | 389–432 | intensely devastating |
| T10 | >121 | >436 | super |

Note that the scale is defined in terms of wind speeds in $m s^{-1}$, but that the speeds in $km h^{-1}$ are rounded conversions, and thus appear discontinuous.

**towering** A *refraction effect that causes distant objects to appear elongated vertically. Although there is normally a decrease in air density with height, when this is far more rapid than normal (i.e. there is a sharp increase in temperature with height), rays of light from the top of the object undergo considerably greater refraction than those from the bottom, and appear to come from higher in the sky. Sometimes incorrectly called *looming. The opposite effect is known as *stooping. *See also* MIRAGE.

**towering cumulus** *See* CONGESTUS.

**Tower of the Winds** An octagonal building at Athens, still existing, erected at some time in the period 100–50 BC by Andronicus of Cyrrhus; known originally as the *Horlogion* (Timepiece). The building was surmounted by a weathervane, with a pointer that would indicate one of the eight sides, each of which carried a sculpture personifying the wind in question. Several sides also carried sundials, and research suggests that the interior contained a water-driven device, more complex than a simple clepsydra (water-clock), that rotated continuously to indicate the current aspect of the heavens.

**trace (of rain)** In *synoptic meteorology, this term is used to indicate either that the measured amount is less than 0.05 mm and the observer is confident that this has not been left from a previous measurement; or that the *rain gauge is empty even though *precipitation is known to have fallen.

**tracer** A substance or property that may be used to track the motions of a fluid such as air or water. A fluorescent material, fluorescein, is commonly used in tracing water movements, for example. Wet-bulb *potential temperature is a useful indicator of the presence of specific *air masses.

**trade winds (trades)** The north-easterly (in the northern hemisphere) and south-easterly (in the southern) winds that flow from the *subtropical anticyclones at *c*.30–40° N and S towards the equatorial region, where they converge at the *Intertropical Convergence Zone. Because the air originates in subsidence, the poleward and eastern zones within the trade-wind belt tend to experience fine weather. The winds are particularly steady between *c*.10–20° N and S. Closer to the Equator and in the western, oceanic portions of the zones, the winds acquire more moisture, giving rise to more cloudy and showery conditions. Note that the name derives from the phrase 'to blow trade', i.e. in a constant course or fashion, not from any association with shipping or commercial trade. Also known as the *equatorial easterlies or **tropical easterlies.** *See also* ANTI-TRADES; EQUATORIAL WESTERLIES; HADLEY CELL; HARMATTAN.

**tramontana** [from Italian *tramontano*: 'from the mountains'] A northerly or north-easterly wind over north-eastern Spain, south-eastern France, the Balearic Islands, the western Mediterranean, and Italy, which is normally cold and dry. Also French *tramontane.*

**transient luminous event (TLE)** A term sometimes applied to *blue jets, *elves, and *sprites.

**trans-ionospheric pulse pair (TIPP)** A distinctive pair of intense VHF pulses, lasting a few microseconds, first detected by the ALEXIS satellite. It was suspected that their origin lay in *thunderstorms and this was subsequently confirmed by instrumentation on *FORTE. They originate in intense inter-cloud *lightning discharges, with the second pulse being the reflection of the first pulse from the Earth's surface, slightly delayed by the additional distance travelled.

**translucidus (tr)** A *cloud variety: cloud that (quite apart from any breaks) is thin enough to be translucent and to show the position of the Sun or Moon. May be applied to *altocumulus, *altostratus, *stratocumulus, or *stratus.

**transmissometer** A device for automatically measuring visibility. Most modern instruments consist of a light source and a separate detector at an appropriate distance. In another version, backscatter from a source is used to determine the particulate content of the atmosphere and hence the visibility.

**transparency** The measure of the degree to which a specific medium permits radiation to be transmitted through it. *Absorption, *reflection, and *scattering may all decrease a medium's transparency.

**transpiration** The process in which the leaves of plants lose *water vapour to the atmosphere. The term is sometimes applied to the overall process by which water is extracted from the soil by the roots, transported through the plant, and finally lost by *evaporation. *See also* EVAPOTRANSPIRATION; GUTTATION.

**treeline** The absolute limit in latitude or altitude beyond which no trees occur, not even in stunted or dwarf form. *See* TIMBERLINE.

**tree ring** The layer of wood laid down by trees during a single year, and particularly characteristic of certain species mainly found in the northern *temperate zone. The cells produced during spring have large (radial) diameters and thin walls. Later in the year, the cell walls thicken and the radial dimensions become smaller. The sharp change where late and early growth adjoin produces the visible ring. Tree rings (also known as **annual rings**) and their variations are the basis of *dendrochronology and *dendroclimatology.

**trifluoromethyl sulphur pentafluoride ($SF_5CF_3$)** A recently discovered pollutant that is an extremely active *greenhouse gas. It has an exceptionally high heat capacity (*see* SPECIFIC HEAT), some 18 000 times that of *carbon dioxide, and may have an atmospheric lifetime of thousands of years. Its origin is currently unknown, but it may be created when sulphur hexafluoride ($SF_6$), an insulator used in the electricity supply industry, reacts with the perfluorinated carbon compounds that are also commonly employed.

**triple point** **1.** The unique temperature at which all three phases (solid, liquid, and gaseous) of a substance exist in equilibrium. For pure water, the triple point is defined as 273.16 K or 0.01 °C. (Note that this is fractionally higher than the freezing or melting point, 273.15 K or 0 °C.) **2.** The point at which warm, cold, and occluded *fronts meet in a *depression. **3.** In the *tropics (2), the point at which three air masses are in contact and where the formation of tropical cyclones (*see feature*) is favoured.

**tritium** The radioactive *isotope of *hydrogen, $^3H$, the nucleus of which consists of one proton and two neutrons. It is present in trace amounts in natural sources of hydrogen, such as sea-water, with an abundance of *c.*$10^{-18}$ of that of normal hydrogen, $^1H$.

**TRMM** *See* TROPICAL RAINFALL MEASUREMENT MISSION.

**tropical air** An *air mass whose *source region is at middle latitudes, normally one of the *subtropical anticyclones. *Continental tropical air (cT) arises over the arid and desert regions and is hot and very dry. *Maritime tropical air (mT) is, however, the dominant air mass at these latitudes, and originates over tropical and subtropical oceans. It is very warm and very humid. *See also* EQUATORIAL AIR; POLAR AIR.

**Tropical Atmosphere–Ocean Array (TAO)** An array of nearly seventy moored buoys that spans the equatorial Pacific. The array is designed to monitor meteorological and oceanographical variables, with a view to understanding and predicting climatic variations in the region, particularly the El Niño Southern Oscillation (*see feature*). Sometimes (confusingly) abbreviated TOA.

**tropical climate** A climatic zone typically found in the equatorial or tropical zone and characterized by high temperatures throughout the year (i.e. with no marked 'winter' season), generally high humidity, and high precipitation, although the latter may occur in a distinct rainy season. *Köppen classification Af and Am (rainforest) and Aw and As (*savanna).

**tropical continental air** *See* CONTINENTAL TROPICAL AIR.

**tropical depression** In the *tropics or *subtropics, an organized, but non-frontal, system of convective clouds around a low-pressure centre, with closed *isobars and circulation. These systems tend to occur where there is *convergence, especially along the *Intertropical Convergence Zone. The maximum sustained (1-minute mean) wind speeds are less than $18\,m\,s^{-1}$ (approximately force 7 on the *Beaufort scale). Although most do not develop further, some may intensify sufficiently to be called a *tropical storm, and even subsequently evolve into a tropical cyclone (*see feature*).

**tropical disturbance** A system of organized convection that occurs in the *tropics or *subtropics. It is associated with a weak area of low pressure, light winds, some *precipitation, and extensive cloud cover, but is non-frontal. Such an initial disturbance may, if conditions are appropriate, eventually develop into a tropical cyclone (*see feature*), the next stage being a *tropical depression.

**tropical maritime air** *See* MARITIME TROPICAL AIR.

**Tropical Ocean Global Atmosphere (TOGA)** A major international programme for the study of the tropical oceans and their interactions with the atmosphere. One primary objective was to develop an ocean-observing system to study these large-scale interactions on both a seasonal and interannual time-scale. This has been achieved with the deployment of the *Tropical Atmosphere–Ocean Array of buoys in the equatorial Pacific.

**Tropical Rainfall Measurement Mission (TRMM)** A joint NASA and Japanese Space Exploration Agency satellite mission launched on 27 November 1997 to study the *hydrological cycle, particularly over the tropics. Notable for the discovery of distinct differences in the reflective and heat-retaining properties of clouds forming over land and over the oceans, and for being the first satellite to carry the *CERES instrument, scheduled to be operated from a series of five satellites to provide a 15-year database for climate studies. Similar observations to those made from TRMM are currently undertaken by the *Aqua and *Terra satellites.

**tropical storm** A distinct, well-organized circulation around a low-pressure core, with maximum sustained (1-minute mean) wind speeds of $18–32\,m\,s^{-1}$ (34–62 knots). The curved cloud bands are distinctly visible in satellite imagery. A tropical storm develops from a *tropical depression and may, if

## tropical cyclone

A *synoptic-scale, non-frontal, low-pressure system in the *tropics or *subtropics, with maximum sustained wind speeds (1-minute mean) greater than 33 $m s^{-1}$ (62 knots, *c*.120 $km h^{-1}$). The central pressure often reaches 950 hPa or lower (the record being 870 hPa in Typhoon Tip, 12 October 1979). The typical structure consists of bands of deep convective cells (*cumulonimbus clouds) that spiral in towards a clear central *eye, where air is descending from upper levels. Surrounding the eye is the *eye wall, which is the site of the most intense *convection and *precipitation and the highest wind speeds. The cumulonimbus towers typically reach *c*.12 km (40 000 ft) or more, and spread out as a *cirrus shield above the system, which may have a diameter of 650 km or more. (Hurricane Gilbert reached 3500 km in September 1988.)

Various conditions must be satisfied for a tropical cyclone to form; it generally originates as a *tropical disturbance and progresses through the stages of a *tropical depression and *tropical storm. The sea-surface temperature must exceed 27 °C, there must be little vertical *wind shear throughout the *troposphere, but there must be upper-level *divergence. The system must also be far enough from the Equator for the Coriolis acceleration (*see* CORIOLIS EFFECT) to operate. In general, tropical cyclones originate about 5–10° of latitude poleward of the *equatorial trough in every ocean, although only one (Catarina in March 2004) has ever been observed in the South Atlantic. A specific feature is that a tropical cyclone consists of a *vortex with a warm core, unlike other disturbances, which have cold cores. The heating arises from the release of vast amounts of *latent heat in the cumulonimbus towers in the spiral bands. The strong updraughts and downdraughts often give rise to subsidiary tornadoes (*see feature*), and the intense precipitation provides a significant fraction of total annual rainfall in many regions.

Although their tracks may be very erratic, systems typically move westwards and slightly towards the poles at speeds of *c*.10 knots. If they reach latitudes 20–30° N or S, they often show dramatic *recurvature, when the track rapidly swings round to north-east or south-east in the northern and southern hemispheres, respectively. Systems typically decay when they reach land or move over cooler waters that deprive them of their heat source. They may then become, or merge with, *extratropical cyclones (*depressions).

Tropical cyclones are known by various names: **cyclone** in the Indian Ocean and off eastern Australia; **hurricane** in the North Atlantic and eastern Pacific; and **typhoon** in the western Pacific. *See also feature* HURRICANES KATRINA, RITA, AND WILMA.

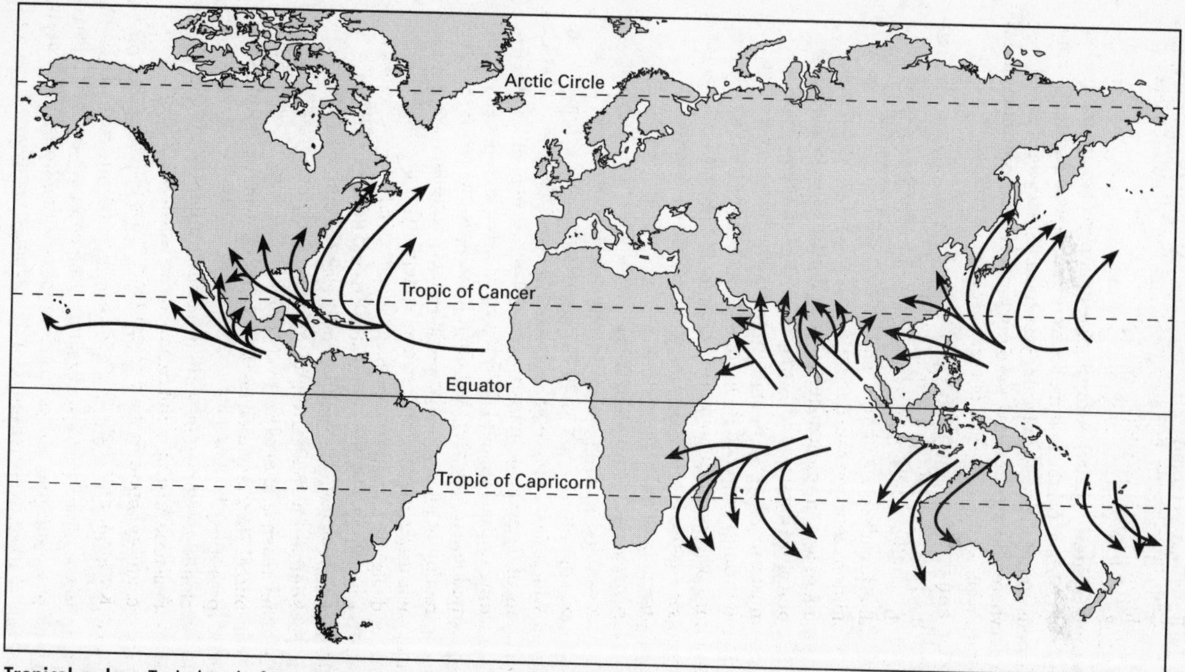


**Tropical cyclone** Typical tracks for tropical cyclones, not all of which occur in any one year. Note the strong *recurvature towards the northeast and southeast when the tracks move away from the tropics.

conditions are favourable, develop into a tropical cyclone (*see feature*). A system may be described as a **moderate** (18–25 m $s^{-1}$) or **severe** (26–32 m $s^{-1}$) **tropical storm**, force 8–9 and 10–11, respectively, on the *Beaufort scale. Sometimes termed a **tropical revolving storm**.

**tropical wave (easterly wave)** A large-scale shallow trough, more marked on upper-air than surface charts, that moves from east to west within the zone covered by the equatorial easterlies (*trade winds), and causes a marked increase in cloud and heavy *showers. An early precursor of tropical cyclones (*see feature*).

**tropics** **1.** The two lines of latitude at which the Sun is directly overhead at the *solstices: the Tropic of Cancer and the Tropic of Capricorn. **2.** The climatic region that lies between these two latitudes, sometimes called the **equatorial zone**.

**tropopause** The boundary between the *troposphere and the overlying *stratosphere, normally defined by a sudden decline in the *lapse rate to 2 deg C $km^{-1}$ or less. It is thus an *inversion. Its height, which varies with season, ranges between 14–18 km over the equatorial zone and 5–8 km over the poles (where it may be indistinct in winter). Because the tropical tropopause is so much higher, its temperature is generally much colder than the corresponding value over the polar regions.

The tropopause is by no means smooth and continuous. Abrupt changes in level and discontinuities occur at approximately latitudes 30° N and S, where there is a major temperature gradient in the upper troposphere, and also at *fronts (particularly near the *Polar Front). *Jet streams are generally located in these regions, which is also where most of the interchange of air between the troposphere and stratosphere takes place. If, above the tropopause, the lapse rate becomes 3 deg C $km^{-1}$ (or more) in an intervening layer, a second tropopause may exist at a higher level, giving rise to a double tropopause. Multiple instances also occur.

**troposphere** [Greek: 'sphere of change'] The lowermost layer of the *atmosphere, extending from the surface to the *tropopause at *c.*7 km altitude near the poles to 14–18 km at the Equator. Originally named by *Teisserenc de Bort in 1902. Most clouds, significant weather systems, and *water vapour are confined to the troposphere, which has an overall positive *lapse rate (i.e. temperature declines with height) up to the tropopause. There is extensive vertical mixing within the layer, unlike in the overlying *stratosphere. *See* ATMOSPHERE, STRUCTURE OF.

**trough** An extension of an area of low atmospheric *pressure, shown on a *synoptic chart by *isobars that are concave towards the centre of the low, and where the air flow exhibits *cyclonic curvature. The counterpart of a *ridge. A trough is always associated with a *front, but non-frontal troughs also occur and are usually marked by increased *cloudiness and *precipitation.

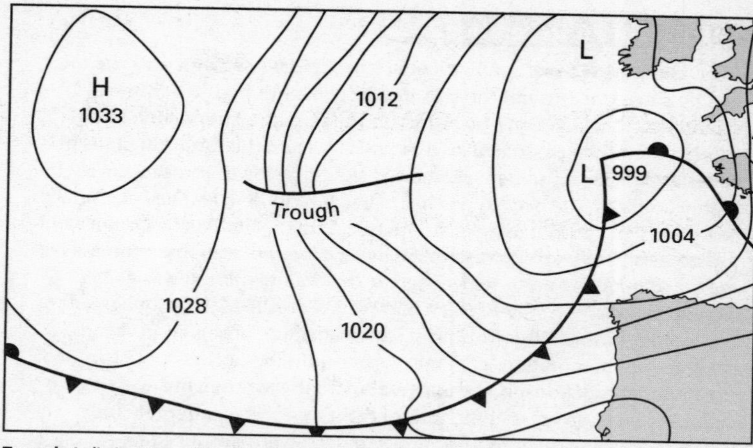


**Trough** A distinct trough line developing in the circulation behind the primary depression, which will give enhanced precipitation, and may eventually develop into a *secondary depression.

**trough line** The central axis of a trough.

**trowal** [from 'trough of warm air aloft'] A primarily Canadian term for a line on a *synoptic chart that marks the projection on to the Earth's surface of the base of the wedge of warm air aloft that is found at an *occlusion. Passage of a trowal normally results in changes in cloud type, *precipitation, and *pressure tendency.

**tsunameter** An installation for detecting *tsunami waves in the open ocean. An installation consists of a pressure sensor located on the sea-floor (typically at a depth of *c.*5 000 m) which, on detecting a tsunami wave, sends an acoustic signal to a nearby moored surface buoy, which relays the signal, via a satellite link, to the monitoring centre. A minimum of five installations are required to identify a tsunami wave, but a larger number (thirteen) are needed to give a positive identification within 30 minutes.

**tsunami** [Japanese: 'harbour wave'] A destructive wave (or more frequently, a series of waves) often many metres high, produced by violent motion of the sea-floor. The latter is itself caused by sudden displacement on a geological fault (accompanied by an earthquake), a landslide (either underwater or affecting the shoreline), or volcanic activity (through faulting or slumping of the sea-floor or, more directly, violent explosions and the effects of pyroclastic flows on the sea). Also known as a **seismic sea wave**, but the commonly used term 'tidal wave' is a misnomer, because tides are not involved.

Fault-generated tsunamis may propagate thousands of kilometres across ocean basins at velocities as high as 800 km $h^{-1}$. Wavelengths are often 150 km or more, but in mid-ocean the height may be negligible, perhaps no more than

## Tsunami of 26 December 2004

This *tsunami, with its death toll of at least 225 000, was not only the most deadly since the eruption of Krakatau in 1883, but highly significant scientifically. The rupture occurred on the Sumatra subduction zone, where the Indian tectonic plate descends beneath the Eurasian plate, and which was believed to be locked or so slow-moving that rupture was highly unlikely. The occurrence has called attention to similar, potentially hazardous locations in the North Atlantic, Indian, and Pacific Oceans and the necessity for warning systems to cover those areas. Some estimates of the accompanying earthquake suggest that its magnitude was as high as 9.3, the largest since the Alaskan earthquake of 1964. The propagation of the waves across all the world's ocean basins was observed in detail by satellite instrumentation, generally confirming the accuracy of current computer simulations, but also revealing that wave-heights and travel distances were greatest along mid-ocean ridges. The extremely high waves (exceeding 30 m in northern Sumatra) were produced by displacement of the sea-floor by as much as 8 m in places, extending for 1 200 km along the plate boundary. The rupture occurred beneath deep water, and the fault continued to slip after the initial seismic event, increasing the amount of water displaced and the consequent size of the waves. A magnitude-8.7 shock farther south on the same fault on 28 March 2005 which occurred below shallow water produced much smaller (although still destructive) waves with a run-up of *c.*4 m.

30–50 cm. As with *storm surges, they may grow extremely high when they encounter a shoreline, particularly if the latter is gently shelving. Their effects may be modified by *seiches that they generate, and which may persist for days.

The greatest tsunami ever recorded occurred in Lituya Bay, Alaska, on 9 July 1958, when a landslide caused a wave 530 m high on the opposite side of the fjord. The volcanic explosion and pyroclastic flows of Krakatau in 1883 sent tsunamis *c.*30–40 m high over the shores of Java and Sumatra, resulting in over 36000 deaths. It is now recognized that earthquake-induced slumping of offshore sediments may cause exceptionally destructive, local tsunamis, such as the *c.*14 m waves that hit part of Papua New Guinea on 17 July 1998.

Because it is ringed by tectonically active regions, the Pacific Ocean is particularly subject to tsunamis. Nations around the ocean basin participate in the Pacific Tsunami Warning System, which has its headquarters at Ewa Beach in Hawaii. Following the disastrous tsunami of 26 December 2004 (*see feature*), the initial stages of a warning system for the Indian-Ocean basin have been established. A future system may be implemented for the North Atlantic, but risks in the South Atlantic are considered to be essentially non-existent. Advance warnings are feasible only for tsunamis that propagate across ocean basins. There appears to be no practicable method of predicting tsunamis caused by nearby underwater slumping.

**tuba (tub)** [Latin: 'trumpet'] A *supplementary cloud feature consisting of a column or cone of cloud that descends from the base of a *cumulonimbus or (occasionally) *cumulus *congestus cloud. It is the central core of a vortex in which the reduced pressure causes *condensation to occur. Commonly known as a **funnel cloud** when not in contact with the surface, it may become a true *landspout or *waterspout if it does touch down.

**tundra** [from Finnish *tunturi*: 'treeless heights'] Treeless and generally low-lying plains, poleward of the *treeline, and underlain by *permafrost. Largely confined to the northern hemisphere, where the tundra occupies wide areas of the Eurasian and North American land masses. In mountainous areas, the region above the treeline is often known as **alpine tundra**, from its similarities in vegetation and general climate.

**turbidity** A measure of the degree to which an imperfectly transparent medium (such as the atmosphere) causes *attenuation of incoming solar radiation, through *absorption and *scattering. In atmospheric research, turbidity is normally taken to include the effects of *aerosols and *particulates and to exclude scattering by air molecules, i.e. *Rayleigh scattering.

**turbidity current** A deep-water, oceanic density current consisting of a dense water–sediment mixture, created by slumping of accumulated sediments. Such currents normally originate at the top of underwater canyons or other steep slopes, and are sometimes triggered by earthquakes. They give rise to coarse-grained deposits (**turbidites**), resembling shallow-water sediments, but at great depths.

**turbulence** Essentially random fluctuations in the velocities of portions of a fluid flow. The study of turbulence is a major problem in hydrodynamics and often requires the use of the most powerful supercomputers. In meteorology, it is particularly associated with the airflow over mountains, motion in the *planetary boundary layer, and phenomena such as *clear-air turbulence.

**twilight** The period of intermediate illumination before sunrise and after sunset, when the sky is not completely dark. The duration of twilight is strongly dependent on latitude and the season of the year. Three forms of twilight are defined as beginning and ending when the centre of the Sun is a specific distance below the horizon: **civil twilight**, less than 6°; **nautical twilight**, between 6° and 12°; **astronomical twilight**, between 12° and 18°.

**twilight arch** The segment of coloured light that lies above the western horizon after the Sun has set. Although often poorly defined, it is initially almost semicircular with reddish and yellow tints, but goes through numerous changes of shape and coloration, until only a narrow segment—sometimes known as the **bright segment** or **afterglow**—remains above the horizon. A similar region and comparable changes are seen in the east before sunrise. *See also* PURPLE LIGHT.

**twilight flash** An alternative name for *twilight glow.

**twilight glow** A strong enhancement of *airglow emission (primarily of the *sodium D lines at 589.3 nm) that occurs at sunrise and sunset. It may be used to determine the distribution of sodium in the upper atmosphere.

**twinkling** *See* SCINTILLATION.

**twister** *See feature* TORNADO.

**type** **1.** A term used for weather with certain characteristic features, basically defined by the pressure pattern that prevails over the area of interest. The description may be in terms of pressure (anticyclonic type, cyclonic type) or wind (westerly, easterly, etc.). **2.** A specific form of *noctilucent cloud.

**typhoon** [from Chinese *tai fun*: 'great wind'] The name given to a tropical cyclone (*see feature*) in the western North Pacific. Because of the extremely long *fetch over the warm Pacific Ocean, typhoons tend to be more intense, with lower central pressures and higher wind speeds, than tropical cyclones in other parts of the world. The largest and strongest are sometimes termed **supertyphoons**, of which Typhoon Tip is a good example. It reached a diameter of 1 680 km and had the lowest atmospheric pressure ever recorded, 870 hPa on 12 October 1979, when west of Guam.

# U

**UARS** *See* Upper Atmosphere Research Satellite.

**UKCIP** *See* UK Climate Impacts Programme.

**UK Climate Impacts Programme (UKCIP)** A programme established by the British government in 1997 to assist both public and private organizations to assess the impact of climatic change (*see feature*) and to initiate appropriate responses.

**ULDB** *See* ultra-long duration balloon.

**Ulloa's circle** *See* Bouguer's halo.

**ultra-long duration balloon (ULDB; superpressure balloons, superballoons)** A new form of long-duration, high-altitude balloon for meteorological and astronomical research. The balloons are designed to withstand daytime heating and maintain an overpressure at night. The volume of the balloon thus remains unchanging, resulting in a constant altitude. Flight duration in the upper levels of the *stratosphere is expected to be several months, so the balloons may replace satellites for certain research programmes.

**ultrasonic anemometer (sonic anemometer)** A form of *anemometer that uses several ultrasonic emitters/receivers to monitor the flow of air in the small area, a few centimetres across, between the transducers. Such anemometers are capable of determining vertical as well as horizontal motion, but are currently too expensive to be widely deployed at *synoptic or *climatological stations.

**ultraviolet radiation (UV)** The portion of the electromagnetic spectrum that lies between *c*.4 and *c*.400 nm. In physics, it is conventionally divided into four regions: near (300–400 nm); middle (200–300 nm); far (100–200 nm); and extreme (4–100 nm). Based on its effects on biological materials, a slightly different division is frequently used: **UV-A** (315–400 nm); **UV-B** (280–315 nm); and **UV-C** (100–280 nm). The last of these never reaches the Earth's surface. Significant amounts of damaging UV-B penetrate to the ground only with the growth of an *ozone hole.

**uncinus (unc)** [Latin: 'hooked'] A *cloud variety: *cirrus with a distinct hook or comma shape. The *generating head is not rounded (*cumuliform), but is either simple or in the form of a small tuft.

**undersun** *See* SUBSUN.

**undular bore** An atmospheric phenomenon, resembling a river *bore where a gravity wave propagates away from an initial disturbance, such as a line of thunderstorms. They appear to occur most commonly at an *inversion, when radiative cooling has created a stable layer near the ground. Typical wavelengths are *c.*8 km, and speeds of 16–80 km $h^{-1}$ have been recorded. As the waves travel overhead there may be a reversal of wind directions at the surface. Undular bores are suspected of influencing other atmospheric events (including thunderstorms) and, most notably, one is thought to have intensified a tornado (*see feature*) on the outskirts of Birmingham, Alabama in April 1998, causing it to become an F5 tornado (on the *Fujita Scale) and create extensive damage. *See also* MORNING GLORY.

**undulatus (un)** A *cloud variety: a layer of cloud that shows parallel undulations. The cloud may consist of either separate elements (*altocumulus, *cirrocumulus, *stratocumulus) or continuous sheets (*altocumulus, *cirrostratus). Occasionally more than one set of undulations may be present.

**Universal Time (UT)** The name by which Greenwich Mean Time is known when it is used for scientific purposes. The term was introduced in 1928. As originally defined, GMT (and thus UT) was based on the rotation of the Earth, but this was found to be non-uniform and therefore unsuitable for precision measurements. Several forms of UT are used for specific purposes, but the most common (and the one broadcast as time signals) is Coordinated Universal Time (UTC), derived by the comparison of atomic clocks kept at various observatories.

**upbank thaw** A thaw that occurs at high levels in a valley or on hills or mountains before thawing occurs lower down the slope. This frequently happens when there is warm air aloft, as when a *warm front is encroaching, or when there is forced descent (and *adiabatic warming) of air from middle levels.

**updraught** An ascending current of air occurring in a relatively restricted region of a *cumulonimbus cloud, *multicell or *supercell storm, tropical cyclone (*see feature*), or similar system. Updraughts originate when more general ascending currents in *thermals become concentrated and organized into a single flow, often in response to the development of a strong, cold *downdraught. When the latter hits the ground and fans out ahead of a storm, it may lift a significant amount of air into the developing system. Updraughts may also originate as a compensating flow to a violent downdraught, and this mechanism is thought to be important in the formation of *landspouts and *waterspouts. In supercell storms, the updraught develops into a large rotating *mesocyclone, which may then give rise to tornadoes (*see feature*).

**upper air (upper atmosphere)** Commonly used term, with no official definition, for the atmospheric layers above the lower *troposphere. Generally regarded as being the layers above 850 hPa.

**upper-air observations** *Synoptic observations from higher levels of the atmosphere. There is a worldwide network of about 500 stations that release *radiosondes twice daily at 00:00 and 12:00 GMT. These ascend at a predetermined rate to altitudes of 30 000 m or more. Observations of *pressure, *temperature, and *relative humidity are transmitted back to the surface station, which may track the balloons by radar or *radio-theodolite. Some stations release radiosondes that use *GPS or *LORAN signals for position finding. In either case, the tracking information is used to determine wind speeds and directions throughout the ascent. In future, radiosonde observations may be supplemented by those from *aerosondes.

**Upper Atmosphere Research Satellite (UARS)** A NASA satellite, active from 1991 to 2005, devoted to monitoring *ozone and various chemical compounds found in the *ozone layer, temperatures and winds in the *stratosphere, and energy input from the Sun. Atmospheric composition measurements are being continued by *Aura, and solar energy studies by the *SORCE satellite.

**upper cold front** A *cold front that is produced when the descending branch of the *cold conveyor belt (*see* DRY INTRUSION) has a greater eastwards motion than the *warm conveyor belt, which it overruns, creating an upper cold front (a *katafront), which lies ahead of the surface cold front. This leads to increased *instability, convective activity (often producing thunderstorms), and heavy *precipitation. In many cases, now known to be relatively common, this gives rise to the distinct, and important, form known as a *split front.

**upper-level chart** Any *synoptic chart that is produced for an *isobaric surface above 1000 hPa (which may be considered to represent surface conditions).

**upper-level rainband** A broad band of convective rain found in a *depression at middle and upper levels of the *troposphere. It is associated with convective cells in the *warm conveyor belt and is embedded in the widespread *precipitation on the *warm front.

**upper-level ridge (upper ridge)** A ridge of high pressure which is more marked at upper levels than at the surface. *See also* OMEGA HIGH.

**upper-level trough (upper trough)** A *trough of low pressure which is more pronounced at upper levels than at the surface.

**upshear** Opposite to the direction of maximum vertical *wind shear, generally relative to the main convective circulation (i.e. *updraught). When wind shear is present, *cumulus clouds, for example, exhibit **upshear tilt**.

Back-building *cumulonimbus *anvils extend on the upshear side. *See also* DOWNSHEAR.

**upslope fog** A type of *fog that occurs on the windward slopes of high ground. It forms when stable air is subject to *orographic lifting, being forced to rise, cools by *adiabatic expansion, and eventually reaches the *dew point. It may be distinguished from *hill fog.

**upwelling** The movement of oceanic water up to the surface from deeper in the water column. Upwelling often occurs when the prevailing wind is offshore and may then bring nutrient-rich *bottom water to the surface. This occurs, for example, under normal conditions, on the west coast of South America.

**UV, UV-A, UV-B, UV-C** *See* ULTRAVIOLET RADIATION.

# V

**valley wind** A localized *anabatic wind that flows up a valley and the surrounding slopes in daytime. It is strongest under warm, calm conditions (i.e. when a *gradient wind is absent) and tends to be weaker than the corresponding night-time *katabatic wind.

**vapour** A substance that is in a gaseous state at a temperature below its *critical point, and which may therefore be liquefied by sufficient pressure. In meteorology, the predominant such substance is *water vapour. Other similar vapours are *carbon dioxide and *sulphur dioxide, but their concentrations in the atmosphere are so low that they do not exist as liquids at the temperatures and pressures that occur.

**vapour pressure** The pressure exerted by a vapour. All liquid and solid substances give off vapour from their surfaces, in the process known as *evaporation, and this consists of the atoms (or molecules) that have sufficiently high *kinetic energies to escape. These gaseous particles then exert a vapour pressure. In an enclosed system, an equilibrium will be reached between the number of particles that enter the gaseous state and those that strike the surface of the material and re-enter it. This equilibrium will occur at a specific vapour pressure, known as the **saturation vapour pressure** (or **equilibrium vapour pressure**), which is dependent on only the temperature and the substance involved. In meteorology, vapour pressure is taken to be the portion of atmospheric *pressure that is exerted by *water vapour alone.

**varve** A pair of layers of fine and coarse silt, representing an annual cycle in the size of material deposited at different seasons. The term was originally applied to sediments found in glacial lakes, but was subsequently extended to any such annual lake deposits. Varves may be employed for absolute dating when used in conjunction with *carbon dating or *dendrochronology, or for relative dating at earlier epochs.

**vault** *See* BOUNDED WEAK ECHO REGION.

**veer** Of the wind: a clockwise change in direction (e.g. south-east to south to south-west), the opposite of *back. In the northern hemisphere the wind veers with the passage of a *front. Note that this term applies in both hemispheres. It is incorrect to use it (as sometimes occurs) to imply anticlockwise change in the southern hemisphere.

**VEI** *See* VOLCANIC EXPLOSIVITY INDEX.

**veil 1.** A form of *aurora, consisting of a very tenuous glow, lacking any structure, which may appear over a wide area of sky and as a background to other forms. **2.** The simplest type of *noctilucent cloud (Type I), consisting of thin, tenuous cloud, sometimes with faint fibrous structure. This form often precedes the main display, which usually exhibits more definite types, such as *bands (2), *billows (2), and *whirls (2).

**velum (vel)** [Latin: 'veil'] An *accessory cloud that consists of a relatively thin layer or veil of considerable horizontal extent that is draped over or pierced by *cumuliform clouds. It occurs with *cumulus or *cumulonimbus.

**vendaval (or vendavales)** A strong, blustery south-west wind in the Strait of Gibraltar and on the southern and south-eastern coasts of Spain. It occurs with *depressions from late autumn to early spring, and is often accompanied by violent *squalls and *thunder. It may be particularly strong in the Straits of Gibraltar because of the funnelling effect between the Atlas Mountains and the Spanish highlands. *See also* LEVANTER.

**veranillo** [Spanish: 'little summer'] On the Pacific coast of Central America and Mexico, the short period of dry weather that occurs in the middle of the summer rainy season.

**verano** [Spanish: 'summer'] In Central America (including Mexico), the long dry season that occurs in winter, approximately between November and April.

**verglas** *See* GLAZE.

**vertebratus (ve)** A *cloud variety, in which the cloud appears like ribs, vertebrae, or the bones of a fish (Plate 3). Mainly applied to *cirrus.

**vertical visibility** The vertical distance above the observer at which a dark object may be detected during daylight. The distance may be determined from observation of a *pilot balloon. *See also* HORIZONTAL VISIBILITY; OBLIQUE VISIBILITY.

**Very Short-Range Forecasting (VSRF)** The preparation of weather forecasts up to 12 hours ahead. The domain may be taken as overlapping with and extending *nowcasting.

**virazon** The extremely strong southwesterly *sea-breeze encountered on the Pacific coast, west of the Andean mountain chain, particularly in Chile. *See also* TERRAL.

**virga (vir)** [Latin: 'rod'] A *supplementary cloud feature that consists of a trail of *precipitation that does not reach the ground. Also known as a **fallstreak.** Virga are particularly associated with *altocumulus, *altostratus, *cirrocumulus, *cumulonimbus, *cumulus, *nimbostratus, and *stratocumulus. *Cirrus cloud, particularly cirrus *uncinus, may also be regarded as virga, although not normally recorded as such.

**visibility** The greatest distance at which an object with specific characteristics is visible to the naked eye. Ideally, the distant objects used in daytime should be black and visible against the sky. At night, lights of known intensity at known distances are used. These may be assessed with a *visibility meter. Such a visual process is now being superseded by the use of an instrument known as a *transmissometer to determine atmospheric transparency. *See also* METEOROLOGICAL OPTICAL RANGE.

**visibility meter** An instrument for determining *visibility, in particular a device that an observer uses to look at lights of known brightness at specific distances, through a graduated filter. Now generally replaced by some form of *transmissometer.

**visible infrared spin scan radiometer (VISSR)** A scanning *radiometer that is the principal instrument on the *GMS series of satellites and provides hourly, full-disk images at visible and infrared wavelengths.

**visible spectrum** The portion of the overall electromagnetic spectrum that is visible to humans. It is generally taken to run from *c.*400 nm in the violet to 700 nm in the red, although there is considerable variation between individuals, and sensitivity also depends on the intensity of the light.

**VISSR** *See* VISIBLE INFRARED SPIN SCAN RADIOMETER.

**Viviani, Vincenzo** (b. Florence, 5 April 1622; d. Florence, 22 September 1703) Italian mathematician, a pupil of Galileo *Galilei, who assisted *Torricelli in constructing the first *barometer. Together with Alfonso Borelli, he greatly improved on earlier work by Pierre Gassendi and obtained the first accurate value for the speed of sound in the atmosphere.

**VOC** *See* VOLATILE ORGANIC COMPOUNDS.

**vog** Volcanic *smog: a term originally used in Hawaii, but which has subsequently been employed elsewhere. Both gaseous emissions (such as *sulphur dioxide) and *volcanic dust may be involved, either combined with a normal *mist or *fog or else in the form of a dry *photochemical haze.

**volatile organic compounds (VOC)** A blanket term for a whole range of organic compounds of low molecular weight that are of great concern as pollutants, and which (particularly through their interaction with vehicle emissions) are responsible for much of the *ozone present at low levels. They include many solvents; propane (a common spray propellant now that *chlorofluorocarbons are banned and no longer used); compounds released during loading and unloading of petrol tankers; and substances such as formaldehyde (methanal) given off by many materials used in building construction. The European Union has set a target of an 8 per cent reduction of

VOCs by 2010, and intends to reduce emissions even further in subsequent years.

**volcanic dust** Fine particles ejected by volcanoes, the effects of which depend upon whether the eruption is powerful enough to send material into the *stratosphere, and also on the volcano's geographical position. Following an eruption, dust spreads around the globe, and that from an equatorial eruption tends to spread both north and south to extratropical zones. At middle or higher latitudes, however, the dust does not spread across the Equator into the other hemisphere.

With weak eruptions, material in the *troposphere is normally removed by *precipitation within days or weeks and any effects are generally confined to the region around the volcano. When material is injected into the stratosphere by powerful eruptions, it may persist for months or even years. By absorbing incident radiation from the Sun, it causes warming of the stratosphere and cooling of the underlying troposphere. Major episodes of volcanic activity (especially those that have produced the major volcanic provinces) have been proposed as the causes of abnormally cold years (such as 1816, known as the 'year without a summer' following the eruption of Tambora in 1815), past climatic change (*see feature*), and as triggers for the onset of *ice ages.

The dust causes optical phenomena, notably *Bishop's ring, *blue moon, and *purple light. The finest particles also act as *nuclei for sulphate *aerosols. *See also* SULPHUR DIOXIDE.

**Volcanic Explosivity Index (VEI)** A measure of the interactions between volcanic activity and climate, introduced by Newhall and Self in 1982, and based on data for 6 000 historic eruptions. It replaced the earlier *Dust Veil Index.

**VOLMET** A spoken radio forecast given in a fixed format and at specific times of the day.

**vortex** Technically, any fluid flow that possesses *vorticity or any flow that has closed *streamlines around a centre. Although in meteorology the term may be applied to phenomena on many scales, the most common usages are in the description of tornadoes (*see feature*) and similar *whirls, tropical cyclones (*see feature*), and in connection with the closed circulation around the Earth's poles (*see* POLAR VORTEX).

**vorticity** A measure of the rotation of an element in a fluid flow, in particular of air, and thus of importance in many atmospheric processes. In *synoptic-scale systems such as *anticyclones, *depressions, and tropical cyclones (*see feature*) the most significant form is rotation about a vertical axis. **Relative** (or **local**) **vorticity** is defined in a system of coordinates that is fixed to the surface, and includes *curvature and *wind shear. **Absolute vorticity** takes the rotation of the Earth into account and is thus zero at the Equator and reaches a maximum at the poles. By convention, the relative vorticity present in a *cyclone is defined as positive and that in an *anticyclone as negative.

# W

**wadi** A watercourse in North Africa and Arabia that is dry for most of the year. It may be the site of *flash floods during the rainy season.

**Walker, Sir Gilbert Thomas** (b. Rochdale, England, 14 June 1868; d. Coulsdon, England, 4 November 1958) British mathematician and meteorologist, who became Director-General of Observatories in the Indian Meteorological Service. Aware of the extreme importance of the *monsoon, he began extensive statistical studies of correlations with worldwide meteorological data. In 1922–3 he published his findings on the *Southern Oscillation. His work was not of practical use at the time, but the significance of his results has subsequently been fully appreciated. The *Walker Circulation, which he first described, is named after him.

**Walker Circulation** The large-scale, vertical, *zonal circulation in the tropics. The convection cells (**Walker cells**) are primarily driven by the *pressure gradients produced by air rising over the warmer areas of the oceans and continents, and subsiding over the cooler regions. First identified by *Walker in 1922–3 from the inverse correlation between rainfall over India and Java and pressure over the eastern Pacific Ocean. Variations in the Walker Circulation form part of the El Niño Southern Oscillation (*see feature*).

**wall cloud** A small-scale, persistent lowering from the *rain-free base of a *thunderstorm, particularly of a *supercell thunderstorm, and normally found on the inflow side of the storm (to south or south-west, in the northern hemisphere). Wall clouds range between 1 and 8km in diameter and often display rapid upward motion. In addition, some exhibit *cyclonic rotation, and these often develop into tornadoes (*see feature*). Also known as a **lowering**. The term is occasionally used instead of *eye wall.

**warm conveyor belt (WCB)** A relatively narrow flow of air that transfers significant amounts of heat, moisture, and momentum from lower latitudes towards the poles. In a northern-hemisphere *depression, for example, it flows northwards roughly parallel to the *cold front, and rises up the *warm front, producing cloud and often giving significant amounts of *precipitation. It tends to split, with a secondary conveyor belt closer to the depression centre, contributing to the *cloud head. The main warm conveyor belt finally turns towards the east and south-east, spreading a shield of *cirrus ahead of the low-pressure system.

**warm front** A *front between two *air masses, where the warmer air is advancing, and replacing the cold air at the surface. (When three air masses are involved, the warmest having been lifted away from the surface, the front is more correctly described as a warm *occlusion.) The approach of a warm front is always marked by a rapid decline in pressure. Its passage is usually accompanied by a *veer in the wind direction (in the northern hemisphere), and a rise in temperature and *dew point. Warm fronts have a typical slope of 1:100 to 1:150—roughly half that of a *cold front.

**warm-front wave** An uncommon type of *secondary depression that forms on the *warm front far ahead (1000–1 500km) of the parent *depression. Such secondaries generally move rapidly away eastwards or south-eastwards, and, although normally shallow, may cause prolonged *precipitation along the warm front.

**warm-occlusion depression** A *secondary depression that forms at the *triple point where warm and cold *fronts combine to create a warm *occlusion. Such depressions are more active than *cold-occlusion depressions and often deepen and move away from the primary as *breakaway depressions.

**warm pocket (warm pool)** An area of warm air surrounded by closed *isotherms and isolated from the main mass of warm air further towards the Equator. At the surface, such a region may exhibit instability, whereas at middle levels it tends to be stable.

**warm rain** An informal term for rain that has originated by *coalescence in clouds whose tops do not reach the freezing level. It is distinguished from *cold rain, which arises through the *Bergeron process.

**warm ridge** A *ridge in which the temperature is higher than in the immediately surrounding area.

**warm sector** The roughly triangular region that lies between the *warm and *cold fronts early in the lifetime of a *depression, where a warm *air mass is present at the surface, between the two areas of colder air. The warm sector narrows as the depression ages and the cold front catches up with the warm front, often overtaking it to produce an *occlusion. Depressions are sometimes known as **warm-sector depressions**. *See figure opposite.*

**warning** Notification that hazardous weather is imminent or has actually been reported. It is an indication that action should be taken immediately to protect lives and property.

**washout** The removal of solid particles, such as *dust, from the air through capture by *precipitation and deposition on the surface. *See* COLOURED RAIN; FALLOUT.

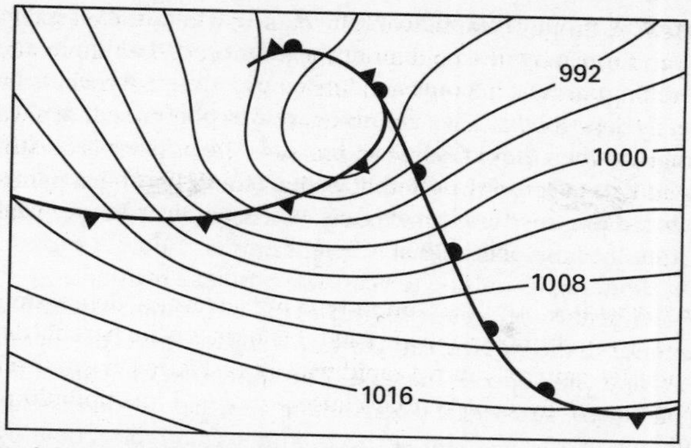


**Warm sector** In many cases, the warm sector may be extremely broad, even when the *depression has started to occlude, as here.

**watch** A notification that the conditions exist for a particular form of hazardous weather. It is the first stage of an alert, indicating the need for precautionary planning, initial preparations, and taking steps to ensure that any further information will be received.

**water** Dihydrogen oxide ($H_2O$), a substance that readily exists in solid, liquid, and gaseous phases on the Earth's surface and in its atmosphere. The exchanges of *latent heat involved in its various phase transitions are of fundamental importance in the atmospheric, oceanic, and terrestrial heat budgets, as well as being involved in many individual atmospheric processes. Pure water has its greatest density at 4 °C, its *freezing point at 0 °C, and *boiling point at 100 °C (all at *standard atmospheric pressure). *See also* DEPOSITION; EVAPORATION; SUBLIMATION.

**water cloud** A cloud that is everywhere above 0 °C and contains no ice crystals, thus consisting of water droplets only.

**watermelon snow** A colloquial term for snow that has a pale red tint caused by the presence of the single-celled alga *Chlamydomonas nivalis*. It is generally found in harsh environments at altitudes of 2 500 metres and above.

**watershed** **1.** The high ground or ridge separating the catchment areas drained by two adjacent river systems. **2.** In North American usage, the term is sometimes used for the region drained by a river system.

**water sky** In the polar regions, a darker appearance to the underside of a low cloud layer above open water, as distinct from the appearance above ice or snow. *See also* ICE BLINK; SNOW BLINK.

**water smoke** *See* STEAM FOG.

**waterspout** A rapidly rotating column of air that occurs over a river, lake or sea. Although some may be true tornadoes (*see feature*) that happen to cross a stretch of water, most waterspouts are created by a different mechanism, which also applies to *landspouts. They are much weaker phenomena and tend to dissipate rapidly when they cross on to dry land. The initial stage is marked by a **dark spot** where a strong downdraught from a parent *cumulus *congestus or *cumulonimbus reaches the water surface. Subsequently, a *funnel cloud (*tuba) descends from the base of the cloud. There is a spiral inflow at surface level, but the rotating *updraughts outside the core are normally invisible.

Beyond a critical value (about 80 km $h^{-1}$) the circulation intensifies and the funnel descends to the surface, throwing up a ring of spray. It generally shrinks in diameter, and wind speeds increase. The spray ring may become a distinct cylinder (the **bush**), superficially resembling a tornado's *debris cloud. The average lifetime of a waterspout is *c.*15 minutes, with typical diameters of 15–50 m. Heights may reach 1 000 m, but the dimensions and form are very variable, with multiple spouts often reported.

Waterspouts are most common over tropical and subtropical waters, but have been reliably reported as far north as Scotland.

**water table** The highest level attained by ground water that is free to seek its own level, i.e. water that is not confined under pressure beneath an impermeable layer of rock, such as that found in artesian basins. The water table reaches the surface at springs and in lakes, rivers, and streams.

**water vapour** The gaseous phase of water, which is the most variable component of the atmosphere at all levels (*see* AIR, COMPOSITION OF) in both space and time. It has a number of important effects: it absorbs incoming solar radiation more strongly than any other atmospheric component; it is an important *greenhouse gas; and it has a strong influence on atmospheric *stability. In addition, it enters the atmosphere through the processes of *evaporation and *sublimation, two processes that require a great deal of energy, which is stored as *latent heat. This energy is transported to different parts of the globe and released in the reverse processes of *condensation and *deposition. Similarly, the release of latent heat when water vapour condenses into cloud droplets or freezes into *ice crystals assists *convection, and is an extremely important source of energy in tropical cyclones (*see feature*), *thunderstorms, and other systems. It is, for example, responsible for the difference between the *saturated adiabatic and *dry adiabatic lapse rates.

The amount of water vapour in the air is determined with a *hygrometer, and is measured in terms of *humidity (*see* HUMIDITY MIXING RATIO; RELATIVE HUMIDITY; SATURATION; VAPOUR PRESSURE). *See also* DEW POINT; FROST POINT; SPECIFIC HEAT.

**wave clouds** Clouds that form at the crests of *lee waves created by wind flow over hilly or mountainous terrain, and which may either be in the form of relatively isolated *lenticular clouds or appear as long, smooth lines of cloud approximately parallel to the line of peaks. Depending upon conditions, wave clouds may be found at great distances downwind of the source of the waves.

**wave cyclone (wave depression)** A *depression that forms at the site of a wave-like buckle on a *front. As the wave moves along the front, the latter splits into distinct *warm and *cold fronts. *See figure overleaf. See also* POLAR-AIR DEPRESSION; SECONDARY DEPRESSION.

**wave motion** An oscillation in the motion of particles of a fluid, often with a greater or lesser degree of periodicity, and which may remain stationary or propagate downstream. In the atmosphere, waves may involve vertical or horizontal motion, or both. Waves may arise from a flow across a physical obstacle (such as airflow across a mountain range), or from instabilities along the boundary between fluids of differing density or temperature (as in the formation of a *wave cyclone on a *quasi-stationary front). *See* GRAVITY WAVE; LEE WAVE; KELVIN WAVE; KELVIN–HELMHOLTZ INSTABILITY; PLANETARY WAVE; ROSSBY WAVE.

**wave recorder** An instrument for recording ocean waves, normally designed to register the waves produced by the wind, but in some forms of instrument (*tsunameters) specifically constructed to record *tsunamis.

**WCB** *See* WARM CONVEYOR BELT.

**WCP** *See* WORLD CLIMATE PROGRAMME.

**WCRP** *See* *WORLD CLIMATE RESEARCH PROGRAMME.

**weakening** A decrease with time of the overall *pressure gradient around the centre of a system, such as an *anticyclone or *depression. The converse is *intensification.

**weather** The state of the atmosphere, particularly with regard to its immediate effects upon human affairs, plants, animals, and, to a lesser extent, upon inanimate objects and processes. Of particular significance are short-term changes in various parameters—such as *temperature, *humidity, wind (*see feature*), *pressure, etc.—which occur over periods of minutes to a few days or weeks, as compared with the longer-term fluctuations that define the *climate.

**weather lore** Folk sayings and rhymes, based partly on empirical experience but more largely on false correlations and superstition, which purport to predict forthcoming weather. In general, those that are related to current conditions may be valid in the short term (i.e. over the next few hours), and have a certain scientific basis. Items of lore that purport to predict conditions for forthcoming months or seasons may be shown to have no real validity.

**weather map** *See* SYNOPTIC CHART.

**weather minima** The meteorological criteria, as defined by aircraft operators, which must be met for a specific aircraft to use a particular airport.

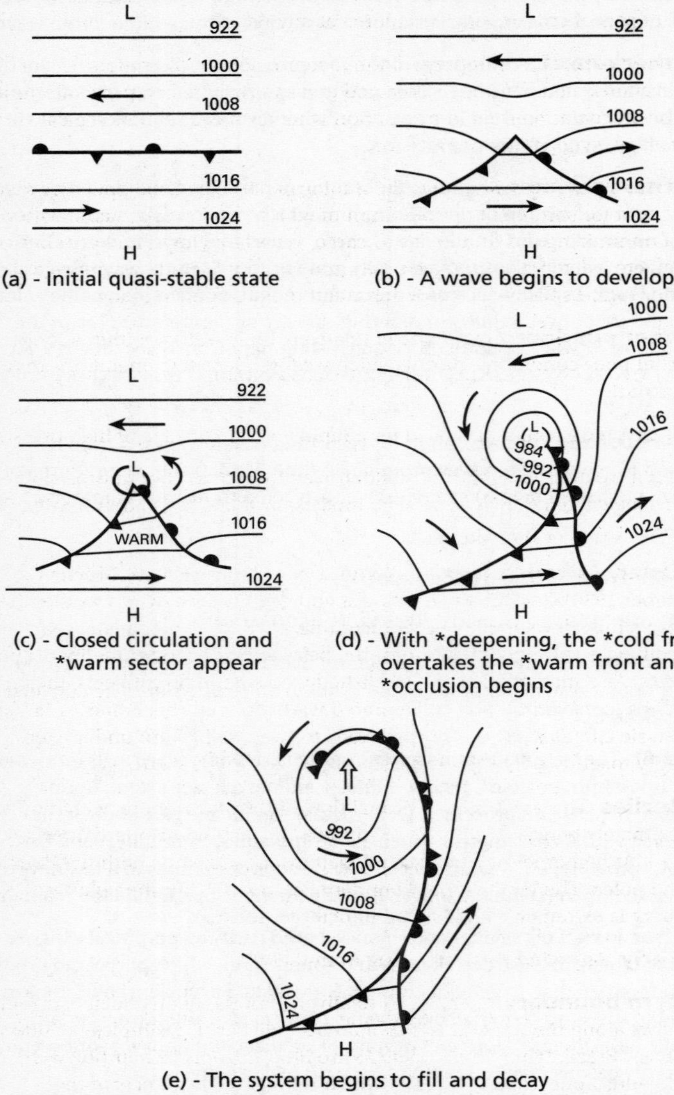


(a) - Initial quasi-stable state

(b) - A wave begins to develop

(c) - Closed circulation and *warm sector appear

(d) - With *deepening, the *cold front overtakes the *warm front and *occlusion begins

(e) - The system begins to fill and decay

**Wave cyclone** Development and evolution of a typical wave cyclone on the Polar Front. High pressure (H) and warmer air lie to the south. Arrows represent the *geostrophic flow.

Because of differences in aircraft performance and instrumentation, various different sets of criteria may be in force at any one time.

**weather report** A summary of the meteorological parameters (i.e. of *observations) that have been obtained at a specific site at a particular time. The specific data reported depend upon whether it is a *climatological, *rainfall, or *synoptic station.

**weather routing** The provision of information about the optimum route to be taken for shipping, the relocation of oil rigs, and similar tasks, with the aim of minimizing risk of damage to cargo, vessels, or tows, and economizing on fuel, etc. Factors that may be taken into account include wave height and direction, wind speeds, ice conditions, and the occurrence of fog.

**weather window** An interval of relatively fine weather within a longer period of poor conditions, during which weather-sensitive operations may be carried out.

**wedge** A term sometimes used for a rapidly moving *ridge of high pressure.

**Wefax** ['weather facsimile'] A standardized format for the transmission of meteorological data, such as satellite images, by analogue, rather than digital, methods.

**Wegener, Alfred Lothar** (b. Berlin, 1 November 1880; d. Greenland, November 1930) German meteorologist and geophysicist, who was the first to propose a fully developed theory of *continental drift, which, many years after his death, was subsequently to form the basis for the modern theory of *plate tectonics. As a meteorologist he published works on many subjects, including tornadoes (*see feature*), and collaborated with *Köppen (his father-in-law) on prehistoric climates. His four expeditions to Greenland were undertaken primarily to study polar *air masses.

**westerlies** The predominant *zonal flow in the atmosphere, which occurs approximately between latitudes 35° and 65° in each hemisphere (at the surface). At higher levels, the westerly belt extends towards both the Equator and the poles. The low-latitude boundary lies at *c*.30°, but the poleward boundary is extremely variable and difficult to define.

**westerly** Any wind that is blowing from the west.

**western boundary current** A relatively narrow, high-velocity current that flows along the western edges of an oceanic basin. Examples include the *Agulhas, *Brazil, *Gulf Stream, and *Kuro Shio Currents. Their poleward motion adds anticyclonic *vorticity which reinforces the effects of the prevailing winds, resulting in their being considerably stronger and faster than the corresponding eastern boundary currents.

**West Wind Drift** *See* ANTARCTIC CIRCUMPOLAR CURRENT.

**wet adiabatic** *See* SATURATED ADIABATIC LAPSE RATE.

**wet air** A term for the conditions prevailing when a cold, dry *air mass is replaced by one that is warm and either *saturated or nearly saturated. Air in contact with the surface or objects is cooled to the *dew point, often giving rise to significant quantities of *condensation.

**wet-bulb depression** The amount by which the wet-bulb reading in a *psychrometer is less than that shown by the dry bulb.

**wet-bulb potential temperature** The temperature that a parcel of air at any level would have if, starting at the *wet-bulb temperature, it were brought at the *saturated adiabatic lapse rate to the standard pressure of 1000 hPa. It may be determined simply from an *aerological diagram. (*See also* POTENTIAL TEMPERATURE.) This property of an air parcel is of particular value, because it is effectively conserved (i.e. does not change) through *condensation, *evaporation, and both dry adiabatic and saturated *adiabatic temperature changes. It may thus be used as a *tracer of particular *air masses.

**wet-bulb temperature** The temperature a parcel of air would have if it were cooled *adiabatically (at constant pressure) to *saturation by the evaporation of water into it, with the *latent heat being supplied by the parcel. This is essentially the same as the temperature indicated by a wet-bulb thermometer.

**wet-bulb thermometer** The thermometer in a *psychrometer that is covered by a muslin sleeve moistened with distilled water, and which records the *wet-bulb temperature.

**wet season** *See* RAINY SEASON.

**whirl** **1.** A general term for various forms of rapidly rotating vortex, covering a wide range of strengths and wind speeds, including *devils, *gustnadoes, *landspouts, tornadoes (*see feature*), and *waterspouts. Various methods of formation occur. The weakest whirls arise from the funnelling effects of local topography (as in snow devils and water devils); vigorous convection from intense heating that creates a *superadiabatic lapse rate initiates some *dust devils; strong downdraughts are involved in gustnadoes, landspouts, and waterspouts; and a wide range of factors combine to produce violent tornadoes. **2.** A form of *noctilucent cloud (classified as Type IV), consisting of arcs or (rarely) rings of cloud with dark centres.

**whirling psychrometer (also known as a sling psychrometer)** A *psychrometer in which the pair of thermometers are mounted in a frame that may be whirled in the air (like a football rattle) to obtain rapid measurement of the dry- and *wet-bulb temperatures.

**whirlwind** A term popularly (and indiscriminately) applied to any form of *whirl (1), *gustnado, *landspout, *waterspout, or tornado (*see feature*).

**whistler** A radio disturbance that is audible as a repeated set of falling tones. It arises when an electromagnetic pulse created by a *lightning discharge travels along the lines of the Earth's magnetic field to the conjugate point in the opposite hemisphere, where the pulse is reflected backwards. It may complete several round trips before its intensity becomes too low to be detectable. The higher frequencies travel faster than the lower ones, and are thus detected earlier, leading to the audible descending tone.

**white-out** A condition that occurs in polar regions, when all visual clues to direction and distance are lost, leading to a dangerous state of disorientation. When the sky is uniformly *overcast and snow cover is complete, the horizon, clouds, and the surface itself cannot be distinguished. An equal amount of light is received from all directions, so no shadows are cast and all depth perception is lost. Contrary to popular belief, a white-out does not necessarily involve *blizzard conditions, i.e. blowing snow.

**whumpf** [onomatopoeic] A colloquial term for the sudden collapse of weak layers within a snow pack on level terrain. Failure of the lower, fragile layers allows the overlying surface to drop, with a characteristic sound. Although the vertical motion may be small (a few millimetres) it is often sufficient to affect a large area, several hundred metres or even kilometres across, and, if it reaches steep terrain, may trigger an *avalanche.

**williwaw** A sudden *gust of wind that descends from mountains bordering the sea; the term is particularly applied to conditions in the Strait of Magellan and also in the Aleutian Islands.

**will-o'-the-wisp** *See* IGNIS FATUUS.

**willy-willy** **1.** An Australian term for *dust devil. **2.** In early Australian accounts, a commonly used (but now obsolete) term for a tropical cyclone (*see feature*).

**Wilson, Charles Thomson Rees** (b. Glencorse, Scotland, 14 February 1869; d. Carlops, Scotland, 15 November 1959) British meteorologist and physicist. To study clouds, Wilson constructed a chamber that, when full of saturated air, could be expanded, thus modelling the natural expansion and cooling involved. He discovered that *condensation nuclei were essential for forming cloud droplets, and that in the absence of nuclei, charged particles would initiate droplet formation in *supercooled water vapour. The cloud chamber became an important tool in particle physics, and earned Wilson the Nobel Prize in 1927. His later studies concentrated on *lightning and all aspects of *atmospheric electricity, suggesting that the downward transfer of charge in *thunderstorms was balanced by an upward flow of current to the *ionosphere. The discovery of *blue jets, *elves, and *sprites decades later confirmed this theory.

## Wind

The horizontal motion of air across the surface of the Earth, described, by convention, as the direction from which the wind is blowing. Any vertical component is normally much smaller and of lesser importance, except in certain specific instances. By international agreement, wind speeds are measured in *knots (kt), although they are also commonly given in terms of the *Beaufort scale and in physical units, particularly metres per second ($m\,s^{-1}$). The instrument for obtaining wind speeds is known as an *anemometer.

A wind arises as a result of a *pressure gradient, which itself normally originates from a temperature (and thus density) difference between different regions. There is a close relationship between the value of the pressure gradient and the resulting wind velocity. Globally the overall pattern of winds is known as the general circulation (*see feature*). *See figures overleaf.* At large scales, the relationship between pressure and wind speed gives rise to *geostrophic and *gradient winds, and also to *Buys Ballot's law. In the *planetary boundary layer, the changes in wind speed and direction are related to surface *friction and, at higher levels, to the *thermal wind.

On a smaller scale, various local processes operate to create *anabatic and *katabatic winds, *land- and *sea-breezes, and *mountain, *mountain-gap, *ravine, and *valley winds. *See also* CORIOLIS EFFECT; EKMAN EFFECT; JET STREAM; LOCAL AND NAMED WINDS.

**wind arrow (wind-direction shaft)** A symbol used for *station plots on a *synoptic chart to indicate wind direction and strength. The symbol consists of a **shaft**, drawn upwind of the station symbol, its direction representing the wind direction to the nearest 10°. At the outer end of the shaft, **feathers** (also known as **barbs**) or **pennants** are added to indicate wind speed. Modern plots show speeds in multiples of 5 *knots. A feather consists of a short oblique stroke at the far end of the shaft, drawn on the side nearest low pressure (i.e. on the left looking towards the station in the northern hemisphere, and on the right in the southern—*see* BUYS BALLOT'S LAW). A half-length feather indicates 5 knots; a full-length one 10 knots; and a pennant (a small triangular 'flag') is used for 50 knots. Originally a half-length feather indicated a Beaufort force of 1 (*see* BEAUFORT SCALE), and a full feather a Beaufort force of 2. The term 'barb' is the commonly used form in North America.

**windbreak** An artificial barrier to the wind, erected to provide protection to objects on the *leeward side. This may be designed to act directly by reducing the force of the wind (as with barriers to protect exposed roads or *rain gauges), or indirectly by altering the wind flow (as with a *snow fence, which causes snow to be deposited away from the road or railway being protected).

**wind-chill** That portion of a body's heat loss that is caused through cooling by the wind. The calculation of the amount of wind-chill is complex and various formulae are used. For the guidance of the general public, the results may be expressed either as a *wind-chill effective temperature or as the more technical *wind-chill factor.

**wind-chill effective temperature (WET)** An air-temperature index that is an approximation of the loss of *sensible heat from exposed skin through a combination of low temperatures and wind. It is defined, and quoted, as the still-air temperature that would give the same heat loss (*wind-chill factor) as exposure to the ambient air under the prevailing wind conditions.

**wind-chill factor** A measure of the rate at which warm objects are cooled by the wind until they reach the ambient air temperature, measured in watts per square metre ($W\ m^{-2}$). The actual numerical values are given in *forecasts by some meteorological services, such as Environment Canada.

**wind profiler** *See* PROFILER.

**wind rose** A diagram that shows the frequency of occurrence of the wind from various directions at a particular point, normally averaged over a year or more. A common variant also displays the range of wind speeds for each of the directions shown.

**wind shear** A difference in wind velocity at neighbouring points. Three cases may be distinguished: **vertical wind shear**, where the wind speed varies with height; **horizontal** or **lateral wind shear**, where the speed varies with horizontal distance, measured at right angles to the wind direction; and **updraught** or **downdraught shear**, where the vertical component varies with horizontal distance.

Vertical wind shear is involved in *clear-air turbulence, the formation of *billows, and Kelvin–Helmholz waves (*see* KELVIN–HELMHOLTZ INSTABILITY). Lateral wind shear is encountered near *jet streams, for example; and shear caused by vertical motion of the air is a major factor in *downbursts and tornadoes (*see feature*).

**windsonde** A term sometimes used for a *radiosonde, specifically designed for upper-wind determinations, which returns pressure observations only (not temperature or humidity).

**wind stress** The force on the land or sea surface generated by the air moving in the *surface boundary layer.

**wind vane** A device for indicating the direction of the wind with reference to the principal compass points. Indication may be either direct for visual inspection, or remote by means of suitable encoders and read-out devices.

**windward** Towards, or on the side from which the wind is blowing; upwind. *Compare* LEEWARD.

**winter** *See* SEASON.

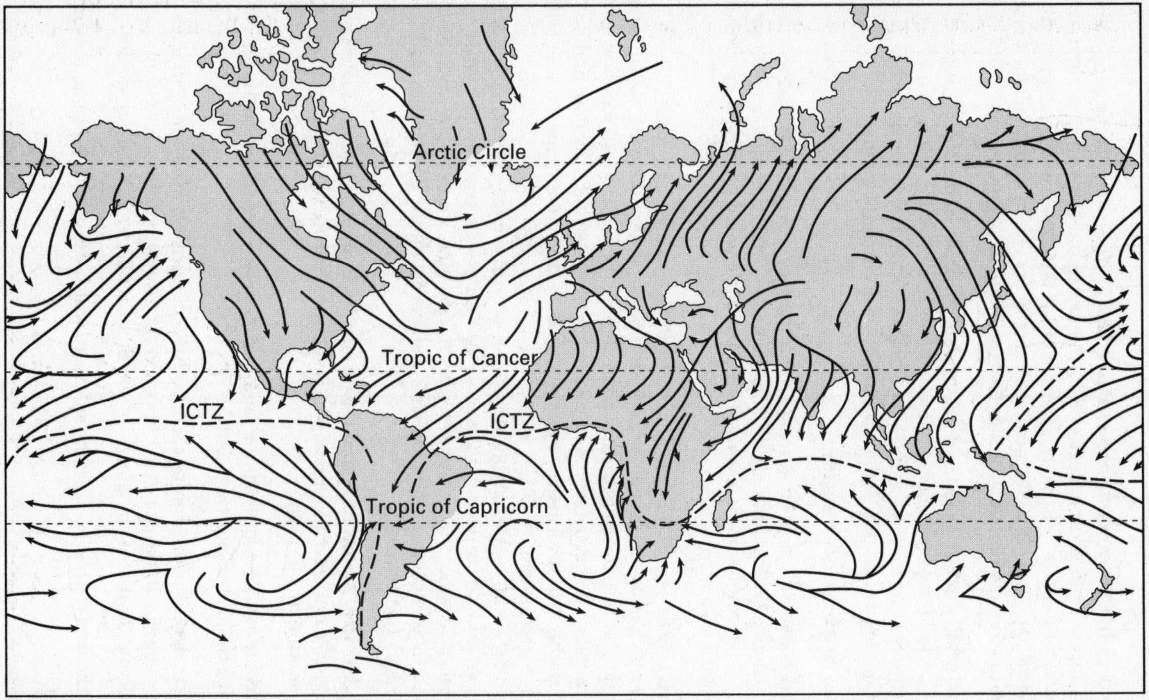


**Wind—typical patterns for January** *Convergence is indicated by bold dashed lines. Note the shift in the position of the *Intertropical Convergence Zone (ICTZ), and the reversal of the *monsoon winds from the July pattern.

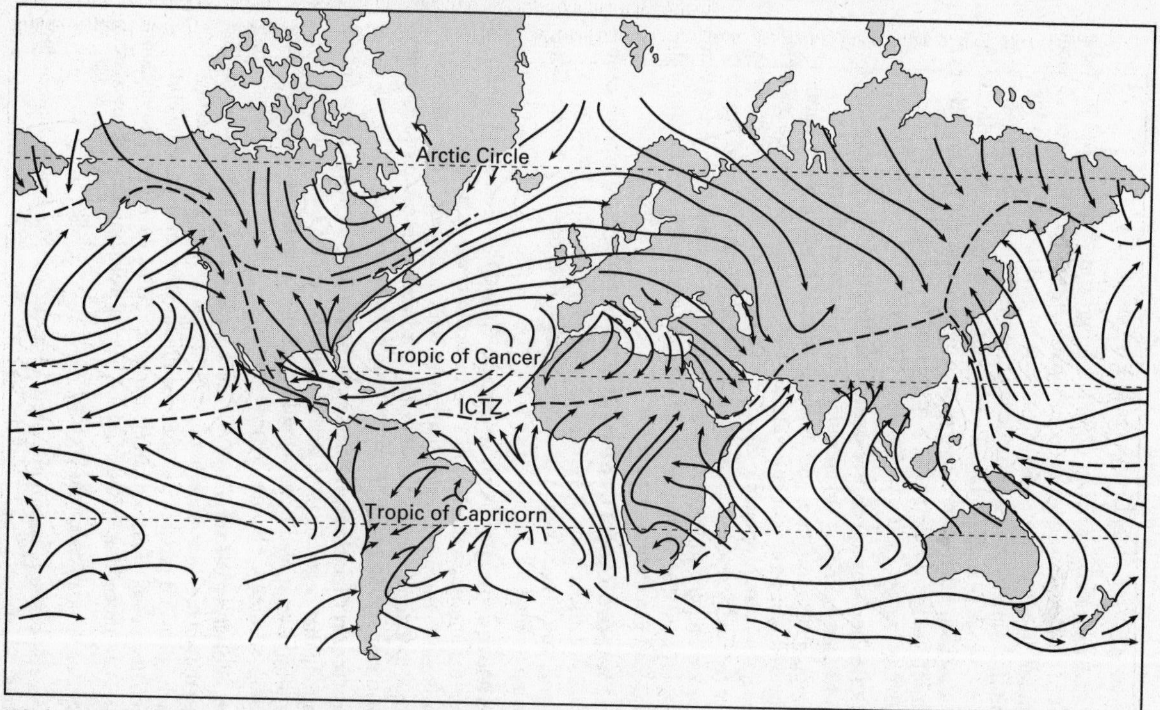


**Wind—typical patterns for July** *Convergence is indicated by bold dashed lines. Note the shift in the position of the *Intertropical Convergence Zone (ICTZ), and the reversal of the *monsoon winds from the January pattern.

**WMC** *See* World Meteorological Centre.

**WMO** *See* World Meteorological Organization.

**WOCE** *See* World Ocean Circulation Experiment.

**World Climate Programme (WCP)** A programme carried out under the auspices of the United Nations and coordinated by the *World Meteorological Organization. There are four basic aims: obtaining data and monitoring climate; improving the use of climatic data for the management of resources, agriculture, urban planning, etc.; research into climatic processes (the *World Climate Research Programme); assessment of the impact of climatic change (*see feature*) and the development of response strategies. *See also* Intergovernmental Panel on Climate Change.

**World Climate Research Programme (WCRP)** A joint programme of the *World Meteorological Organization, the Intergovernmental Oceanographic Commission, and the International Council for Scientific Unions with the basic aim of determining whether *climate may be predicted and the extent of anthropogenic effects. In addition to various lesser projects, there are five main research programmes: *Arctic Climate System Study; *Climate Variability and Predictability Programme; *Global Energy and Water Cycle Experiment; *Stratospheric Processes and Their Role in Climate; and *World Ocean Circulation Experiment.

**World Meteorological Centre (WMC)** One of the three primary centres (Melbourne, Moscow, and Washington) responsible for the collection, processing, and dissemination of meteorological data over the *Global Data-Processing System.

**World Meteorological Organization (WMO)** A specialized agency of the United Nations, established in 1948 to facilitate the exchange of meteorological data worldwide. Apart from standardization, administrative, and other functions, it is responsible for coordinating research programmes and for the operation of the *World Weather Watch.

**World Ocean Circulation Experiment (WOCE)** A major international programme that began in 1990, designed to obtain data for use in the preparation of improved ocean-circulation models, and specifically for the prediction of climatic change (*see feature*).

**World Weather Watch (WWW)** The coordinated, worldwide system that ensures that all member states of the *World Meteorological Organization obtain the data and services that they require for weather forecasting and research. It consists of three main elements: the *Global Observing System, the *Global Data-Processing System, and the *Global Telecommunications System.

**WWW** *See* World Weather Watch.

**WX** An abbreviation, commonly used—especially in the United States—for 'weather'.

# X

**xenon (Xe)** An *inert gas, atomic number 54, that occurs at extremely low concentration ($8.0 \times 10^{-6}$ per cent by volume) in the atmosphere.

**X-ray event** *See* GAMMA-RAY AND X-RAY EVENTS.

# Z

**zastruga** *See* SASTRUGA.

**zenith** **1.** The point on the celestial sphere directly above the observer. **2.** In a looser sense, an area of sky, of indefinite size, immediately above the observer.

**zenith, magnetic** The point on the celestial sphere indicated by a freely suspended magnetic needle and at which the magnetic field lines appear to converge. *See* CORONA (2).

**zephyr** A light, warm breeze. Derived from *Zephyros,* the personification by the ancient Greeks of the warm west wind that was thought to prevail around the time of the summer *solstice.

**zodiacal band** An extremely faint band of light that may be observed in the night-time sky along the *ecliptic and which may merge into the *zodiacal light. It is caused by sunlight scattered by interplanetary dust particles that lie outside the Earth's orbit. *See also* GEGENSCHEIN.

**zodiacal light** A faint, approximately conical area of light with its apex on the *ecliptic that extends upwards from the horizon and is seen shortly before sunrise or after sunset. The visible area actually forms part of an elliptical region centred on the Sun. It is most readily visible around the *equinoxes, when the ecliptic is at the steepest angle to the horizon. It consists of sunlight scattered by fine dust particles in interplanetary space that lie approximately in the Earth's orbital plane.

**zonal circulation** A circulation that occurs essentially in an east–west plane (along a parallel of latitude), flowing in one direction at the surface and in the opposite direction at altitude (*compare* MERIDIONAL CIRCULATION). The principal example is the *Walker Circulation, which forms part of the El Niño Southern Oscillation phenomenon (*see feature*).

**zonal flow** Large-scale airflow in which the motion occurs primarily along parallels of latitude, i.e. east–west or vice versa. *See also* MERIDIONAL CIRCULATION.

**zonal index** A parameter used to specify the strength of *zonal flow within a particular range of latitudes. It may be determined from the mean surface pressure difference at specific latitudes. With a high index the *jet stream and *westerlies lie well towards the pole. With a low zonal index the westerly flow may be completely fragmented into a cellular form, with deep cold *depressions at low latitudes and warm *blocking anticyclones at high latitudes. The variation between high and low indices is referred to as the **index cycle.**

**zonal wave number** A measure of the number of complete waves around the globe at a particular latitude, and usually determined from the contours at a specific *isobaric surface. Typical wave numbers for the northern hemisphere are 4 or 5, indicating the numbers of ridges and troughs found around the Earth. A wave number of 1 is an indication of the eccentricity of the basic flow with respect to the geographic pole.

**zonda** A warm wind in Argentina, either a wintertime westerly *föhn wind that descends the eastern slopes of the Andes, or a moist northerly ahead of a *depression.

**zone of silence** A region around a sound source (e.g. a volcanic explosion) in which none of the sound is audible. Its inner boundary is the limit of surface-propagated sound, and the outer is the distance at which sound waves refracted in the upper atmosphere reach the ground. These boundaries are very approximately 100 and 200 km, respectively, from the source, but differ considerably depending on the atmosphere's temperature (and hence density) profile. They are also often far from circular, and may be affected by local topography.

**Zulu (Z)** The designation for the time zone in which Greenwich Mean Time applies, centred on 0° longitude. For scientific purposes the time for this zone is known as *Universal Time. All meteorological observations worldwide are reported using this time (for ease of comparison and computation), and are commonly quoted with the suffix 'Z'.

# Appendix A: Weather Records

## Drought

**Lowest mean annual rainfall:** 0.1 mm, Pacific coast of Chile

**Driest place:** 14 years without rain, Iquique, Chile (parts of the Chilean Atacama Desert may have passed several decades without rain)

## Hail

**Largest single hailstone:** 1 kg, Gopalganj, Bangladesh, 14 April 1986

**Largest hailstone aggregates:** 3.4 kg, Hyderabad, India, 1939; 4 kg, Yüwu, China, 1902

**Highest death toll:** 246, Moradabad, India, 20 April 1888

## Humidity

**Highest:** average afternoon dewpoint >28.9 °C, Aseb, Eritrea

## Iceberg

**Largest tabular iceberg:** *c.*33 000 km$^2$, off Scott Island, Southern Ocean, November 1956

## Pressure

**Highest surface pressure:** 1085.7 hPa, Tosontsengel, Mongolia, 19 December 2001

**Lowest surface pressure:** 870 hPa, within Typhoon Tip, west of Guam, Pacific Ocean, 12 October 1979

## Rainfall

**Greatest yearly average:** 11 873 mm, Mawsynram, Meghalaya State, India **(estimated):** 13 299.44 mm, Lloro, Colombia

**Lowest yearly average:** 0.5 mm, Quillagua, Chile

**Most in one year:** 26 461 mm, Cherrapunji, India, 1 August 1860 to 31 July 1861

**Most in one day:** 1870 mm, Cilaos, La Réunion, Indian Ocean, 16 March 1952

**Most intense:** 38.1 mm in 1 minute, Barst, Guadeloupe, 26 November 1970

**Most days of rain:** 325 yearly average, Bahía Felix, Chile

## Sea temperature

**Highest recorded:** 37 °C, Persian Gulf

## Snowfall

**Greatest in one year:** 28956mm, Mt Baker Lodge, Washington, winter 1998–9
**Greatest single:** 4800mm, Mt Shasta Ski Bowl, California, 13–19 February 1959
**Greatest daily:** 1930mm, Silver Lake, Colorado, 14–15 April 1921
**Greatest depth:** 11.820m, Mount Ibuki, Japan, 1927
**Heaviest continuous:** 1730mm in 19 hours, Bessus, Savoie, 5–6 April 1969

## Storm surge

**Greatest height:** 12.2m, Hatia Island, Bangladesh, 12 November 1970
**Most deaths:** about 300000 in Bakarganj (now in Bangladesh), November 1970

## Sunshine

**Greatest annual total:** 4300 hours (97% of total possible), eastern Sahara

## Temperature

**Lowest recorded:** –89.2 °C, Vostok, Antarctica, 21 July 1983
**Lowest for a permanently inhabited place:** –68 °C, Oymyakon, Siberia, 1933 (unofficial subsequent record –72 °C)
**Lowest mean annual:** –58 °C, Polus Nedostupnosti (Pole of Inaccessibility), Antarctica
**Highest mean annual:** 34.4 °C, Dallol, Ethiopia, 1960–66
**Highest recorded:** 58 °C, Al'Aziziyah, Libya, 13 September 1922
**Greatest rise:** 27 deg C (from –20 °C to 7 °C) in 2 minutes, Spearfish, South Dakota, 23 January 1943
**Greatest range recorded:** 105 deg C (from –68 °C to 37 °C), Verkhoyansk, Siberia
**Greatest change in one day:** –56 deg C (from 7 °C to –49 °C), Browning, Montana, 23–24 January 1916
**Longest heatwave:** 162 days over 38 °C, Marble Bar, Western Australia, 23 October 1923 to 7 April 1924
**Longest period of stable temperatures:** 1927–35, Garapan, Saipan Island, Northern Marianas, never above 31.4 °C nor below 19.6 °C

## Thunder and lightning

**Highest annual thunderstorm days:** 262, Kampala, Uganda
**Longest measured lightning channel:** 150 km—radar measurements in 1956 over Colorado

## Tornado

**Highest wind speed:** 512 km $h^{-1}$, Oklahoma, 3 May 1999

**Longest track:** 472 km, Illinois–Indiana, 26 May 1917

**Most deaths:** *c.*1300, with 50000 homeless, Shaturia, Bangladesh, 26 April 1989

**Greatest outbreak:** 148 tornadoes, 3–4 April 1974, covering 13 US states, 315 dead, 5484 injured, with 6 F5 tornadoes

## Tropical cyclone

**Largest:** 3500 km, Hurricane Gilbert, Caribbean, September 1988

**Longest duration:** 31 days for 'John', Pacific Ocean, August–September 1994

**Most deaths:** over 78000, Irrawaddy Delta, Burma (Myanmar), Cyclone Nargis, 3 May 2008. Prior to this, the Great Hurricane of 1780 is reputed to have killed more than 22000 people in Barbados and Martinique.

## Waterspout

**Highest recorded:** 1528 m, off Eden, New South Wales, 16 May 1898

## Wave, sea

**Highest:** 34 m, USS Ramapo, mid-Pacific, 6–7 February 1933

## Wind speed (excluding tornadoes)

**Highest recorded gust:** 371 km $h^{-1}$, Mt Washington, New Hampshire, 20 March 1986

**Highest 5-minute speed:** 303 km $h^{-1}$, Mt Washington, New Hampshire, 20 March 1986

**Highest recorded jet-stream speed:** 656 km $h^{-1}$, above South Uist, Outer Hebrides, Scotland, 13 December 1967

**Windiest place:** sustained wind speeds of 67 km $h^{-1}$ throughout year, Commonwealth Bay, George V Coast, Antarctica; record 320 km $h^{-1}$

# Appendix B: British Climate Data

Typical climatic data for various selected sites in the British Isles, the majority airport or aerodrome sites. Temperatures are mean temperature, and average daily maximum and minimum; sunshine and rainfall are average monthly totals. Data for the years 1961–90.

| | J | F | M | A | M | J | J | A | S | O | N | D | Year |
|---|---|---|---|---|---|---|---|---|---|---|---|---|---|
| **Belfast (Aldergrove Airport), Northern Ireland** | | | | | | | | | | | | | |
| Mean °C | 3.8 | 3.9 | 5.4 | 7.5 | 10.2 | 13.2 | 14.7 | 14.4 | 12.4 | 9.9 | 6.0 | 4.6 | 8.8 |
| Max. °C | 6.5 | 6.8 | 8.8 | 11.5 | 14.4 | 17.3 | 18.5 | 18.2 | 15.9 | 13.0 | 8.9 | 7.3 | 12.3 |
| Min. °C | 1.1 | 1.1 | 2.0 | 3.5 | 6.1 | 9.1 | 10.9 | 10.7 | 9.0 | 6.9 | 3.1 | 2.0 | 5.5 |
| Sun. hr | 45.4 | 66.5 | 97.6 | 154.0 | 183.7 | 170.1 | 147.7 | 139.4 | 109.0 | 85.2 | 59.0 | 38.8 | 1296.6 |
| Rain. mm | 86 | 58 | 67 | 53 | 60 | 63 | 64 | 80 | 85 | 88 | 78 | 78 | 860 |
| **Birmingham Airport** | | | | | | | | | | | | | |
| Mean °C | 3.1 | 3.1 | 5.2 | 7.6 | 10.6 | 14.0 | 15.8 | 15.4 | 13.2 | 10.0 | 6.0 | 4.2 | 9.0 |
| Max. °C | 6.0 | 6.2 | 8.9 | 11.9 | 15.3 | 18.8 | 20.6 | 20.1 | 17.6 | 13.8 | 9.2 | 7.1 | 13.0 |
| Min. °C | 0.3 | 0.1 | 1.5 | 3.3 | 6.0 | 9.2 | 11.1 | 10.8 | 8.8 | 6.2 | 2.9 | 1.3 | 5.1 |
| Sun. hr | 49.7 | 60.0 | 101.5 | 129.2 | 178.0 | 186.2 | 181.0 | 166.8 | 134.3 | 97.2 | 64.2 | 46.9 | 1395 |
| Rain. mm | 57 | 48 | 51 | 49 | 56 | 56 | 46 | 66 | 54 | 52 | 59 | 66 | 660 |
| **Cardiff Airport** | | | | | | | | | | | | | |
| Mean °C | 4.4 | 4.2 | 5.9 | 8.0 | 11.1 | 14.1 | 16.0 | 15.9 | 14.0 | 11.1 | 7.3 | 5.4 | 9.8 |
| Max. °C | 6.9 | 6.9 | 9.2 | 11.9 | 15.1 | 18.1 | 20.0 | 19.8 | 17.5 | 14.1 | 10.1 | 8.0 | 13.1 |
| Min. °C | 2.0 | 1.6 | 2.7 | 4.2 | 7.1 | 10.1 | 12.0 | 12.1 | 10.5 | 8.1 | 4.5 | 2.8 | 6.5 |
| Sun. hr | 52.3 | 70.6 | 118.3 | 168.3 | 204.6 | 207.9 | 210.0 | 195.0 | 144.8 | 94.0 | 71.5 | 49.4 | 1586.7 |
| Rain. mm | 91 | 67 | 76 | 57 | 64 | 66 | 74 | 80 | 92 | 96 | 100 | 98 | 961 |
| **Edinburgh Airport** | | | | | | | | | | | | | |
| Mean °C | 3.2 | 3.3 | 5.1 | 7.1 | 9.9 | 13.0 | 14.5 | 14.3 | 12.3 | 9.5 | 5.4 | 3.9 | 8.5 |
| Max. °C | 6.2 | 6.5 | 8.7 | 11.1 | 14.2 | 17.3 | 18.8 | 18.5 | 16.2 | 13.2 | 8.7 | 6.9 | 12.2 |
| Min. °C | 0.3 | 0.0 | 1.5 | 3.1 | 5.7 | 8.7 | 10.3 | 10.2 | 8.4 | 5.9 | 2.1 | 0.9 | 4.8 |
| Sun. hr | 46.2 | 65.7 | 99.4 | 132.9 | 168.0 | 169.7 | 162.4 | 144.7 | 109.8 | 86.6 | 60.9 | 39.1 | 1285.4 |
| Rain. mm | 57 | 42 | 51 | 41 | 51 | 51 | 57 | 65 | 67 | 65 | 63 | 58 | 668 |
| **Eskdalemuir, Dumfries & Galloway** | | | | | | | | | | | | | |
| Mean °C | 1.8 | 1.8 | 3.5 | 5.7 | 8.7 | 11.7 | 13.2 | 12.9 | 10.7 | 8.1 | 4.1 | 2.5 | 7.1 |
| Max. °C | 4.5 | 4.7 | 6.9 | 10.0 | 13.4 | 16.3 | 17.6 | 17.2 | 14.6 | 11.6 | 7.2 | 5.3 | 10.8 |
| Min. °C | −1.0 | −1.2 | 0.1 | 1.4 | 4.0 | 7.1 | 8.8 | 8.7 | 6.9 | 4.7 | 1.0 | −0.2 | 3.4 |
| Sun. hr | 38.7 | 62.6 | 85.9 | 132.2 | 160.8 | 158.4 | 150.9 | 134.9 | 94.8 | 73.8 | 55.7 | 38.6 | 1187.3 |
| Rain. mm | 166 | 110 | 137 | 83 | 95 | 94 | 98 | 125 | 149 | 164 | 152 | 165 | 1538 |
| **Lerwick, Shetland** | | | | | | | | | | | | | |
| Mean °C | 3.2 | 3.1 | 3.8 | 5.2 | 7.6 | 10.0 | 11.4 | 11.8 | 10.1 | 8.2 | 5.1 | 3.9 | 7.0 |
| Max. °C | 5.3 | 5.1 | 6.0 | 7.7 | 10.0 | 12.5 | 13.7 | 14.1 | 12.3 | 10.2 | 7.2 | 6.0 | 9.2 |
| Min. °C | 1.2 | 1.1 | 1.6 | 2.7 | 5.2 | 7.5 | 9.1 | 9.5 | 8.0 | 6.3 | 3.1 | 1.8 | 4.8 |
| Sun. hr | 22.0 | 52.5 | 85.3 | 132.4 | 147.1 | 150.9 | 123.3 | 124.0 | 94.4 | 60.5 | 31.2 | 13.4 | 1037.0 |
| Rain. mm | 131 | 92 | 114 | 72 | 62 | 60 | 62 | 79 | 118 | 134 | 143 | 142 | 1209 |
| **London (Gatwick Airport)** | | | | | | | | | | | | | |
| Mean °C | 3.5 | 3.8 | 5.7 | 8.0 | 11.3 | 14.4 | 16.5 | 16.1 | 13.8 | 10.7 | 6.4 | 4.5 | 9.6 |
| Max. °C | 6.7 | 7.1 | 9.9 | 12.6 | 16.3 | 19.6 | 21.7 | 21.4 | 18.8 | 15.0 | 10.1 | 7.7 | 13.9 |
| Min. °C | 0.4 | 0.5 | 1.5 | 3.4 | 6.3 | 9.3 | 11.3 | 10.9 | 8.8 | 6.4 | 2.8 | 1.3 | 5.2 |
| Sun. hr | 52.4 | 71.3 | 113.4 | 153.0 | 204.3 | 204.3 | 204.5 | 195.3 | 148.1 | 110.5 | 69.3 | 47.8 | 1574.2 |
| Rain. mm | 78 | 51 | 61 | 54 | 55 | 57 | 45 | 56 | 68 | 73 | 77 | 79 | 754 |
| **Plymouth** | | | | | | | | | | | | | |
| Mean °C | 5.9 | 5.7 | 6.9 | 8.8 | 11.5 | 14.3 | 16.1 | 16.0 | 14.3 | 11.9 | 8.5 | 7.0 | 10.6 |
| Max. °C | 8.5 | 8.3 | 9.8 | 12.0 | 14.8 | 17.6 | 19.5 | 19.3 | 17.5 | 14.7 | 11.3 | 9.6 | 13.6 |
| Min. °C | 3.4 | 3.2 | 4.1 | 5.5 | 8.3 | 11.0 | 12.8 | 12.8 | 11.2 | 9.2 | 5.7 | 4.4 | 7.6 |
| Sun. hr | 54.8 | 76.8 | 125.8 | 180.5 | 218.0 | 217.8 | 220.3 | 200.4 | 156.8 | 107.6 | 74.7 | 50.9 | 1684.4 |
| Rain. mm | 114 | 92 | 87 | 59 | 61 | 57 | 55 | 69 | 76 | 95 | 101 | 116 | 982 |
| **Waddington, Lincolnshire** | | | | | | | | | | | | | |
| Mean °C | 3.2 | 3.3 | 5.4 | 7.5 | 10.8 | 13.9 | 15.8 | 15.7 | 13.6 | 10.3 | 6.0 | 3.9 | 9.1 |
| Max. °C | 5.8 | 6.1 | 8.9 | 11.4 | 15.2 | 18.4 | 20.3 | 20.1 | 17.7 | 13.8 | 8.9 | 6.5 | 12.8 |
| Min. °C | 0.6 | 0.5 | 1.9 | 3.6 | 6.5 | 9.5 | 11.4 | 11.4 | 9.5 | 6.8 | 3.2 | 1.4 | 5.5 |
| Sun. hr | 54.7 | 67.4 | 109.2 | 143.5 | 198.2 | 196.6 | 186.5 | 175.5 | 137.8 | 103.7 | 65.4 | 49.9 | 1488.4 |
| Rain. mm | 49 | 38 | 49 | 47 | 50 | 53 | 53 | 63 | 47 | 47 | 55 | 53 | 604 |

# Appendix C: World Climate Data

A selection of sites, representing various *Köppen classification types. London in the preceding table (British Climate Data) may be taken as representative of type Cb, a maritime, west-coast climate. Maximum and minimum temperatures, and sunshine hours are daily averages. Rainfall is average monthly total. Humidity is average humidity at noon.

| | J | F | M | A | M | J | J | A | S | O | N | D |
|---|---|---|---|---|---|---|---|---|---|---|---|---|
| **Moscow (Db – continental, cool summer)** | | | | | | | | | | | | |
| Max. °C | –9 | –6 | 0 | 10 | 19 | 21 | 23 | 22 | 16 | 9 | 2 | –5 |
| Min. °C | –16 | –14 | –8 | 1 | 8 | 11 | 13 | 12 | 7 | 3 | –3 | –10 |
| Sun. hr | 1 | 3 | 4 | 5 | 8 | 9 | 9 | 8 | 6 | 3 | 1 | 0 |
| Rain. mm | 39 | 38 | 36 | 37 | 53 | 58 | 88 | 71 | 58 | 45 | 47 | 54 |
| Hum. % | 77 | 66 | 64 | 54 | 43 | 47 | 54 | 55 | 59 | 67 | 79 | 83 |
| **New York (Da – continental, hot summer)** | | | | | | | | | | | | |
| Max. °C | 4 | 4 | 9 | 15 | 21 | 26 | 28 | 27 | 24 | 18 | 12 | 6 |
| Min. °C | –3 | –2 | 1 | 6 | 12 | 17 | 20 | 19 | 16 | 10 | 4 | –1 |
| Sun. hr | 5 | 6 | 7 | 7 | 8 | 10 | 10 | 9 | 8 | 7 | 6 | 5 |
| Rain. mm | 89 | 74 | 104 | 89 | 91 | 86 | 102 | 119 | 89 | 84 | 89 | 84 |
| Hum. % | 61 | 58 | 56 | 55 | 56 | 58 | 57 | 60 | 60 | 59 | 60 | 60 |
| **Rome (Cs – Mediterranean)** | | | | | | | | | | | | |
| Max. °C | 11 | 13 | 15 | 19 | 23 | 28 | 30 | 30 | 26 | 22 | 16 | 13 |
| Min. °C | 5 | 5 | 7 | 10 | 13 | 17 | 20 | 19 | 17 | 13 | 9 | 6 |
| Sun. hr | 4 | 4 | 6 | 7 | 8 | 9 | 11 | 10 | 8 | 6 | 4 | 4 |
| Rain. mm | 71 | 62 | 57 | 51 | 46 | 37 | 15 | 21 | 63 | 99 | 129 | 93 |
| Hum. % | 68 | 64 | 56 | 54 | 54 | 48 | 42 | 43 | 50 | 59 | 66 | 70 |
| **Cairo (BW – desert)** | | | | | | | | | | | | |
| Max. °C | 19 | 21 | 24 | 28 | 32 | 35 | 35 | 35 | 33 | 30 | 26 | 21 |
| Min. °C | 9 | 9 | 12 | 14 | 18 | 20 | 22 | 22 | 20 | 18 | 14 | 10 |
| Sun. hr | 7 | 8 | 9 | 10 | 10 | 12 | 12 | 11 | 10 | 9 | 8 | 6 |
| Rain. mm | 4 | 4 | 3 | 1 | 2 | 0 | 0 | 0 | 1 | 1 | 3 | 7 |
| Hum. % | 43 | 39 | 33 | 28 | 25 | 27 | 31 | 35 | 37 | 36 | 42 | 46 |
| **Dar-es-Salam (Aw – savanna, winter dry season)** | | | | | | | | | | | | |
| Max. °C | 31 | 32 | 32 | 31 | 30 | 29 | 29 | 29 | 30 | 30 | 31 | 31 |
| Min. °C | 24 | 24 | 23 | 23 | 21 | 19 | 19 | 18 | 19 | 20 | 22 | 23 |
| Sun. hr | 8 | 7 | 7 | 5 | 7 | 7 | 7 | 9 | 9 | 9 | 8 | 8 |
| Rain. mm | 61 | 81 | 115 | 286 | 183 | 31 | 29 | 26 | 32 | 62 | 104 | 95 |
| Hum. % | 69 | 68 | 70 | 75 | 68 | 60 | 58 | 57 | 58 | 61 | 66 | 69 |
| **Hong Kong (Ca – humid subtropical)** | | | | | | | | | | | | |
| Max. °C | 18 | 17 | 19 | 24 | 28 | 29 | 31 | 31 | 29 | 27 | 23 | 20 |
| Min. °C | 13 | 13 | 16 | 19 | 23 | 26 | 26 | 26 | 25 | 23 | 18 | 15 |
| Sun. hr | 5 | 4 | 3 | 4 | 5 | 5 | 8 | 6 | 6 | 7 | 7 | 6 |
| Rain. mm | 33 | 46 | 74 | 137 | 292 | 394 | 381 | 367 | 257 | 114 | 43 | 31 |
| Hum. % | 66 | 73 | 74 | 77 | 78 | 77 | 77 | 77 | 72 | 63 | 60 | 63 |
| **Freetown (Am – tropical monsoon)** | | | | | | | | | | | | |
| Max. °C | 29 | 30 | 30 | 30 | 30 | 30 | 28 | 28 | 28 | 29 | 29 | 29 |
| Min. °C | 24 | 24 | 25 | 25 | 25 | 24 | 23 | 23 | 23 | 23 | 24 | 24 |
| Sun. hr | 8 | 8 | 8 | 7 | 6 | 5 | 3 | 2 | 4 | 6 | 7 | 7 |
| Rain. mm | 11 | 8 | 18 | 59 | 160 | 351 | 845 | 835 | 566 | 262 | 123 | 41 |
| Hum. % | 70 | 69 | 71 | 72 | 75 | 78 | 81 | 82 | 81 | 77 | 76 | 72 |

# Appendix D: Conversion Tables

## Knots to $km h^{-1}$ and $m s^{-1}$

| Knots | $km h^{-1}$ | $m s^{-1}$ | Knots | $km h^{-1}$ | $m s^{-1}$ |
|---|---|---|---|---|---|
| 1 | 1.852 | 0.514 | 70 | 129.6 | 36.0 |
| 2 | 3.7 | 1.0 | 75 | 138.9 | 38.6 |
| 3 | 5.6 | 1.5 | 80 | 148.2 | 41.2 |
| 4 | 7.4 | 2.1 | 85 | 157.4 | 43.7 |
| 5 | 9.3 | 2.6 | 90 | 166.7 | 46.3 |
| 6 | 11.1 | 3.1 | 95 | 176.0 | 48.9 |
| 7 | 13.0 | 3.6 | 100 | 185.2 | 51.4 |
| 8 | 14.8 | 4.1 | 105 | 194.5 | 54.0 |
| 9 | 16.7 | 4.6 | 110 | 203.7 | 56.6 |
| 10 | 18.5 | 5.1 | 115 | 213.0 | 59.2 |
| 15 | 27.8 | 7.7 | 120 | 222.2 | 61.7 |
| 20 | 37.0 | 10.3 | 125 | 231.5 | 64.3 |
| 25 | 46.3 | 12.7 | 130 | 240.8 | 66.9 |
| 30 | 55.6 | 15.4 | 135 | 250.0 | 69.5 |
| 35 | 64.8 | 18.0 | 140 | 259.3 | 72.0 |
| 40 | 74.1 | 20.6 | 145 | 268.5 | 74.6 |
| 45 | 83.3 | 23.2 | 150 | 277.8 | 77.2 |
| 50 | 92.6 | 25.7 | 155 | 287.1 | 79.7 |
| 55 | 101.9 | 28.3 | 160 | 296.3 | 82.3 |
| 60 | 111.1 | 30.9 | 165 | 305.6 | 84.9 |
| 65 | 120.4 | 33.4 | 170 | 314.8 | 87.5 |

## Approximate Altitude Equivalents

| Feet | m | km |
|---|---|---|
| 5000 | 1524 | 1.5 |
| 6500 | 1981 | 2.0 |
| 10000 | 3048 | 3.0 |
| 15000 | 4572 | 4.6 |
| 20000 | 6096 | 6.1 |
| 25000 | 7620 | 7.6 |
| 30000 | 9144 | 9.1 |
| 35000 | 10668 | 10.7 |
| 40000 | 12192 | 12.2 |
| 45000 | 13716 | 13.7 |
| 50000 | 15240 | 15.2 |

## Atmospheric Pressure

| hPa | Hg | hPa | Hg | hPa | Hg |
|---|---|---|---|---|---|
| 960 | 28.35 | 990 | 29.23 | 1020 | 30.12 |
| 961 | 28.38 | 991 | 29.26 | 1021 | 30.15 |
| 962 | 28.41 | 992 | 29.29 | 1022 | 30.18 |
| 963 | 28.44 | 993 | 29.32 | 1023 | 30.21 |
| 964 | 28.47 | 994 | 29.35 | 1024 | 30.24 |
| 965 | 28.50 | 995 | 29.38 | 1025 | 30.27 |
| 966 | 28.53 | 996 | 29.41 | 1026 | 30.30 |
| 967 | 28.56 | 997 | 29.44 | 1027 | 30.33 |
| 968 | 28.59 | 998 | 29.47 | 1028 | 30.36 |
| 969 | 28.61 | 999 | 29.50 | 1029 | 30.39 |
| 970 | 28.64 | 1000 | 29.53 | 1030 | 30.42 |
| 971 | 28.67 | 1001 | 29.56 | 1031 | 30.45 |
| 972 | 28.70 | 1002 | 29.59 | 1032 | 30.47 |
| 973 | 28.73 | 1003 | 29.62 | 1033 | 30.50 |
| 974 | 28.76 | 1004 | 29.65 | 1034 | 30.53 |
| 975 | 28.79 | 1005 | 29.68 | 1035 | 30.56 |
| 976 | 28.82 | 1006 | 29.71 | 1036 | 30.59 |
| 977 | 28.85 | 1007 | 29.74 | 1037 | 30.62 |
| 978 | 28.88 | 1008 | 29.77 | 1038 | 30.65 |
| 979 | 28.91 | 1009 | 29.80 | 1039 | 30.68 |
| 980 | 28.94 | 1010 | 29.83 | 1040 | 30.71 |
| 981 | 28.97 | 1011 | 29.85 | 1041 | 30.74 |
| 982 | 29.00 | 1012 | 29.88 | 1042 | 30.77 |
| 983 | 29.03 | 1013 | 29.91 | 1043 | 30.80 |
| 984 | 29.06 | 1014 | 29.94 | 1044 | 30.83 |
| 985 | 29.09 | 1015 | 29.97 | 1045 | 30.86 |
| 986 | 29.12 | 1016 | 30.00 | 1046 | 30.89 |
| 987 | 29.15 | 1017 | 30.03 | 1047 | 30.92 |
| 988 | 29.18 | 1018 | 30.06 | 1048 | 30.95 |
| 989 | 29.21 | 1019 | 30.09 | 1049 | 30.98 |

# Appendix D: Conversion Tables

## Temperature

| °F | °C | °F | °C | °C | °F | °C | °F |
|---|---|---|---|---|---|---|---|
| 100 | 37.8 | 50 | 10.0 | 50 | 122.0 | 0 | 32.0 |
| 99 | 37.2 | 49 | 9.4 | 49 | 120.2 | –1 | 30.2 |
| 98 | 36.7 | 48 | 8.9 | 48 | 118.4 | –2 | 28.4 |
| 97 | 36.1 | 47 | 8.3 | 47 | 116.6 | –3 | 26.6 |
| 96 | 35.6 | 46 | 7.8 | 46 | 114.8 | –4 | 24.8 |
| 95 | 35.0 | 45 | 7.2 | 45 | 113.0 | –5 | 23.0 |
| 94 | 34.4 | 44 | 6.7 | 44 | 111.2 | –6 | 21.2 |
| 93 | 33.9 | 43 | 6.1 | 43 | 109.4 | –7 | 19.4 |
| 92 | 33.3 | 42 | 5.6 | 42 | 107.6 | –8 | 17.6 |
| 91 | 32.8 | 41 | 5.0 | 41 | 105.8 | –9 | 15.8 |
| 90 | 32.2 | 40 | 4.4 | 40 | 104.0 | –10 | 14.0 |
| 89 | 31.7 | 39 | 3.9 | 39 | 102.2 | –11 | 12.2 |
| 88 | 31.1 | 38 | 3.3 | 38 | 100.4 | –12 | 10.4 |
| 87 | 30.6 | 37 | 2.8 | 37 | 98.6 | –13 | 8.6 |
| 86 | 30.0 | 36 | 2.2 | 36 | 96.8 | –14 | 6.8 |
| 85 | 29.4 | 35 | 1.7 | 35 | 95.0 | –15 | 5.0 |
| 84 | 28.9 | 34 | 1.1 | 34 | 93.2 | –16 | 3.2 |
| 83 | 28.3 | 33 | 0.6 | 33 | 91.4 | –17 | 1.4 |
| 82 | 27.8 | 32 | 0.0 | 32 | 89.6 | –18 | –0.4 |
| 81 | 27.2 | 31 | –0.5 | 31 | 87.8 | –19 | –2.2 |
| 80 | 26.7 | 30 | –1.1 | 30 | 86.0 | –20 | –4.0 |
| 79 | 26.1 | 29 | –1.7 | 29 | 84.2 | –21 | –5.8 |
| 78 | 25.6 | 28 | –2.2 | 28 | 82.4 | –22 | –7.6 |
| 77 | 25.0 | 27 | –2.8 | 27 | 80.6 | –23 | –9.4 |
| 76 | 24.4 | 26 | –3.3 | 26 | 78.8 | –24 | –11.2 |
| 75 | 23.9 | 25 | –3.9 | 25 | 77.0 | –25 | –13.0 |
| 74 | 23.3 | 24 | –4.4 | 24 | 75.2 | –26 | –14.8 |
| 73 | 22.8 | 23 | –5.0 | 23 | 73.4 | –27 | –16.6 |
| 72 | 22.2 | 22 | –5.6 | 22 | 71.6 | –28 | –18.4 |
| 71 | 21.7 | 21 | –6.1 | 21 | 69.8 | –29 | –20.2 |
| 70 | 21.1 | 20 | –6.7 | 20 | 68.0 | –30 | –22.0 |
| 69 | 20.6 | 19 | –7.2 | 19 | 66.2 | –31 | –23.8 |
| 68 | 20.0 | 18 | –7.8 | 18 | 64.4 | –32 | –25.6 |
| 67 | 19.4 | 17 | –8.3 | 17 | 62.6 | –33 | –27.4 |
| 66 | 18.9 | 16 | –8.9 | 16 | 60.8 | –34 | –29.2 |
| 65 | 18.3 | 15 | –9.4 | 15 | 59.0 | –35 | –31.0 |
| 64 | 17.8 | 14 | –10.0 | 14 | 57.2 | –36 | –32.8 |
| 63 | 17.2 | 13 | –10.6 | 13 | 55.4 | –37 | –34.6 |
| 62 | 16.7 | 12 | –11.1 | 12 | 53.6 | –38 | –36.4 |
| 61 | 16.1 | 11 | –11.7 | 11 | 51.8 | –39 | –38.2 |
| 60 | 15.6 | 10 | –12.2 | 10 | 50.0 | –40 | –40.0 |
| 59 | 15.0 | 9 | –12.8 | 9 | 48.2 | –41 | –41.8 |
| 58 | 14.4 | 8 | –13.3 | 8 | 46.4 | –42 | –43.6 |
| 57 | 13.9 | 7 | –13.9 | 7 | 44.6 | –43 | –45.4 |
| 56 | 13.3 | 6 | –14.4 | 6 | 42.8 | –44 | –47.2 |
| 55 | 12.8 | 5 | –15.0 | 5 | 41.0 | –45 | –49.0 |
| 54 | 12.2 | 4 | –15.6 | 4 | 39.2 | –46 | –50.8 |
| 53 | 11.7 | 3 | –16.1 | 3 | 37.4 | –47 | –52.6 |
| 52 | 11.1 | 2 | –16.7 | 2 | 35.6 | –48 | –54.4 |
| 51 | 10.6 | 1 | –17.2 | 1 | 33.8 | –49 | –56.2 |

# Further Reference

Glickman, T. S. (Managing Ed.), *Glossary of Meteorology*, 2nd edn, American Meteorological Society, Boston, 2000

Greer, I. W. (ed.), *Glossary of Weather and Climate*, American Meteorological Society, Boston, 1996

Lewis, R. P. W. (ed.), *Meteorological Glossary*, 6th edn, HMSO, London, 1991

Pearce, E. A., and Smith, C. G., *Hutchinson World Weather Guide*, 5th edn, Helicon, 2000

Schneider, S.H. (ed.), *Encyclopedia of Climate and Weather*, Oxford University Press, New York, 1996

World Meteorological Organization, *International Cloud Atlas*, Vol. I, WMO, Geneva, 1975

World Meteorological Organization, *International Cloud Atlas*, Vol. II, WMO, Geneva, 1987

World Meteorological Organization, *International Meteorological Vocabulary*, 2nd edn, WMO, Geneva, 1992

# Internet Sources

SEE WEB LINKS

This is a web-linked dictionary. To access the websites listed below, go to the dictionary's web page at http://www.oup.com/uk/reference/resources/weather, click on the Web links in the Resources section and click straight through to the relevant websites.

## Current weather

- AccuWeather.com's homepage showing US forecasts.
- AccuWeather.com's UK forecasts.
- BBC Weather homepage.
- CNN Weather homepage.
- Intellicast homepage with US weather maps.
- ITV Weather homepage.
- Unisys Weather homepage, with data provided by the US National Weather Service.
- Homepage of the UK Meteorological Office.
- The Weather Channel homepage.
- German website showing UK station plots.

## General information

- Homepage of HurricaneZone.net with global hurricane information.
- Homepage of the US National Climate Data Centre.
- Information on extreme weather by the US National Climate Data Centre.
- Homepage of the US National Hurricane Center.
- University of Reading's Department of Meteorology homepage.
- Forecasts and supplementary weather information relevant to the British Isles.
- Unisys Atlantic tropical storm tracking by year.
- World climate data (historical weather averages) from a wide range of sources.

## Meteorological offices, agencies, and organizations

- Homepage of the Bureau of Meteorology, Australia.
- Homepage of the Meteorological Service of Canada (MSC).
- Homepage of the European Centre for Medium-Range Weather Forecasts.
- Homepage of the European Meteorological Satellite Organization.
- Homepage of the Intergovernmental Panel on Climate Change.
- Homepage of the National Oceanic and Atmospheric Administration (NOAA).
- Homepage of NOAA's Climate Prediction Center.
- Homepage of the National Weather Service (NWS).
- Homepage of the UK Meteorological Office.
- UK forecasts provided by the UK Meteorological Office.
- Homepage of the Meteorological Office Hadley Centre for Climate Change.
- UK map showing up-to-date surface observations.
- Homepage of the World Meteorological Organization.

## Satellite agencies

- Homepage of the European Space Agency.
- Homepage of the European organization for the Exploitation of Meteorological Satellites (Eumetsat).

- Homepage of the National Aeronautics and Space Administration (NASA).
- Homepage of the National Environmental Satellite, Data and Information Service (NESDIS).

## Societies

- Homepage of the American Meteorological Society.
- Homepage of the Australian Meteorological and Oceanographic Society.
- Homepage of the Canadian Meteorological and Oceanographic Society.
- Homepage of the European Meteorological Society.
- Homepage of the Irish Meteorological Society.
- Homepage of the National Weather Association, USA.
- Homepage of the New Zealand Meteorological Society.
- Homepage of the Royal Meteorological Society.
- Tornado and Storm Research Organization (TORRO).

# Oxford Paperback Reference

**The Concise Oxford Dictionary of Quotations**
Edited by Elizabeth Knowles

Based on the highly acclaimed *Oxford Dictionary of Quotations*, this paperback edition maintains its extensive coverage of literary and historical quotations, and contains completely up-to-date material. A fascinating read and an essential reference tool.

**The Oxford Dictionary of Humorous Quotations**
Edited by Ned Sherrin

From the sharply witty to the downright hilarious, this sparkling collection will appeal to all senses of humour.

**Quotations by Subject**
Edited by Susan Ratcliffe

A collection of over 7,000 quotations, arranged thematically for easy look-up. Covers an enormous range of nearly 600 themes from 'The Internet' to 'Parliament'.

**The Concise Oxford Dictionary of Phrase and Fable**
Edited by Elizabeth Knowles

Provides a wealth of fascinating and informative detail for over 10,000 phrases and allusions used in English today. Find out about anything from the 'Trojan horse' to 'ground zero'.